Industrial Process Improvement by Automation and Robotics

Industrial Process Improvement by Automation and Robotics

Editors

Raul D. S. G. Campilho
Francisco J. G. Silva

MDPI

Basel • Beijing • Wuhan • Barcelona • Belgrade • Novi Sad • Cluj • Manchester

Editors

Raul D. S. G. Campilho
Mechanical Engineering
ISEP–School of Engineering
Porto
Portugal

Francisco J. G. Silva
Mechanical Engineering
ISEP–School of Engineering
Porto
Portugal

Editorial Office
MDPI
St. Alban-Anlage 66
4052 Basel, Switzerland

This is a reprint of articles from the Special Issue published online in the open access journal *Machines* (ISSN 2075-1702) (available at: www.mdpi.com/journal/machines/special_issues/ indusctrial_process_automation).

For citation purposes, cite each article independently as indicated on the article page online and as indicated below:

Lastname, A.A.; Lastname, B.B. Article Title. *Journal Name* **Year**, *Volume Number*, Page Range.

ISBN 978-3-0365-9467-5 (Hbk)
ISBN 978-3-0365-9466-8 (PDF)
doi.org/10.3390/books978-3-0365-9466-8

Contents

About the Editors

Raul D. S. G. Campilho

Raul Duarte Salgueiral Gomes Campilho is an Assistant Professor with Habilitation at ISEP—School of Engineering and a researcher in INEGI. His main research interests are composite materials, numerical modelling, and industrial process improvement by automation and robotics.

Francisco J. G. Silva

Francisco José Gomes da Silva is an Associate Professor with Habilitation at ISEP—School of Engineering and a researcher in INEGI. His main research interests are coatings, machining processes, and industrial process improvement by automation and robotics.

Preface

The pursuit of industrial process improvement has been a continuous task in modern manufacturing and production. The confluence of automation and robotics has not only revolutionized manufacturing processes but has also redefined the benchmarks of efficiency, precision, and adaptability in various industries. This reprint, "Industrial Process Improvement by Automation and Robotics", is a comprehensive collection of research papers, each chapter delving into distinct facets of the integration of automation and robotics to enhance industrial processes across diverse applications. The compilation of fourteen chapters covers a broad spectrum of topics, addressing the latest technological advancements and innovative methodologies employed in industrial automation.

In Chapter 1, the Special Issue editorial sets the stage for this collection, exploring the foundational concepts, current advancements, and future prospects in this field. It serves as an introductory overview, laying the groundwork for the subsequent chapters. Chapter 2 delves into the design of a spiral double-cutting machine tailored for an automotive bowden cable assembly line, showcasing how specialized equipment can streamline production in the automotive sector. In Chapter 3, a pioneering concept of a full-automatic equipment for car seats is discussed, emphasizing high productivity and adaptability in the manufacturing process. Chapter 4 introduces a method for measuring workpiece form deviations using machine vision, contributing to enhanced quality control and accuracy in production. Chapters 5, 6, and 7 shed light on innovative techniques for improving automation systems, refining robot positioning accuracy, and employing cutting-edge methodologies like GAN-BPNN for surface roughness measurement and calibration of industrial robots. The subsequent chapters explore a range of subjects from process simulation and optimization of arc welding robot workstations to collaborative workplaces involving humans and robots, brain–computer interface applications in assembly tasks, and decision support methods for dynamic production planning. The two last chapters focus on critical issues like environmental risk assessment and management in the context of Industry 4.0 and the challenges and potential research paths in Industry 5.0, emphasizing the need for sustainability and technological advancements.

The primary aim of this reprimt is to serve as a compendium for researchers, engineers, industrial practitioners, and academicians actively involved or interested in the domains of automation, robotics, and industrial process improvement. The multidisciplinary approach of these chapters caters to a diverse audience seeking in-depth knowledge, innovative strategies, and insightful perspectives in this evolving field.

Raul D. S. G. Campilho and Francisco J. G. Silva
Editors

machines   MDPI

Editorial

Industrial Process Improvement by Automation and Robotics

Raul D. S. G. Campilho [1,2,*] and **Francisco J. G. Silva [1,2]**

1   ISEP—School of Engineering, Polytechnic of Porto, R. Dr. António Bernardino de Almeida, 431, 4200-072 Porto, Portugal; fgs@isep.ipp.pt

2   INEGI—Institute of Science and Innovation in Mechanical and Industrial Engineering, Pólo FEUP, Rua Dr. Roberto Frias, 400, 4200-465 Porto, Portugal

*   Correspondence: rds@isep.ipp.pt; Tel.: +351-939-526-892

check for updates

Citation: Campilho, R.D.S.G.; Silva, F.J.G. Industrial Process Improvement by Automation and Robotics. *Machines* **2023**, *11*, 1011. https://doi.org/10.3390/machines11111011

Received: 30 October 2023
Accepted: 3 November 2023
Published: 6 November 2023

Automation and robotics have revolutionized industrial processes, making them more efficient, precise, and flexible. The integration of automation and robotics into manufacturing and production has been a pivotal driver of industrial advancements [1,2]. The ability to improve quality, reduce human error, and increase production speed has made these concepts indispensable for various industries [3]. Moreover, automation and robotics are becoming particularly relevant in the era of Industry 4.0, where smart manufacturing and mechatronics play a crucial role [4]. In this Editorial, the state of the art in automation and robotics, their applications, current limitations, and future perspectives within the context of improvements in the industrial process are explored.

Automation involves the use of various control/sensor systems and actuators to operate machinery, reducing the need for human intervention [5]. Automation can be as simple as a thermostat regulating room temperature or as complex as a fully automated assembly line [6]. The primary goal is to enhance efficiency and productivity while minimizing errors [7]. This approach has found its place in a variety of industries, with the automotive sector emerging as a major catalyst for the advancement of automation systems, driven by the pursuit of heightened productivity and enhanced flexibility [8,9]. Groover [10] outlines several key factors that prompt businesses to embrace process automation, including increased productivity, reduced production costs, improved part quality, shorter delivery times, the execution of tasks that are impractical for manual labor, the prevention of non-automation-related expenses, and the reduction in or elimination of manual operations. Moreover, automation enables line operators to transition to a more supervisory role, relieving them of monotonous, repetitive, and labor-intensive work, while simultaneously ensuring the company's competitiveness [11]. One of the most significant advancements in automation is the implementation in the principles of Industry 4.0. Industry 4.0 represents the fourth industrial revolution and is characterized by the integration of digital technologies into industrial processes [12]. This concept includes the use of the Internet of Things (IoT), artificial intelligence (AI), and big data analytics. In Industry 4.0, machines communicate and make decisions independently, leading to what is often referred to as the "smart factory" [13]. This principle results in enhanced efficiency and productivity and reduced downtime by the application of predictive maintenance supported on data analytics, machine learning, and the IoT [14]. Automation and Industry 4.0 have allowed for more efficient production, quicker decision making, and improved resource allocation. With real-time data analysis and optimization, companies can minimize waste, reduce energy consumption, and increase the quality of their products [15].

Robotics goes beyond automation by introducing physical machines that can perform tasks with a high degree of autonomy. These machines are equipped with sensor and actuator systems that enable them to interact with their environment [16]. Robotics plays a significant role in flexible production, particularly when tasks require precision and adaptability [17]. Robotics has become increasingly prevalent across diverse industrial applications [18,19]. While various definitions of robots exist, the ISO 8373 standard characterizes a robot as a reprogrammable and multifunctional manipulator, controlled in position, with one or multiple degrees of freedom, capable of manipulating objects using programmed movements to execute various functions [20]. Industrial robots typically comprise three key components: the manipulator (robot), a controller, and a user interface (programming console). The robot is equipped with sensors and actuators, which are the senses and muscles of robotic systems. Sensors provide data about the robot's surroundings, including information about temperature, humidity, light, and object detection [21]. Actuators are responsible for converting digital instructions into physical movement. Advances in sensor technology, such as Light Detection and Ranging (LiDAR), cameras, and ultrasonic sensors, have improved robots' ability to navigate and interact with their environment [22,23]. Similarly, actuator advancements, like advanced servo motors, enable robots to perform tasks with greater accuracy and agility [24]. The integration of sensor and actuator technologies has had a profound impact on the field of robotics. Robots are now capable of performing complex tasks like pick-and-place operations, assembly, and even intricate surgical procedures [25]. They can work alongside humans in collaborative settings, which is particularly beneficial in manufacturing environments [26]. Presently, industrial robots play a pivotal role in automotive production lines, with the application of robots in this industry gaining substantial traction in recent years. This move toward robotization facilitates the assembly of different vehicles on a shared production line, resulting in reduced production costs for small- and medium-scale operations compared to dedicated automation or manual labor-based assembly lines [16,27]. The primary drivers for integrating robots in industries are the need to operate in hazardous environments, the execution of repetitive tasks, the management of intricate handling processes, and the maintenance of continuous operation [28,29]. The current state of the art in robot development, which includes control systems and sensor technologies, ensures the safe utilization of these systems in production and assembly lines [30,31], and this safety extends to collaborative operations, combining the productivity attributes of robots with the improved cognitive and decision-making skills of human operators, thereby enhancing overall manufacturing and assembly efficiency [32].

Currently, the applications of automation and robotics are widespread in industry and society in general. The most prominent applications are as follows:

- Manufacturing: One of the primary applications of automation and robotics is in manufacturing. Automated assembly lines have become the main assurance of companies' competitiveness [33,34], producing a wide range of products, from consumer electronics to automobiles. Robots can handle repetitive and hazardous tasks with precision and consistency. Their application ensures that defects are minimized, resulting in higher-quality products [35].
- Healthcare: In the healthcare sector, robotics has seen major advances. Robotic surgery, for instance, has become more common, allowing for minimally invasive procedures to be performed with high precision [36]. Robots can also assist in patient care, such as in the delivery of medications or in the rehabilitation of patients [37].
- Logistics and warehousing: E-commerce and the demand for rapid order fulfillment have led to the adoption of robotics in logistics and warehousing. Automated guided vehicles (AGVs) and drones are used for material handling and order picking. This procedure speeds up the process and reduces the risk of errors in inventory management [38].
- Agriculture Robots are used for tasks like planting, harvesting, and monitoring crops. These machines can work uninterruptedly, improving the efficiency of farming operations [39]. The integration of automation and robotics in agriculture is essential to meet the growing global food demand [40].
- Service and entertainment: Robotic technology has also found its way into the service and entertainment industries. Robots are used as receptionists, guides in museums, and even as companions for the elderly [41]. Entertainment robots, like those used

in theme parks, enhance visitor experiences and provide a unique form of entertainment [42].

Despite the major advances and breakthroughs in automation and robotics, which has led to the most diverse applications, as described, limitations persist related to these technologies that need to be addressed, such as:

- High initial investment: The initial cost of implementing automation and robotics systems can be substantial. Small- and medium-sized enterprises (SMEs) may find it challenging to invest in this technology, hindering its widespread adoption [43].
- Complexity and integration: Integrating automation and robotics into existing systems can be complex. It requires a deep understanding of the specific needs of the industry and often involves custom solutions. This complexity can be a barrier for many businesses [44].
- Workforce disruption: The fear of job displacement remains a concern. While automation and robotics can improve efficiency and productivity, they may also lead to job displacement. It is crucial to manage this transition by upskilling the workforce and focusing on roles that complement automation rather than firing the line operators that previously accomplished the repetitive tasks [11].
- Safety: Ensuring the safety of workers and humans when robots operate in shared spaces is of utmost importance. Safety standards and risk assessment procedures must be in place to prevent accidents and injuries [45].
- Lack of standardization: The lack of standardized interfaces and communication protocols can hinder the interoperability of different automation and robotics systems [46]. Standardization efforts are ongoing, but more progress is needed to achieve seamless integration [47].

As technology and scientific knowledge continue to evolve, the future of automation and robotics holds promising opportunities and prospects for future research:

- Human–robot collaboration: Collaborative robots, or "cobots," are becoming increasingly applied on the factory floor. These robots work alongside humans, enhancing productivity in complex tasks [48]. Future developments in this area will focus on improving the ease of programming and the flexibility of these systems [49].
- AI and machine learning: Advancements in AI and machine learning will lead to more intelligent and adaptable robots that will be capable of learning from their experiences and continuously improving their performance [50].
- Interconnected systems: The integration of robotics and automation with Industry 4.0 principles will lead to more interconnected systems [51], leading to higher efficiency and productivity, reduced downtime, as well as improved resource allocation [52].
- Accessibility: Efforts are being made to reduce the cost and complexity of adopting automation and robotics. As a result, the technology will be more accessible to a broader range of industries, including SMEs [53].
- Sustainability: The concept of sustainability will be a key focus in the future. Robots and automated systems can play a crucial role in reducing waste and energy consumption [54]. Sustainable practices will become an integral part of automation and robotics design [55].

In conclusion, automation and robotics have significantly impacted industrial processes by enhancing efficiency, precision, and flexibility. The integration of these technologies into manufacturing, healthcare, logistics, agriculture, and other sectors has brought about numerous benefits. However, challenges such as high initial costs, complexity, workforce disruption, safety concerns, and lack of standardization still prevent a more widespread use of these technologies. Nonetheless, the future of automation and robotics is promising, with major research and improvement areas being identified. It is the aim of this Special Issue to document the main developments in this field and bring new prospects for further development in automation and robotics.

Conflicts of Interest: The authors declare no conflict of interest.

References

1. Dzedzickis, A.; Subačiūtė-Žemaitienė, J.; Šutinys, E.; Samukaitė-Bubnienė, U.; Bučinskas, V. Advanced applications of industrial robotics: New trends and possibilities. *Appl. Sci.* **2022**, *12*, 135. [CrossRef]
2. Saxena, A.; Chaturvedi, R. Advancement of industrial automation in integration with robotics. In Proceedings of the 2021 International Conference on Simulation, Automation & Smart Manufacturing (SASM), Mathura, India, 20–21 August 2021; pp. 1–6.
3. Javaid, M.; Haleem, A.; Singh, R.P.; Suman, R. Substantial capabilities of robotics in enhancing industry 4.0 implementation. *Cogn. Robot.* **2021**, *1*, 58–75. [CrossRef]
4. Liagkou, V.; Stylios, C.; Pappa, L.; Petunin, A. Challenges and opportunities in industry 4.0 for mechatronics, artificial intelligence and cybernetics. *Electronics* **2021**, *10*, 2001. [CrossRef]
5. Gupta, A.; Arora, S.K. *Industrial Automation and Robotics*; Laxmi Publications: New Delhi, India, 2009.
6. Santos, P.M.M.; Campilho, R.D.S.G.; Silva, F.J.G. A new concept of full-automated equipment for the manufacture of shirt collars and cuffs. *Robot. Comput. Integr. Manuf.* **2021**, *67*, 102023. [CrossRef]
7. Sousa, V.F.C.; Silva, F.J.G.d.; Campilho, R.D.S.G.; Pinto, A.G.; Ferreira, L.P.; Martins, N. Developing a novel fully automated concept to produce bowden cables for the automotive industry. *Machines* **2022**, *10*, 290. [CrossRef]
8. Santos, R.F.L.; Silva, F.J.G.; Gouveia, R.M.; Campilho, R.D.S.G.; Pereira, M.T.; Ferreira, L.P. The improvement of an APEX machine involved in the tire manufacturing process. *Procedia Manuf.* **2018**, *17*, 571–578. [CrossRef]
9. Nunes, D.M.; Campilho, R.; Silva, F.J.G. Design of a transfer system for the automotive industry. *Proc. Inst. Mech. Eng. Part E J. Process Mech. Eng.* **2022**, *236*, 2044–2055. [CrossRef]
10. Groover, M.P. *Automation, Production Systems, and Computer-Integrated Manufacturing*; Prentice Hall: Hoboken, NJ, USA, 2007.
11. Araújo, W.F.S.; Silva, F.J.G.; Campilho, R.D.S.G.; Matos, J.A. Manufacturing cushions and suspension mats for vehicle seats: A novel cell concept. *Int. J. Adv. Manuf. Technol.* **2017**, *90*, 1539–1545. [CrossRef]
12. Culot, G.; Nassimbeni, G.; Orzes, G.; Sartor, M. Behind the definition of Industry 4.0: Analysis and open questions. *Int. J. Prod. Econ.* **2020**, *226*, 107617. [CrossRef]
13. Osterrieder, P.; Budde, L.; Friedli, T. The smart factory as a key construct of industry 4.0: A systematic literature review. *Int. J. Prod. Econ.* **2020**, *221*, 107476. [CrossRef]
14. Pech, M.; Vrchota, J.; Bednář, J. Predictive maintenance and intelligent sensors in smart factory: Review. *Sensors* **2021**, *21*, 1470. [CrossRef]
15. Jagtap, S.; Garcia-Garcia, G.; Rahimifard, S. Optimisation of the resource efficiency of food manufacturing via the Internet of Things. *Comput. Ind.* **2021**, *127*, 103397. [CrossRef]
16. Ogenyi, U.E.; Liu, J.; Yang, C.; Ju, Z.; Liu, H. Physical human–robot collaboration: Robotic systems, learning methods, collaborative strategies, sensors, and actuators. *IEEE Trans. Cybern.* **2021**, *51*, 1888–1901. [CrossRef] [PubMed]
17. Michalos, G.; Makris, S.; Spiliotopoulos, J.; Misios, I.; Tsarouchi, P.; Chryssolouris, G. ROBO-PARTNER: Seamless human-robot cooperation for intelligent, flexible and safe operations in the assembly factories of the future. *Procedia CIRP* **2014**, *23*, 71–76. [CrossRef]
18. Neythalath, N.; Søndergaard, A.; Bærentzen, J.A. Adaptive robotic manufacturing using higher order knowledge systems. *Autom. Constr.* **2021**, *127*, 103702. [CrossRef]
19. Francesco, P.; Paolo, G.G. AURA: An example of collaborative robot for automotive and general industry applications. *Procedia Manuf.* **2017**, *11*, 338–345. [CrossRef]
20. Virk, G.S.; Moon, S.; Gelin, R. ISO standards for service robots. In *Advances in Mobile Robotics*; World Scientific: Toh Tuck, Singapure, 2008; Volume 8, pp. 133–138.
21. Seung-Ho, B.; Jae-Han, P.; Jaehan, K.; Kyung-Wook, P.; Moon-Hong, B. Building a smart home environment for service robots based on RFID and sensor networks. In Proceedings of the 2007 International Conference on Control, Automation and Systems, Seoul, Republic of Korea, 17–20 October 2007; pp. 1078–1082.
22. Alatise, M.B.; Hancke, G.P. A review on challenges of autonomous mobile robot and sensor fusion methods. *IEEE Access* **2020**, *8*, 39830–39846. [CrossRef]
23. Tran, T.Q.; Becker, A.; Grzechca, D. Environment mapping using sensor fusion of 2D laser scanner and 3D ultrasonic sensor for a real mobile robot. *Sensors* **2021**, *21*, 3184. [CrossRef]
24. Suwoyo, H.; Hidayat, T.; Jia-nan, F. A transformable wheel-legged mobile robot. *Int. J. Eng. Contin.* **2022**, *2*, 27–39. [CrossRef]
25. Moghaddam, M.; Nof, S.Y. Parallelism of pick-and-place operations by multi-gripper robotic arms. *Robot. Comput. Integr. Manuf.* **2016**, *42*, 135–146. [CrossRef]
26. Conti, C.J.; Varde, A.S.; Wang, W. Human-robot collaboration with commonsense reasoning in smart manufacturing contexts. *IEEE Trans. Autom. Sci. Eng.* **2022**, *19*, 1784–1797. [CrossRef]
27. Villani, V.; Pini, F.; Leali, F.; Secchi, C. Survey on human–robot collaboration in industrial settings: Safety, intuitive interfaces and applications. *Mechatronics* **2018**, *55*, 248–266. [CrossRef]
28. Ore, F.; Vemula, B.R.; Hanson, L.; Wiktorsson, M. Human—Industrial robot collaboration: Application of simulation software for workstation optimisation. *Procedia CIRP* **2016**, *44*, 181–186. [CrossRef]

29. Pedrocchi, N.; Vicentini, F.; Matteo, M.; Tosatti, L.M. Safe human-robot cooperation in an industrial environment. *Int. J. Adv. Robot. Syst.* **2013**, *10*, 27. [CrossRef]
30. Siciliano, B.; Khatib, O. *Springer Handbook of Robotics*; Springer: Berlin/Heidelberg, Germany, 2016.
31. Brogårdh, T. Present and future robot control development—An industrial perspective. *Annu. Rev. Control* **2007**, *31*, 69–79. [CrossRef]
32. Gopinath, V.; Ore, F.; Johansen, K. Safe assembly cell layout through risk assessment—An application with hand guided industrial robot. *Procedia CIRP* **2017**, *63*, 430–435. [CrossRef]
33. Costa, R.J.S.; Silva, F.J.G.; Campilho, R.D.S.G. A novel concept of agile assembly machine for sets applied in the automotive industry. *Int. J. Adv. Manuf. Technol.* **2017**, *91*, 4043–4054. [CrossRef]
34. Barbosa, A.F.G.; Campilho, R.D.S.G.; Silva, F.J.G.; Sánchez-Arce, I.J.; Prakash, C.; Buddhi, D. Design of a spiral double-cutting machine for an automotive bowden cable assembly line. *Machines* **2022**, *10*, 811. [CrossRef]
35. Pereira, J.A.P.; Campilho, R.D.S.G.; Silva, F.J.G.; Sánchez-Arce, I.J. Robotized cell design for part assembly in the automotive industry. *Proc. Inst. Mech. Eng. Part C* **2022**, *236*, 8807–8822. [CrossRef]
36. Peters, B.S.; Armijo, P.R.; Krause, C.; Choudhury, S.A.; Oleynikov, D. Review of emerging surgical robotic technology. *Surg. Endosc.* **2018**, *32*, 1636–1655. [CrossRef]
37. Kyrarini, M.; Lygerakis, F.; Rajavenkatanarayanan, A.; Sevastopoulos, C.; Nambiappan, H.R.; Chaitanya, K.K.; Babu, A.R.; Mathew, J.; Makedon, F. A survey of robots in healthcare. *Technologies* **2021**, *9*, 8. [CrossRef]
38. Boysen, N.; de Koster, R.; Weidinger, F. Warehousing in the e-commerce era: A survey. *Eur. J. Oper. Res.* **2019**, *277*, 396–411. [CrossRef]
39. Aravind, K.R.; Raja, P.; Pérez-Ruiz, M. Task-based agricultural mobile robots in arable farming: A review. *Span. J. Agric. Res.* **2017**, *15*, e02R01. [CrossRef]
40. Grieve, B.D.; Duckett, T.; Collison, M.; Boyd, L.; West, J.; Yin, H.; Arvin, F.; Pearson, S. The challenges posed by global broadacre crops in delivering smart agri-robotic solutions: A fundamental rethink is required. *Glob. Food Secur.* **2019**, *23*, 116–124. [CrossRef]
41. Sharkey, A.; Wood, N.; Aminuddin, R. Robot companions for children and older people. In *Designing Robots, Designing Humans*; Hasse, C., Sondergaard, D.M., Eds.; Routledge: Abingdon, UK, 2019.
42. Milman, A.; Tasci, A.; Zhang, T. Perceived robotic server qualities and functions explaining customer loyalty in the theme park context. *Int. J. Contemp. Hosp. Manag.* **2020**, *32*, 3895–3923. [CrossRef]
43. Barton, M.; Budjac, R.; Tanuska, P.; Gaspar, G.; Schreiber, P. Identification overview of industry 4.0 essential attributes and resource-limited embedded artificial-intelligence-of-things devices for small and medium-sized enterprises. *Appl. Sci.* **2022**, *12*, 5672. [CrossRef]
44. Syed, R.; Suriadi, S.; Adams, M.; Bandara, W.; Leemans, S.J.J.; Ouyang, C.; ter Hofstede, A.H.M.; van de Weerd, I.; Wynn, M.T.; Reijers, H.A. Robotic process automation: Contemporary themes and challenges. *Comput. Ind.* **2020**, *115*, 103162. [CrossRef]
45. Nikolakis, N.; Maratos, V.; Makris, S. A cyber physical system (CPS) approach for safe human-robot collaboration in a shared workplace. *Robot. Comput. Integr. Manuf.* **2019**, *56*, 233–243. [CrossRef]
46. Hazra, A.; Adhikari, M.; Amgoth, T.; Srirama, S.N. A comprehensive survey on interoperability for iiot: Taxonomy, standards, and future directions. *ACM Comput. Surv.* **2021**, *55*, 1–35. [CrossRef]
47. Wickel, N.; Vossel, M.; Yilmaz, O.; Rademacher, K.; Janß, A. Integration of a surgical robotic arm to the connected operating room via ISO IEEE 11073 SDC. *Int. J. Comput. Assist. Radiol. Surg.* **2023**, *18*, 1639–1648. [CrossRef]
48. Simões, A.C.; Pinto, A.; Santos, J.; Pinheiro, S.; Romero, D. Designing human-robot collaboration (HRC) workspaces in industrial settings: A systematic literature review. *J. Manuf. Syst.* **2022**, *62*, 28–43. [CrossRef]
49. Liu, L.; Guo, F.; Zou, Z.; Duffy, V.G. Application, development and future opportunities of collaborative robots (cobots) in manufacturing: A literature review. *Int. J. Hum.-Comput. Interact.* 2022; *in press*.
50. Mohebbi, A. Human-robot interaction in rehabilitation and assistance: A review. *Curr. Robot. Rep.* **2020**, *1*, 131–144. [CrossRef]
51. Goel, R.; Gupta, P. Robotics and Industry 4.0. In *A Roadmap to Industry 4.0: Smart Production, Sharp Business and Sustainable Development*; Nayyar, A., Kumar, A., Eds.; Springer International Publishing: Berlin/Heidelberg, Germany, 2020; pp. 157–169.
52. Jamwal, A.; Agrawal, R.; Sharma, M.; Giallanza, A. Industry 4.0 technologies for manufacturing sustainability: A systematic review and future research directions. *Appl. Sci.* **2021**, *11*, 5725. [CrossRef]
53. Ingaldi, M.; Ulewicz, R. Problems with the Implementation of Industry 4.0 in Enterprises from the SME Sector. *Sustainability* **2020**, *12*, 217. [CrossRef]
54. Sarc, R.; Curtis, A.; Kandlbauer, L.; Khodier, K.; Lorber, K.E.; Pomberger, R. Digitalisation and intelligent robotics in value chain of circular economy oriented waste management—A review. *Waste Manag.* **2019**, *95*, 476–492. [CrossRef]
55. Almusaed, A.; Yitmen, I.; Almssad, A. Reviewing and integrating aec practices into industry 6.0: Strategies for smart and sustainable future-built environments. *Sustainability* **2023**, *15*, 13464. [CrossRef]

machines

MDPI

Article

Design of a Spiral Double-Cutting Machine for an Automotive Bowden Cable Assembly Line

André F. G. Barbosa [1], Raul D. S. G. Campilho [1,2,*], Francisco J. G. Silva [1,2], Isidro J. Sánchez-Arce [2], Chander Prakash [3] and Dharam Buddhi [4]

1   Departamento de Engenharia Mecânica, Instituto Superior de Engenharia do Porto,
    Instituto Politécnico do Porto, R. Dr. António Bernardino de Almeida, 431, 4200-072 Porto, Portugal
2   INEGI-Pólo FEUP, Rua Dr. Roberto Frias, 400, 4200-465 Porto, Portugal
3   School of Mechanical Engineering, Lovely Professional University, Phagwara 144001, India
4   Division of Research & Innovation, Uttaranchal University, Dehradun 248007, India
*   Correspondence: raulcampilho@gmail.com; Tel.: +351-939526892; Fax: +351-228321159

Abstract: The manufacture of automotive components requires innovative technologies and equipment. Due to the competitiveness in the sector, the implementation of automatic and robotic equipment has been vital in its development to produce the largest number of products in the shortest amount of time. Automation leads to a significant reduction in defects and enables mass production and standardization of the final product. This work was based on the need of an automotive components' company to increase the rate of spiral cable cutting, used as protection for Bowden (control) cables. Currently, this component, used in automotive systems, is processed with simple cutting machines and cleaning machines. Based on the design science research (DSR) methodology, this work aims to develop a machine capable of performing the cutting and cleaning of two spiral cables simultaneously and automatically. The development of this machine was based on existing machines, and the biggest challenge was the implementation of a double-cutting system. The designed machine met the initial requirements, such as enabling the simultaneous cut of two spirals, being fully automatic, doubling the output over the current solution, and fully complying with the current legislation.

Keywords: automotive component industry; spiral cable; automation; cutting system; productivity

Citation: Barbosa, A.F.G.; Campilho, R.D.S.G.; Silva, F.J.G.; Sánchez-Arce, I.J.; Prakash, C.; Buddhi, D. Design of a Spiral Double-Cutting Machine for an Automotive Bowden Cable Assembly Line. *Machines* **2022**, *10*, 811. https://doi.org/10.3390/machines10090811

Academic Editor: Fugui Xie

Received: 2 August 2022
Accepted: 14 September 2022
Published: 15 September 2022

1. Introduction

Besides its economic importance, the automotive industry also contributes to technological innovation [1]. The need to remote control different functions within a vehicle is often satisfied using control cables. These functions vary from safety to comfort equipment. Applications of control cables are found in many fields of engineering such as robotics [2], and the automotive industry represents one of the most common and generalized applications [3]. Control cables, also known as Bowden cables, transmit linear motion and power by the relative displacement of cable or wire inside a spiral. Both of these components are flexible, which allows for several applications [4]. In addition, the efficiency of control cables is around 97% [5]. All these factors contribute to their large applicability. Control cables are composed of different components, apart from the inner cable and the spiral. Actually, depending on the application, the control cables may require special features such as grommets, low-friction coatings, anti-corrosive coatings, purpose-made cable ends [6], and supports or end connectors [7]. They are commonly employed to connect the control levers to latches such as those in engine and luggage compartments and door mechanisms. Some of their essential applications are for controlling the accelerator, clutch, and parking brakes through pedals and levers. In addition, they can be applied to comfort devices such as external mirrors and seats. Control cables should also have adequate strength to perform the required tasks and may be lubricated to reduce friction [8]. Since control cables are used to control critical components of vehicles, some standards regarding their materials, testing procedures, and production techniques have been published. For example, ISO standard 2408:2017 [9] specifies requirements for producing, testing, and marking steel cables, which are used in control cables. The inner cable, which is the one transferring motion and force, is made of braided steel wires forming a wire rope [4], with characteristics and types that are provided by Oberg et al. [10]. The cable sheath, also known as the cable housing, is made of a combination of materials, with emphasis on the steel spiral, which is arranged in a way that is strong in compression [4]. The spiral is often covered with plastic to protect it and its inners from the environment. In its interior, the spiral may have an antifriction coating or an inner tube to minimize the friction between the cable and spiral, which also prevents damage to the inner cable due to abrasion [11]. The spiral is a structural part of the control cable [4], hence, it is strong and difficult to cut. Furthermore, vehicle models even from the same manufacturer present different configurations of control cables, making it more difficult to automate the process. Consequently, the production and assembly of control cables often rely on several manual tasks, although some operations can be eased through jigs and fixtures [12].

The production of control cables was addressed as a case study by Moreira et al. [13]. The automated production of two different models of control cable was one of the requirements. The production of these cables was separated into three stages: preparation, injection, and assembly. The requirements were fulfilled by using integrated fully automated processes within a manufacturing cell [14,15]. Moreover, an integrated tooling system was implemented, allowing the production of the two desired models and leaving room for future expansions. The tooling had a computerized identification system, ensuring that the right tooling for the production run is installed in the cell. The concept was successfully implemented. Depending on the application, the inner cable may be subject to other operations prior to its assembly into the control cable set. An automated manufacturing cell was developed by Martins et al. [16], in which the inner cable was prepared for the zamak injection of its terminals and subsequently injected. In addition, the cable was cut-to-length and strength-tested. The manufacturing cell allowed for the production of three different models of control cable, each with different features. Some applications of control cables require lubrication, which is accomplished during the assembly process. In this regard, Ribeiro et al. [17] designed and implemented a system to lubricate control cables. The developed system allowed to maximize the use of the delivered grease regardless of the container type. Furthermore, a system of grease injectors, as well as the accompanying logistics, were developed and implemented. This also included an air detection system to minimize the mixture of grease-air, which negatively affects the lubricant properties. The proposed system led to efficient use of the grease supply, improved the lubrication of cables, and favored ergonomics and logistics on the factory floor. Subsequently, Vieira et al. [7] developed and implemented an automatic process to inject plastic ends into the spiral. The process consisted of the design of a production cell, where the plastic cap ends are injected. The cell was self-contained, allowing for its introduction within a production line of control cables. Although the work reported by Moreira et al. [13] is related to control cables, its focus was on the integration within a manufacturing cell, and, thus, the individual production steps were not fully described or addressed. Similarly, Martins et al. [16] described the overall process and the important steps, but the detail of the cutting process was omitted.

The cutting, deburring, and cleaning processes for the spiral, although mentioned in the literature, are not described for this particular component due to its specific nature. Nevertheless, the spiral is a metallic component, and, hence, cutting processes suitable for metal pipes and stock are applicable [11]. For example, Li et al. [18] performed a review of the state-of-the-art cutting processes and the effect of tool geometry on surface finish, wear, and heating. Nonetheless, the large number of variables involved in this type of process indicates the need for extensive testing for the process of interest due to operational concerns [19]. For example, the plastic coating on the spirals may require special attention during the cutting process to prevent heat damage [20] or catching fire.

This combination of materials, metal, and plastic is suitable to be cut using mechanical or abrasive means [21], providing that cooling is used. In this regard, cooling has to be environmentally and product-friendly. Cooling and lubricating during machining and cutting have different alternatives, which also depend on the type of material being processed. Nevertheless, a current trend is to minimize the amount of coolant used by improving the tool geometry and cutting parameters through simulations and experimentation, as reviewed by Mohamed Akeel et al. [22]. In addition, the available tests in the literature mostly pertain to flat substrates, whilst hollow substrates may present a challenge. Spirals are hollow and relatively thin, so cutting may be difficult, since thin shapes may deform due to cutting forces [18,23,24], especially for shear cutting (with guillotines), which would not be applicable for spiral processing. Resulting from a survey of industrial equipment, the most used method to cut this component is abrasive disk cutting, which typically presents good efficiency and speed that, together with its reduced cost, make this one of the most used methods in the cable industry. Abrasive grains of aluminum oxide, silicon carbide, or zirconium are recommended [25]. However, due to the friction that is generated between the disk and the part, the heat produced in this zone grows exponentially, rapidly expanding to the entire system. Therefore, it is necessary to study this situation in advance, in order to avoid damaging the system. Diamond disk cutting can be an alternative that greatly reduces pollution generation and improve performance with a respective increase in service life, but with a higher cost [26]. Other cutting techniques such as abrasive waterjet [27] or laser cutting [28] have been explored for metallic and composite materials with success, although the parameters must be optimized for each application. Abrasive waterjet cutting in particular involves a good edge sharpness and low initial investment and avoids the heat-affected zone [29], but for spiral cutting can promote the water ingress to the spiral. Laser cutting leads to reduced contamination of the workpiece, good dimensional accuracy, and negligible warping [30]. However, heat generation can cause a problem for plastic components. Plasma cutting seems less suitable, since it is less precise and uses more energy than laser cutting, although it cuts bigger thicknesses [31]. Despite the several processes available to perform the task, their applicability is case-dependent, being necessary to model and experiment with a given material and equipment to maximize performance and the product's quality.

As previously described, amongst the several components of control cables, automated production of spirals ready for assembly has not been studied. This work was based on the need of an automotive components' company to increase the rate of spiral cable cutting, used as protection for Bowden (control) cables. Currently, this component, used in automotive systems, is processed in simple cutting machines and cleaning machines. This work aims to develop a machine capable of performing cutting and cleaning of two spiral cables simultaneously and automatically. The development of this machine was based on existing machines, and the biggest challenge was the implementation of a double-cutting system.

2. Methods

2.1. Selected Methodology

The automatic machine developed in this work for spiral double-cutting followed the design science research (DSR) methodology. This approach is particularly suited for the design and improvement of existing processes, as in the current work, and it involves the detailed study of the existing processes and proposal of new solutions arising from the inputs of different participants in the process, leading to an improvement process that can be used in different disciplines [3]. The DSR methodology is step forward compared to traditional design, in which the interested parties (companies) define the objectives and requirements in a short-term perspective, limited by the available knowledge and productions resources. Some limitations can also be pointed out, such as the requirement for a detailed analysis of the initial process, possible need for multiple iterations, and added time to the project [32].

The DSR methodology can be described by six stages [33]:

- 1—Identification of the problem—Analysis of the current machine, description of limitations and improvement points;
- 2—Objective definition—Establishment of the objectives and requirements to attain the desired result;
- 3—Design and development—Choice of the news concepts to implement, based on the requirements;
- 4—Solution demonstration—Implementation of the concepts defined in the previous stage;
- 5—Solution evaluation—Performance analysis of the solution and verification against the initial objectives/requirements;
- 6—Conclusions—Comparison between the initial and new solutions.

In view of this procedure, Section 2.2 (control cable and initial machine) and Section 2.3 (problem characterization) relate to stage 1 (identification of the problem). Stage 2 (objective definition is accomplished in Section 2.4 (objectives and requirements). The pre-design Section 3.1 of Section 3 partially fulfills stage 3 (design and development), by choosing the cutting concept to implement. It should be emphasized that a similar procedure was followed for the other machine constituents, although in this paper only the main issue to address was explicitly shown. Stages 4 (solution demonstration) and 5 (solution evaluation) are diluted in Section 3.2 (final solution) and Section 3.3 (design process). In these stages, and due to the limitation that the physical construction of the new machine was not possible to accomplish before the idealized solution was made ready for the company, implementation and evaluation of the concepts were carried out virtually using software by the relevant design team and company personnel. Finally, stage 6 (conclusions) is the core of Section 4 (discussion and conclusions), where the initial and new machines are compared, and the improvements are assessed.

2.2. Control Cable and Initial Machine

The automotive components' company produces control cables, many of which are for automotive use. Figure 1 shows an example of control cable. As mentioned before, control cables are composed of different parts depending on the application, with the inner cable and the spiral as their most important constituents.

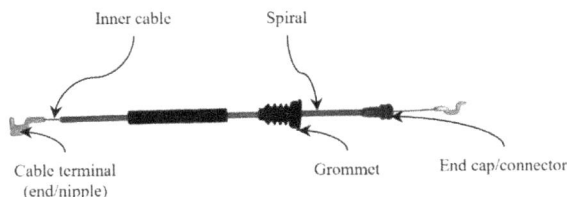

Figure 1. Control cable and its main parts and accessories.

The main materials of the control cable are stainless steel for the inner cable, Zamak 5 alloy (zinc die casting alloy) for the cable terminals, polymers for the outer accessories (including an elastomer for the grommet), and the spiral materials, which are particularly relevant for this work, since these will be subjected to the cutting operation.

- Spiral body-spring steel (medium-carbon, with 0.5–1% C, and main alloy elements 0.15–0.35% Si, 0.6–0.9% Mn, 0.4% Cr, 0.1% Mo, and 0.4% Ni), providing a high yield limit to assure flexibility in the elastic regime. The spiral was heat treated to provide an elastic limit of 650 MPa and tensile strength of 900 MPa.
- Outer coating-used for spiral outer protection and composed of polypropylene applied in the spiral body by wire-coating extrusion, after the spiral reaches room temperature. This polymer is a thermoplastic obtained by chain-growth polymerization from the monomer propylene, with good heat and chemical resistance. The Young's modulus between 1.3 and 1.8 GPa provides the required flexibility without breaking.

- Inner tube-inserted inside the spiral to guide the inner cable during operation with low resistance to motion, made of PA66 (polyamide 66) to provide excellent abrasion resistance and low frictional properties, while retaining acceptable strength and toughness and being resistant to oils, greases, and fuels. The approximate Young's modulus of 3.5 GPa ensures the desired flexibility for this application.

An example of the current machinery is shown in Figure 2. These machines use an abrasive disk to cut the spirals, which are fed by a set of rollers. The cut is performed with a silicon carbide (SiC) disk cooled by using compressed air.

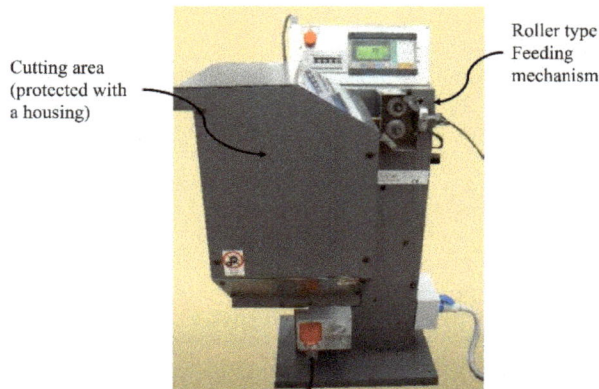

Figure 2. Example of cutting machinery already in use in the facilities.

The current cutting mechanism cuts a spiral at a time. Therefore, this mechanism was used as a starting point for all the improvements. The cutting mechanism possesses a motorized roller type feeder (Figure 3a). In this type of feeder, the raw spiral is compressed between the rollers with enough pressure that the friction between them pushes the spiral into the machine without deforming it. The feeder system pushes the spiral into the cutting area where it is cut using an automatically actuated abrasive disk (Figure 3b). The cutting motion is driven by a servomotor (Figure 4), whilst the cutting disk's wear is compensated by means of sensors, as shown in Figure 3b. The cutting disk is cooled by compressed air directed to it through a nozzle system (Figure 4). Although abrasive disks efficiently cut this type of material, subsequent preparation and cleaning are required due to the burrs left on the surface.

Currently, metal chips and dust are projected during the process, and most of them lie on the floor and machine, despite all the attempts to mitigate this problem. Once the spirals are cut, they are transported to another station where the burrs are removed, and the cuts cleaned for further operations down the line. However, the transport between the cutting station to the cleaning station is done by hand. On the other hand, the cleaning station, as shown in Figure 5, receives the spirals through two conveyor belts, which have an adjustable span to allow for different spiral lengths. The conveyor belt system moves the spirals to an alignment device, ensuring that all the spirals are aligned with the machine. Then, each spiral passes through five stations: (1) reaming, (2) grinding, (3) flaring, (4) heating, and (5) blowing. This process results in cable spirals with equal length and up to standard. Regarding the cleaning station, the conveyor belts have magnetic attachments, which allow for holding the spirals, though they also attract metallic debris. The conveyor belt drive is powered by a servomotor, allowing the precise positioning of each spiral through all the stations.

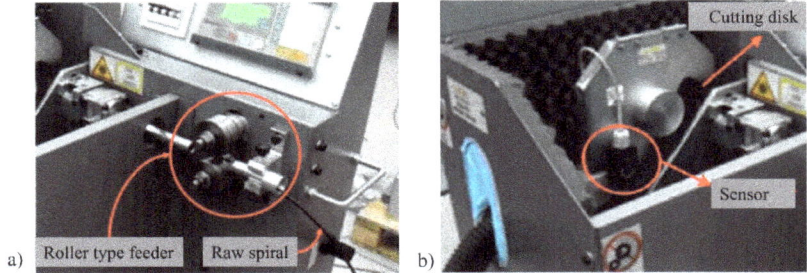

Figure 3. Details of the cutting machine: (**a**) roller type feeder and (**b**) sensor to control disk wear.

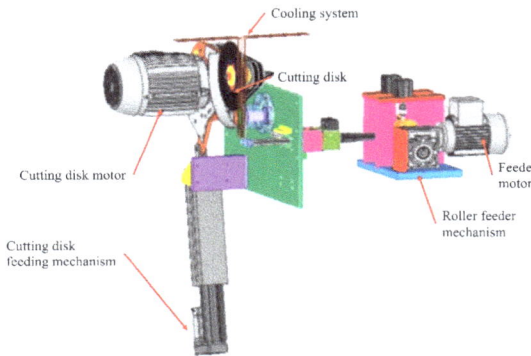

Figure 4. Scheme of the current cutting system.

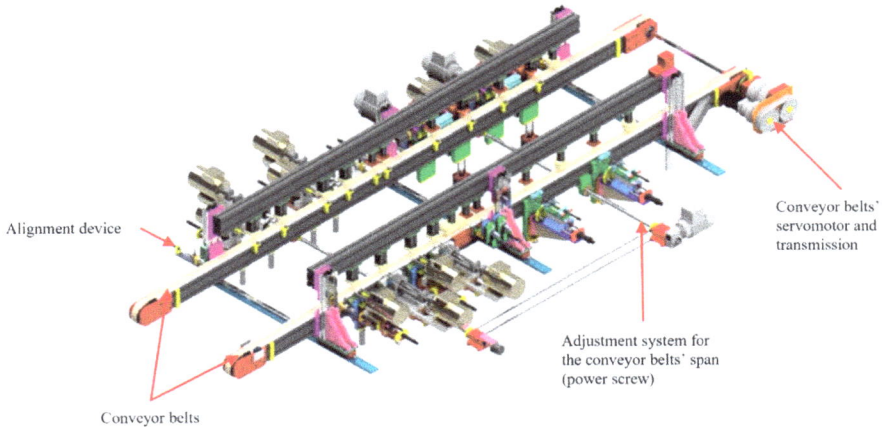

Figure 5. Scheme of the current cleaning/post-processing system.

In summary, the weakest point of the current cutting method is the quality of the final cuts, which require further processes to remove, and associated amount of dust and metallic debris that the process leaves on the machine and spirals. Then, the need for manual transfer between stations limits the production rates. In consequence, there is room for improvement using the current machine as a starting point, which would also increase the productivity of the machinery.

2.3. Problem Characterization

Although the company already has infrastructures to produce the cables, there is the need to optimize the process because, in the current form, the machinery cuts one cable spiral at the time. Once the cable spirals are cut, they are transported to another station where their ends are deburred and cleaned (post-processed). However, the transport and post-processing are still manual tasks. Using single-spiral-cutting machines, given the number of manual tasks, results in low production rates. Although the current process is viable, the results of the cutting process are far from repeatable, affecting the quality. Therefore, it is necessary to improve the process and ensure the final quality of the spirals. In consequence, it is proposed to develop and implement the processes and machinery to allow cutting and post-processing of two spirals simultaneously, and then arrange all of them into a manufacturing cell following safety and ergonomic standards.

2.4. Objectives and Requirements

The main objective is to develop a cutting system able to cut, clean, and deburr two spirals at a time. In addition, the idealized solution should integrate with the current production lines at a minimum cost, while reducing the allocated human resources, assuring the required quality level by the clients, and increasing the production output. As a result of this work, a project should be proposed that can be manufactured, assembled, and used in an industrial environment with significant benefit over the current manually operated single-cutting machine. The raw spirals are supplied wound on spools, and, thus, the system should account for this feature. Then, the raw spirals should be automatically fed into the cutting system. On the other hand, the variety of models produced by the company leads to different cable lengths, varying from 200 mm and 600 mm, so the proposed system should take this into account, expecting an automatic or semi-automatic measurement system. The cutting would proceed in batches of a certain length. Therefore, the length would be adjusted on every batch to be produced. Finally, the system should deburr and clean the inside and outside diameters of the cut spirals as automatically as possible. In addition, the proposed system should take advantage of current equipment and facilities, and the respective implementation should cause minimum disturbances to the production and be affordable.

This project must always consider the fulfilment of the initially imposed requirements. Only compliance with these will guarantee the quality of the products, the necessary production rate, and the proper functioning of the industrial environment in which it will be inserted. In summary, the system's requirements are as follows.

- Automatic spiral feeding system. The raw spiral is wound on a spool; hence being part of the process.
- Cut two spirals simultaneously.
- Automatic or semi-automatic (at minimum) cut to length system. The length ranges from 200 mm to 600 mm. The machine should enable to set the length first and then cut all the batches to length.
- Cutting quality must be guaranteed to facilitate the cleaning process and guarantee the quality of the final product. Thus, the machine must be able to make a clean cut both on the spiral and on the components attached to it, such as the casing or the inner tube.
- Clean and deburr the spirals. The sleeve should come out of the machine ready for use, i.e., it must be ready for a steel cable to be passed inside it, so that it can receive terminals if necessary. For this, it must not have burrs, obstructions in the hole, or damage to the casing.
- Provide a cost-effective solution, which is highly relevant in the competitive automotive components industry, leading to a reduced return-on-investment for currently operating machines.

3. Results

3.1. Pre-Design

The pre-design stage of the current project was carried out to serve as the basis for the project. This study involved all phases of brainstorming, testing, and analyzing of the possible problems and goals to be achieved. The implementation of a double-cutting system creates additional design difficulties that do not exist in a single-cutting system, such as the cooling problem. Thus, different hypotheses emerged to replace the cutting system. After verifying the presented hypotheses, the solutions that would respond to the imposed problems were selected. Due to the experience of using this type of equipment, some changes were also suggested to avoid the additional problems found in the existing machines in the company. The different hypotheses were cutting with an abrasive disk (current solution), laser cutting, waterjet cutting, and diamond disk cutting. Shear cutting was ruled out from the beginning due to the involved deformations arising from the hollow spiral geometry, which prevents the correct functionality of the control cable. The different hypotheses were tested, so that it was possible to choose the most viable one. As for cooling, the possibilities of projecting gases such as carbon dioxide or nitrogen as well as cooling with water and air were considered. The tests carried out for the different cutting techniques considered the use of three types of spirals (Figure 6): laminated spiral without inner tube, push-pull spiral, and braided spiral.

(a)   (b)   (c)

Figure 6. Spiral types: laminated spiral without inner tube (**a**), push-pull spiral (**b**), and braided spiral (**c**).

3.1.1. Laser Cutting

The first solution to replace the current cutting system was laser cutting. Spiral samples were sent to a specialized company in laser cutting (MACSA), which performed tests with an F9100 beam: a 100 W beam with ultra-high speed (UHS), as recommended by the company for the specimen dimensions and materials involved. Sample results for the laminated spiral are shown in Figure 7. Good results were not obtained, as the required cutting time is slightly higher than the current abrasive disk cutting, and the overall cut quality is poor. Actually, although the metal component can be cut with a minor burr, the plastic coating that protects the spiral tends to melt and leaves a poor finish. As shown in the figure, the laser beam ended up damaging the coating even before coming into contact with the spiral cable. Different processing conditions, such as higher power combined with shorter process time, were tested within the limits of current technology and application to a production line, but it was not possible to prevent melting of the plastic coating, which affects the subsequent process operations. Thus, the possibility of laser cutting was ruled out, due to non-compliance with quality requirements.

Figure 7. Laser cutting test results for the laminated spiral without inner tube.

3.1.2. Waterjet Cutting

The second tested solution consisted of waterjet cutting. It was necessary to build a structure that would allow the fixation of the spiral during the cuts. The test setup consisted of a steel fixation jig to hold the wire and hinder spiral deformations during the operation. The tests were carried out at the company JACQUET Portugal, with a pressure of 4000 bar, considering the possibility of using abrasive particles together with the water to facilitate cutting, i.e., tests with and without abrasive were considered. It was concluded that waterjet cutting without abrasive cuts the plastic coating but is not capable of cutting the steel spiral. Figure 8 shows representative examples for a laminated spiral without inner tube (a) and a braided spiral (b), with emphasis on the degradation state of the coating with little to no effect on the metal part.

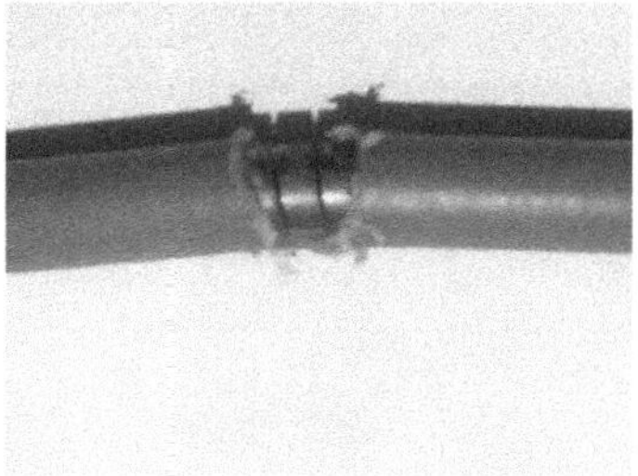

(a)    (b)

Figure 8. Waterjet cutting without abrasive examples for a laminated spiral without inner tube (a) and a braided spiral (b).

On the other hand, the abrasive waterjet cuts the entire cable without difficulty, but, as observed for laser cutting, it results in significant coating degradation. Moreover, in the case of spirals equipped with an internal tube, the cut region becomes clogged with plastic residues, which affect the assembly of the inner cable. Examples are given in Figure 9: a push-pull spiral (a) and a braided spiral (b).

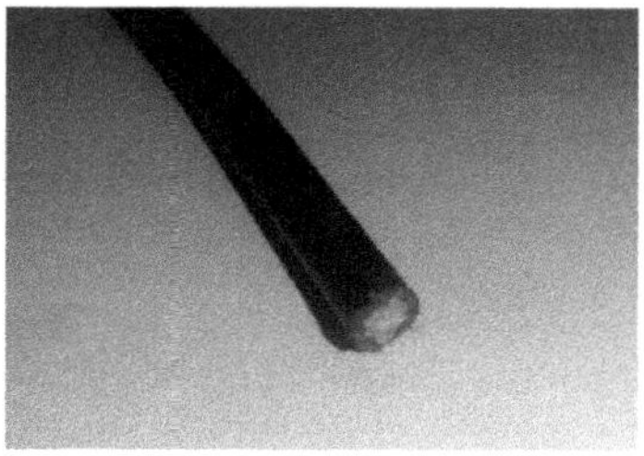

(a) 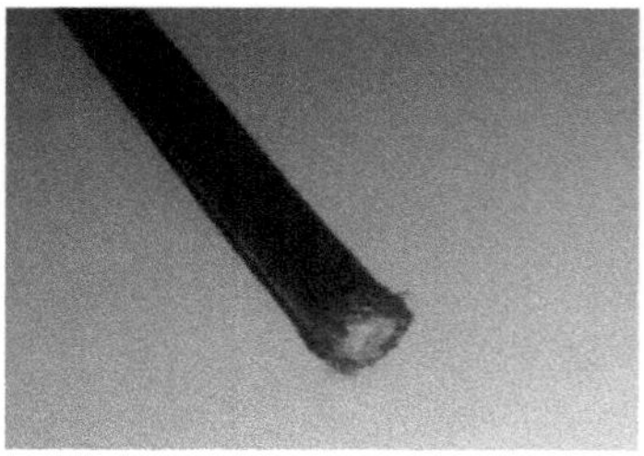   (b)

Figure 9. Waterjet cutting with abrasive examples for a push-pull spiral (a) and a braided spiral (b).

At a later stage, it was also possible to observe that, with time, oxidation of the steel spiral occurs due to water ingress to the interior of the spiral, where it remains and ends up oxidizing the material. It can, thus, be concluded that this process requires the implementation of a system that guarantees the drying of the entire spiral. Thus, by analyzing the obtained results, the possibility of adopting a waterjet cutting system was discarded, since it would not be able to meet the necessary requirements.

3.1.3. Diamond Disk Cutting

The possibility of replacing the abrasive disk with a diamond disk was also tested. For this, a STRUERS Minitom cutting machine was used. Two types of spirals were analyzed: laminated spiral and braided spiral. For the braided spiral, diamond disk-cut samples were collected. For the laminated spiral, samples were collected and cut with diamond and abrasive disks, to compare the results. The samples were sent to a specialized laboratory, where a microscopic scanning analysis was performed. Results of the braided spiral with diamond disk cut are shown in Figure 10 (sample 1). For the laminated spirals, two samples were collected, namely with a diamond disk cut (Figure 11; sample 2) and an abrasive disk cut (Figure 12; sample 3).

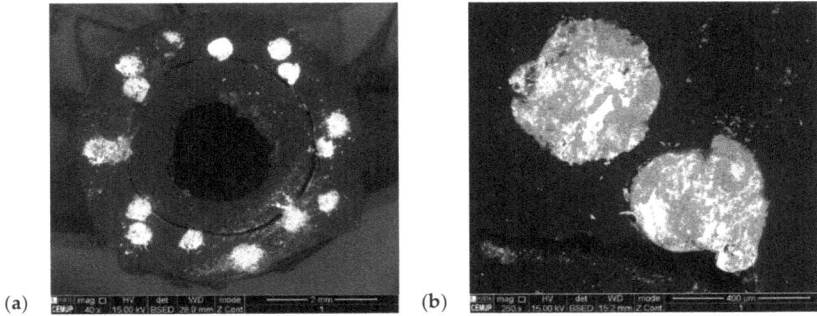

(a) (b)

Figure 10. Microscopic images for the braided spiral with diamond disk cut: magnification of 40× (**a**) and 250× (**b**).

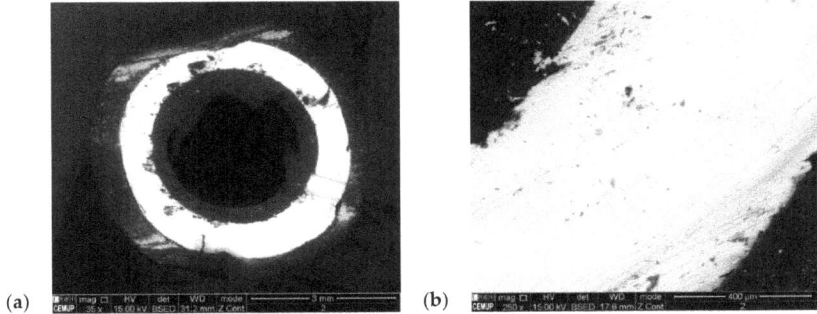

(a) (b)

Figure 11. Microscopic images for the laminated spiral with diamond disk cut: magnification of 40× (**a**) and 250× (**b**).

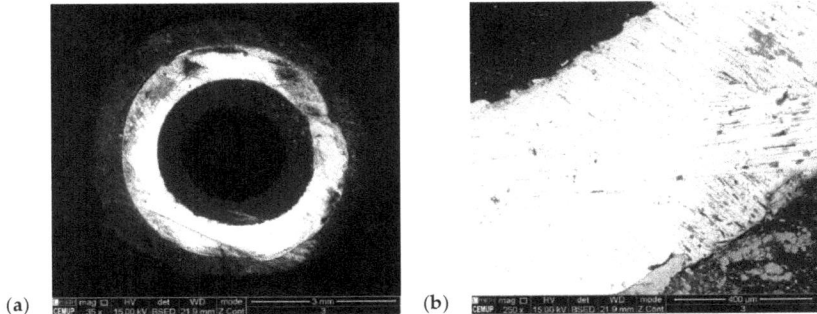

(a) (b)

Figure 12. Microscopic images for the laminated spiral with abrasive disk cut: magnification of 40× (**a**) and 250× (**b**).

Analysis of the results shows that the cut surfaces contain several defects when analyzed microscopically. Sample 1 showed some signs of steel oxidation, due to exposure to air, which is a natural phenomenon in this type of material. In sample 2, there are some defects in the cut of the coating and inner tube. In both samples, the cut surface is scratched due to the action of the rotating disk. In sample 3, however, this defect is more noticeable due to the high roughness of the abrasive disk compared to the diamond disk. Nonetheless, the defects pointed out do not prevent the correct use of the spirals, contrarily to what happened in the laser and waterjet cutting tests. Thus, the implementation of a diamond disk could be a possible solution. However, the diamond disk does not prove to constitute a reason for the reduction in defects, by providing a similar result to the abrasive disk. Moreover, the existing cutting system (abrasive disk) is already proven and put into practice, while the diamond disk solution would need further testing in an industrial environment. On the other hand, it is also necessary to consider the cost of the diamond disk, which is much higher than that of the abrasive disk. The diamond disk also requires specific cutting speeds to avoid rupture, which would require changes in the existing cutting concept. It was, then, assumed that the abrasive disk is sufficient for the work to be carried out. However, it is important to keep the disk at a low enough temperature, so as not to be affected by thermal fatigue.

3.1.4. Abrasive Disk Cutting with Cooling

The hypothesis of keeping the currently used cutting method gathers significant know-how from the company. This method does not require additional testing, as it is validated in real operating conditions. However, it is necessary to consider the heating issue, which is aggravated in a double-cutting system. Cutting disk cooling is, as mentioned, essential due to the heat generated in the cutting disk housing. The temperature is higher compared to a single-spiral-cutting system, and it affects both the surrounding atmosphere and the disk itself. Therefore, it is essential to include a cooling system capable of removing the hot air generated in the cutting area and cooling the cutting disk.

3.1.5. Selection of the Best Idea

The best idea for the cutting process was assessed by the selection table methodology (or Ashby methodology) [34]. Initially, it is necessary to establish the criteria by which each idea should be scored for a comparative evaluation between ideas. After a detailed analysis of machine/production line functionality and required features for the cut spirals, the following criteria were selected.

1. Cutting quality: includes coating degradation, cut section geometry, and possibility of debris accumulated in the spiral, leading to subsequent cleaning operations and increased cost/time.
2. Cutting time: shorter cutting times increase productivity.
3. Process know-how: from the company that requested this machine, to facilitate implementation and operation of the new idea in the company's production lines.
4. Cost: cost to fabricate and implement the cutting system.
5. Heat generation: affects the cut quality and requires proper dissipation systems in the machine.

Next, the relative importance of each criterion (w_i) should be selected. By building a selection table and comparatively ranking each pair of ideas, following the procedure of Nunes et al. [35] and considering criterion 1 as the comparison standard, the following w_i were attained (satisfying $\Sigma w_i = 1$): $w_1 = 0.33$, $w_2 = 0.18$, $w_3 = 0.10$, $w_4 = 0.15$, and $w_5 = 0.24$. The classification of each idea (V_i) is then calculated. With this purpose, a qualitative scale between 1 and 5 was established, in which 1 is the "least favorable" and 5 is the "most favorable" classification. In this process, a comparative evaluation between all ideas is performed to provide a relative hierarchy and reduce uncertainty. The weighted classification of each idea (β_i) is given by $\beta_i = V_i/MV_i$ ($\times 100$), in which MV_i is the highest V_i between all ideas. The idea classification in each criterion ($\Omega_i = w_i \times b_i$) is then calculated to produce the final classification of each idea ($\gamma_i = \Sigma(w_i \times b_i)$), leading to the choice of the highest ranked idea. Table 1 presents the selection matrix for the choice of spiral cutting process, in which the selected criteria were sorted from biggest to smallest w_i in columns, and V_i were defined in view of the aforementioned discussions for each candidate cutting process. By applying this methodology, the current process (abrasive disk cutting with cooling) was selected and is considered in the machine design that follows.

Table 1. Selection matrix for the choice of spiral cutting process.

V_i / β_i	Ω_i	1—Cutting quality		5—Heat Generation		2—Cutting time		4—Cost		3—Process know-how		Final Classification
		ω_1	0.33	ω_5	0.24	ω_2	0.18	ω_4	0.15	ω_3	0.1	γ_i
Laser		3 / 60.0	19.8	4 / 80.0	19.2	4 / 80.0	14.4	2 / 40.0	6.0	1 / 20.0	2.0	61.4
Waterjet		3 / 60.0	19.8	5 / 100.0	24.0	3 / 60.0	10.8	1 / 20.0	3.0	1 / 20.0	2.0	59.6
Diamond disc		5 / 100.0	33.0	2 / 40.0	9.6	5 / 100.0	18.0	3 / 60.0	9.0	4 / 80.0	8.0	77.6
Abrasive disk with cooling		5 / 100.0	33.0	2 / 40.0	9.6	5 / 100.0	18.0	5 / 100.0	15.0	5 / 100.0	10.0	85.6

However, prior to design, for cooling during the abrasive disk cutting process, several hypotheses were considered: with water projection, with compressed air projection, with projection of inert gases, and with fans included in the system. The use of water for disk cooling was a hypothesis discarded early on, due to oxidation of the steel spiral, which is accelerated by the ingress of water, and dirtiness, due to debris coming from the cut in contact with water. The projection of compressed air, as analyzed in the initial project, would become economically unfeasible. For the same reason, the projection of inert gases such as carbon dioxide or nitrogen can also be ruled out. The possibility of inserting hot air exhaust fans present in the cutting disk housing was then considered. With this approach, the temperature in the atmosphere surrounding the working zone of the disk could be reduced. However, the cutting disk would reach very high temperatures, so it is imperative to include another system responsible for its cooling. It was then decided to install a centrifugal fan in the system capable of delivering air to the disk surface without using compressed air. Thus, the solution found for cooling the cutting system involves the inclusion of hot air exhaust fans and a fan with cold air projection directly onto the disk.

3.2. Final Solution

The final machine was developed to comply with the mentioned objectives and requirements, and with the guidelines defined in the pre-design stage. After selecting the cutting and cooling systems, it was necessary to build the machine. The machine can be divided into three main systems, as shown in Figure 13: A-cutting mechanism, B-manipulator, and C-spiral preparation and cleaning system.

These systems interact with each other promoting the desired processing of the spiral. Figure 14 shows the global representation of the machine, including the three main systems and the components that, although not part of the spiral processing, are indispensable in the operation of the machine: 1-spiral coil containers, 2-spiral cut output container, 3-control panel, 4-valve boxes, 5-electrical panel, and 6-safety barrier.

The developed machine makes it possible to feed the spirals from the spiral containers (Figure 14-1). The spirals are removed from the spool and enter the machine through the cutting mechanism (Figure 13-A), where they will be positioned and cut to the desired length (between 200 and 600 mm). The manipulator (Figure 13-B) transports the spirals to the preparation and cleaning system (Figure 13-C). In this system, the necessary processing is carried out at various stations to give the spirals the necessary finish. When this procedure is finished, the spirals fall to an outlet container (Figure 14-2). The machine is operated by the control panel (Figure 14-3), which interacts with the components of the valve boxes (Figure 14-4) and the electrical panel (Figure 14-5), allowing for the driving of the automatic systems of the machine. The safety barrier (Figure 14-6) is a safety component that surrounds all the machine and prevents people from approaching the most dangerous components. To facilitate the analysis of the developed project, different colors were assigned to the components used to promote their distinction. A reference was also given to the parts developed to facilitate a simpler organization of the project. No reference was assigned to standard components, as they already contain the supplier's reference.

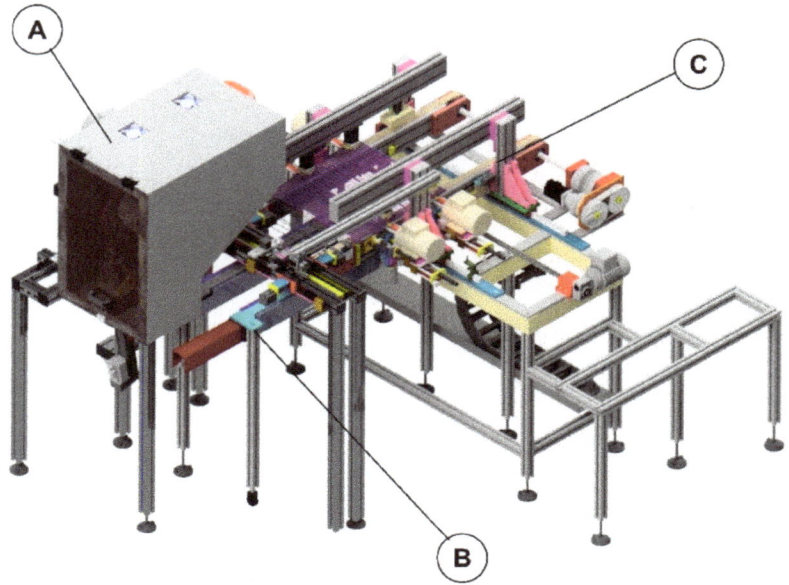

Figure 13. Arrangement of the main systems in the machine.

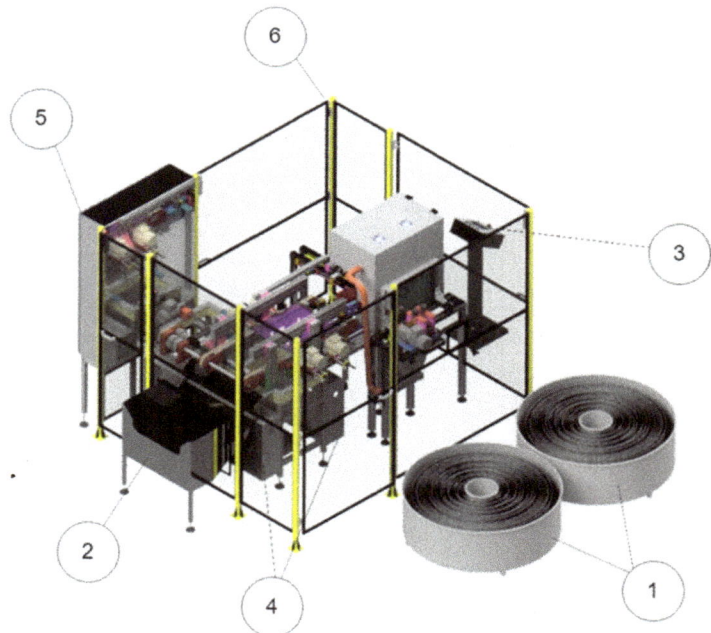

Figure 14. Arrangement of the secondary components in the machine.

3.3. Design Process

3.3.1. Cutting Mechanism

The cutting mechanism (Figure 15) is responsible for dragging/pulling the spirals and cutting them. This mechanism is divided into eight essential systems: (1) drag mechanism, (2) pull system, (3) cutting machine, (4) cutting presser, (5) backstop system, (6) cutting rulers, (7) extraction support system, and (8) detection system.

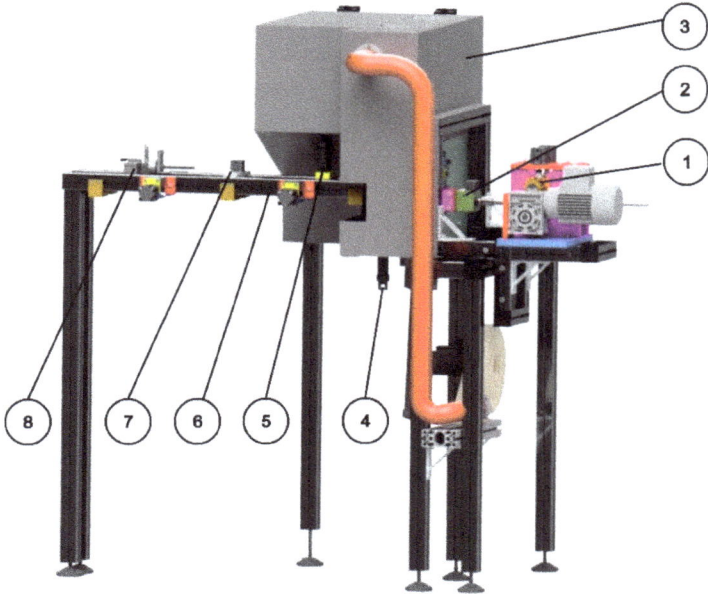

Figure 15. Layout of the cutting mechanism.

The drag mechanism (Figure 15-1) leads the spiral into the cutting machine to the pre-defined position, such that it can be cut correctly. The system is activated by a motor that applies the movement through rollers. For a successful cut, the contact time between the spiral and the cut disk should be reduced, since the disk/spiral interaction highly increases the disk temperature, and defects can appear in the spiral's plastic coating. To reduce the contact time, a pull system was created (Figure 15-2), which promotes the removal of the spiral immediately after the cut. A pneumatic actuator is used to lock the spiral during the cut in the first time. A spring system is used to induce a tensile axial load in the spiral during the cut process and, when it is possible, the spiral recedes. The spirals are cut in the cutting machine (Figure 15-3) using an abrasive disk driven by an electric motor/belt system. The cutting disc used is aluminum oxide with an outside diameter of 180 mm, a thickness of 1 mm, and a grit of 60. Such a disc should operate at cutting speeds below 80 m/s and below 8500 rpm [36]. A servomotor feeds the abrasive disc in the spiral's direction, while the feed rate used was selected upon testing. The servomotor also allows to compensate for the disc's wear. The cutting process leads to the deposition of debris that can affect the normal mechanism operation. Currently, metal chips and dust are projected during the process, and most of them lie on the floor and machine, despite all the attempts to mitigate this problem. To remove the debris from the cutting process, including the metal chips and dust, which comprise one of the main problems pointed out in the initial machine due to floor and machine contamination, a disk protection housing is now used that seals the cutting area and directs the debris downwards (by gravity) towards a suction nozzle, to assure the cleanliness of the process. On the other hand, since the temperature of the cutting disk has an influence on the cut quality, a piping system is used to inject atmospheric air directly on the disk, as represented in Figure 16a. Air pumping is made by a centrifugal fan, and the projection to the disk is accomplished by a copper tube, punctured in predefined locations to project air into the cutting disk and cool it by convection.

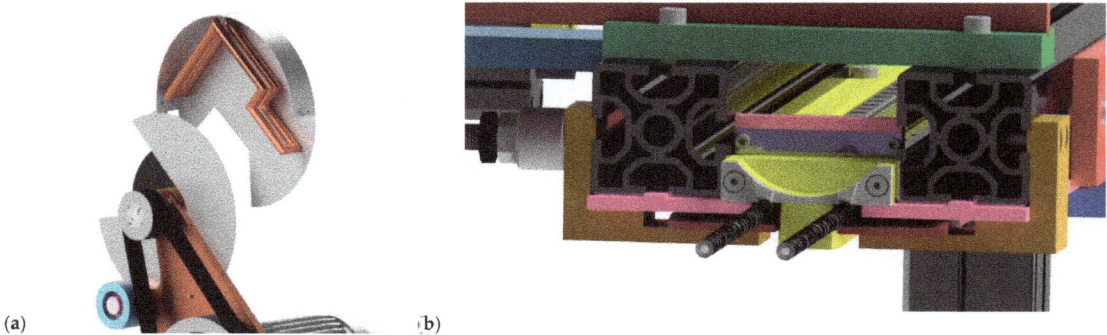

(a)   (b)

Figure 16. Cutting disk cooler system (**a**) and schematic of the cutting rulers (**b**).

To remove the hot air generated at the cutting zone and promote the machine cool down, two fans are used in the protection box of the cutting machine (Figure 15-3), enabling the machine to work continuously, while ensuring quality cuts. The cutting presser (Figure 15-4) aims to prevent bending of the spiral, and it is driven by a pneumatic cylinder that lifts it close to the spiral being cut. The backstop system (Figure 15-5) locks the spiral during cutting. While the spirals are cut, they sit on the cutting rulers (Figure 15-6) until they are expelled from the cutting mechanism. In the final stage of the cutting process, pneumatic actuators drop the spiral to the manipulator, and the second phase of the process is started. Figure 16b represents a pair of spirals in the cutting rulers. Since the spirals can get stuck on the cutting rulers, a pneumatic-driven extraction support system (Figure 15-7) is considered, which pushes the spiral down. The length of the spiral to be cut is defined by the operator, using the detection system adjustment (Figure 15-8). This system uses a sensor that is mechanically activated by a paddle system, when the spiral touches it. This system works with the drag mechanism to define the length of the spiral.

3.3.2. Manipulator

The manipulator is the system responsible for transporting the cut spiral between the cutting mechanism and the spiral preparation and cleaning system. The cutting process causes the deposition of debris. Therefore, the belts of the preparation and cleaning system would quickly become dirty if they were placed under the cutting machine, compromising the correct functioning of the system and causing severe production losses. The manipulator was designed to solve this problem and move the spiral in the X and Y axes, as shown in Figure 17a.

Figure 17b shows the assembled manipulator and spiral transfer system. The horizontal movement (in the X direction) is carried out on sliders that move on linear guides, driven by pneumatic cylinders. The X axis makes the movement between two predefined positions: below the cutting mechanism and over the belts of the spiral preparation/cleaning system. The vertical movement (in the Y direction) is also carried out by two pneumatic cylinders fixed to the sliders using a plate. A spiral attachment component is screwed to the cylinders, where the spirals are supported during the mechanism's operation. A magnet will secure the spirals to this part, preventing them from falling during movement. The Y axis brings the manipulator closer to the cut spiral, to receive it at its ends by two grips, and puts the spiral on the belts when retreating. The described movements work as a cycle and are repeated for each new pair of spirals. The pneumatic cylinders and the linear guides are correctly protected from the cutting debris.

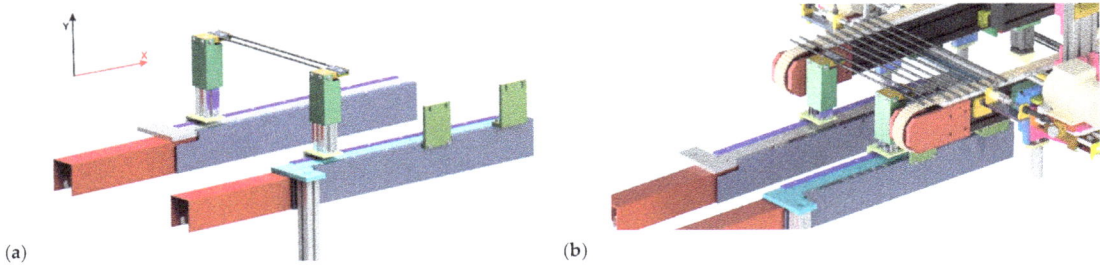

(a)   (b)

Figure 17. Schematic of the manipulator (**a**) and assembled manipulator and spiral transfer system (**b**).

3.3.3. Spiral Preparation and Cleaning System

The cleaning and preparation system is responsible for ensuring a quality finish on the spiral edges. The spiral preparation and cleaning system is based on five mechanisms, represented in Figure 18a: (1) spiral advance system, (2) length tuning system, (3) control positioner, (4) inside and outside cleaning station, and (5) air cleaning station.

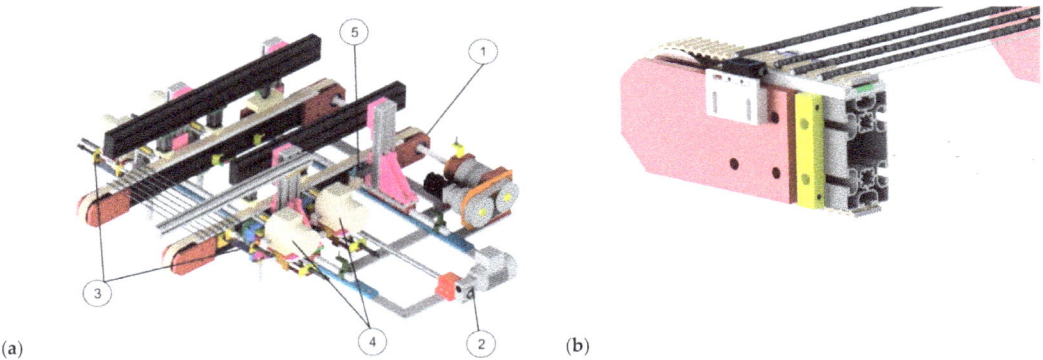

(a)   (b)

Figure 18. Spiral preparation and cleaning system (**a**) and magnet system and photelectric sensor (**b**).

The spiral advance system (Figure 18a-1) is a belt-drive mechanism that transports the spiral along the preparation and cleaning system, handling it through the various stations and pushing it, at the end of the process, into an unloading container. To prevent the spirals from falling out during their movement, a magnet system is used to hold the spiral to the conveyor. A photoelectric sensor (Figure 18b) is also used to define the position of the spiral to each processing station, since the detection corresponds to a known distance, to control the feed of the servomotor actuating the belt drive mechanism. Another sensor is included at each station, to trigger the respective operation when a spiral is detected. The length-tuning system (Figure 18a-2) regulates the cutting length automatically, powered by an electric geared motor, to accommodate batches of different spiral lengths. The system slides on two linear guides that hold all of the spiral preparation and cleaning system. The spirals are placed in the system in a misaligned and random position by the manipulator. To be processed correctly, the spirals need a correct alignment. To correct the existing misalignment, a control positioner is used (Figure 18a-3). The system is pneumatically driven, and it aligns the spirals, as shown in more detail in Figure 19a. The spiral preparation and cleaning system contains four inside and outside cleaning stations (Figure 18a-4). These stations work in pairs to clean both ends of each spiral. Each station consists of two parts: the presser and the cleaning system, as shown in Figure 19b. The presser is used to fix the spiral during processing. The cleaning system has a special tool that processes the spiral inside and outside at the same time.

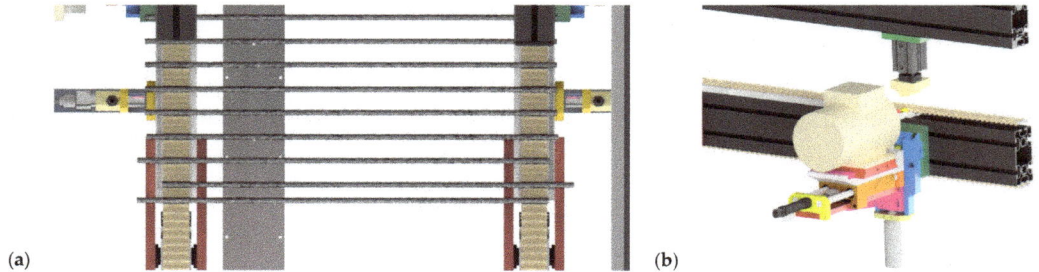

(a)

(b)

Figure 19. Alignment of spirals before and after the control positioner (**a**) and inside and outside cleaning station (b).

To be correctly positioned and concentric with the spirals, the system can be tuned in three axes. This must be done manually by the operator before starting the machine. To reduce the machine cycle time and increase the working time, each inside and outside cleaning station acts on alternating spirals. Thus, each pair of stations performs the cleaning process for the half of the spirals that pass through it, while the other half are processed by the second pair of stations. This concept is described in Figure 20a: the red spirals are processed by the red cleaning station, and the blue ones are processed by the blue cleaning station.

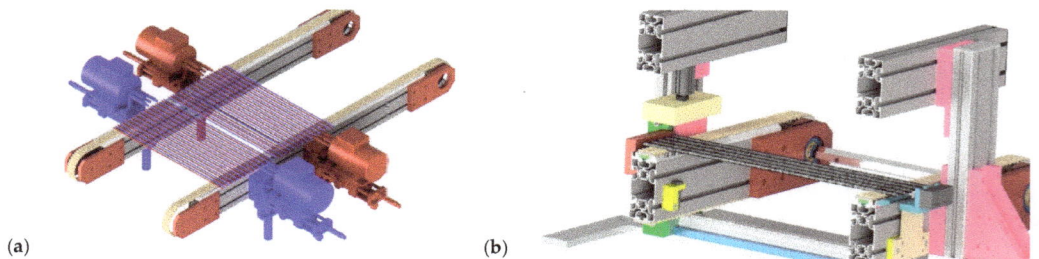

(a)

(b)

Figure 20. Processing of the spirals (**a**) and air cleaning station (**b**).

The air cleaning station (Figure 20b) is the last system through which the spiral passes before leaving the machine (Figure 18a-5). This system aims to unblock the inside of the spiral by injecting compressed air through the interior of the spiral to clean debris. It uses a pressure switch to assess if the spiral is not blocked. If this condition is true, the spiral can be removed from the machine.

3.3.4. Machine Structure

The structure designed to support the cleaning double spiral cutting machine is based on EN AW-6060 aluminum alloy profiles (Figure 21a) due to the long service life, high corrosion resistance, and high strength. The aluminum profiles also can be connected and aligned correctly in a simplified way through universal connectors. Aiming for a higher stiffness in the structure and better positioning between the profiles, aluminum corner connections were implemented, which improve stability (Figure 21b). Leveling feet were used to adjust the height of the mechanisms and correct the misalignment of the floor.

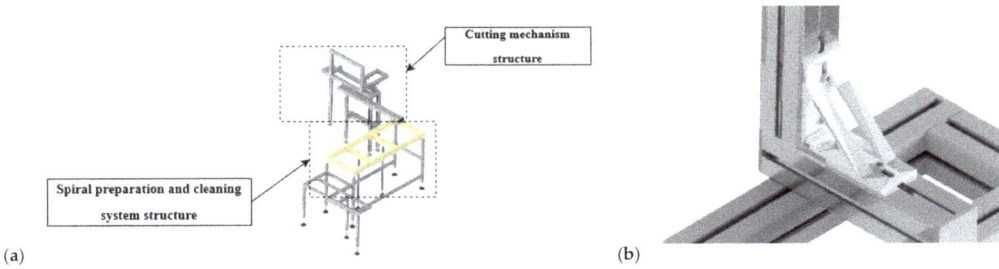

Figure 21. Machine structure (**a**) and corner connections (**b**).

3.3.5. Command System

The command system includes the machine's control panel, the electrical panel, and the solenoid valve boxes. The machine's control panel enables the operator-machine interaction. It contains a human-machine interface (HMI) console that provides information about the systems to the operator and allows the operator to trigger the systems. The HMI communicates with a programmable logic controller (PLC) that controls the electronic and pneumatic components of the entire system. The PLC, signal input and output terminals, power supplies, electrical components, and auxiliary components are fundamental to the operation of the machine. The activation of the pneumatic components in the machine is done by solenoid valves. These components transform an electrical signal into a mechanical response. Design of the control system is not presented, since this task will be accomplished by a specialized company, so it is not part of this work's objectives.

3.3.6. Safety

The double-spiral-cutting machine contains mechanisms that can be dangerous for operators. Thus, it is necessary to create systems capable of guaranteeing the safety of people who circulate in the vicinity of the machine. The operation of the machine must also comply with safety instructions. Aiming to prevent people and objects from approaching the machine mechanisms and protect the machine, a safety barrier was installed (Figure 22).

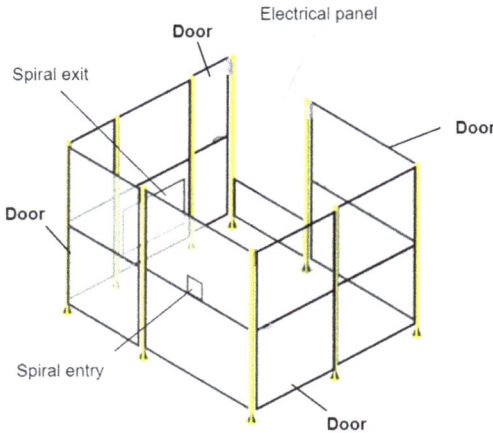

Figure 22. Safety barrier for the machine.

The barrier is built by pillars that guarantee its fixation to the factory floor and by a mesh that seals the system. The mesh must, however, allow the spirals to enter and exit the machine and, therefore, the entrances and exits represented in Figure 22 were opened. Four doors were added to enable access to the interior of the machine, for maintenance, repairs, or changing the cutting disk. In this way, the safety barrier can be closer to the machine and, thus, allow the machine to occupy as little space as possible in the factory, without preventing access to the machine when necessary.

The access doors installed in the barrier are equipped with safety latches that prevent the door from opening when the machine is in operation. As a result, access to the inside of the protection is only possible when there is no danger to the user. The safety latch is bolted to the structural pillar of the safety barrier, and a trigger is applied to the door, as shown in Figure 23a. When the door is closed, the sensor's safety latch locks the trigger, and machine operation is allowed. Once in operation, the safety latch does not allow the door to be opered until it receives information that the machine has stopped.

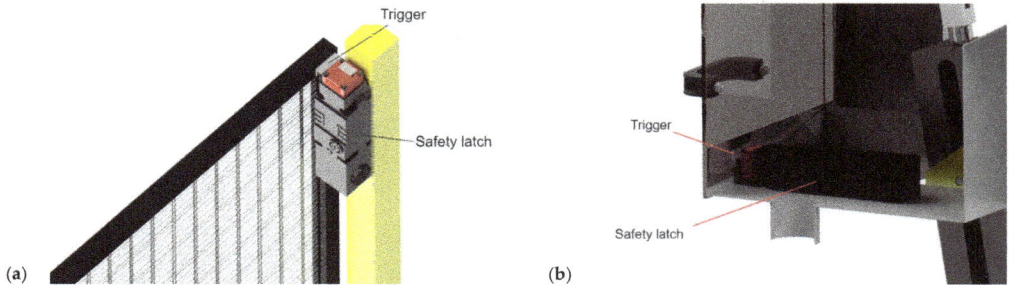

Figure 23. Safety latch for the machine access doors (**a**) and cutting disk housing's safety latch (**b**).

The cutting disk housing has the function of protecting the user during operation. This element is the most dangerous component of the machine, and, therefore, its security system must be studied with greater care. Considering the negligence that may exist by any user of the machine, there is no safety system that prevents the machine from starting when someone is inside. In this sense, it is important to implement an improved safety system for the cutting machine. Thus, the cutting disk housing was also equipped with a safety latch similar to the barrier. This latch prevents the housing from opening when the cutting machine is in operation and the machine from starting while the door is open. Its installation is shown in Figure 23b. The safety latch is fixed to the housing structure and is equipped with an actuator, fixed to the door. The hypothesis of an accident that, for whatever reason, the safety systems have not prevented should also be considered. In this case, there is an emergency button that forces the machine to stop immediately in a controlled manner and as quickly as possible. The emergency button can be deactivated, but the machine is only restarted when the 'START' button is pressed. The emergency button must only be used in case of imminent danger to people or equipment. This should never be activated to turn off the machine.

To prevent accidents from occurring, these safety instructions must be followed.

- The electrical panel must remain closed and prevent access to its interior while the machine is on.
- Components that show movement must not be moved while the machine is on.
- The machine must only be operated by users who are familiar with its operation.
- Safety components must not be removed.
- The functioning of safety components must be checked regularly.
- Maintenance actions must be performed with the machine turned off.

4. Discussion and Conclusions

The search for automatic solutions in industrial equipment is associated with increases in productivity and reduction in production costs, which also favors an increase in product quality. Within this scope, this work used the DSR methodology to propose a new machine design that can be easily integrated in the production line to increase the rate of spiral cutting in a control cable assembly line, by cutting and cleaning two spiral cables simultaneously and automatically, thus doubling the productivity associated to this production stage. The machine developed in this paper included different engineering fields to produce a unique and complex system that meets the defined objectives and requirements. When comparing the initial and final machines, and within the scope of stage 6 (conclusions) of the DSR methodology, a significant simplification can be observed, essentially in the preparation and cleaning system, where all unnecessary components were removed, so the required quality output is assured due to the designed mechanism. The preparation and cleaning system, spiral feed from the spool, and cutting length adjustment system are now automatic. At the structural level, simplifications were also made due to the use of aluminum profiles, which facilitate the assembly of the machine; a smaller machine was also proposed, promoting the efficient use of space and simplified construction. Fabrication costs of the new machine were calculated by the company, including all stages from design to assembly, leading to an estimation of EUR 64,000, which gives a 28% reduction in cost compared to the initial design. Operating costs of the machine diminished, since, in terms of cooling, compressed air was replaced by cooling with a fan, and the dedicated operator used in the initial machine for spiral handling is no longer necessary. Still, downtimes for any industrial equipment represent production losses, and, therefore, interruptions for small maintenance actions are to be avoided. In this aspect, the stops made to clean the cutting debris were eliminated due to the implementation of the manipulator. It is, thus, possible to verify that all the imposed requirements were fulfilled.

As a summary of the accomplished work and degree of novelty, a new and original solution, compared to the current state-of-the-art cutting processes, is proposed to improve the productivity of spiral manipulation, cutting, and cleaning, which can be applied to other companies that work in the same field as well as to other fields of engineering in which similar operations such as feeding, cutting, and cleaning are required. As a results, companies can improve their processes and increase competitiveness. It should be mentioned that, for future works, it is still necessary to design the control system, which was out of the scope of the present work, to be executed by a specialized company, as is the common practice in the company. As for future research directions, aiming for the continuous improvement that is essential in the industry for a company to keep competitive, it would be relevant to perform a detailed line balancing analysis, considering the full production process of control cables, to assess the bottlenecks arising from the production rate increase in the spiral-cutting process, to maximize output. Within this topic, the company would also benefit from moving the automatic component between production stages by an integrated pick-and-place transfer system along the assembly line and a respective automatic artificial vision quality inspection station at its end, leading to quality improvement and labor reduction, with a respective cost advantage. On the other hand, in the advent of the mass production of the proposed machine, it would become increasingly relevant to apply the design toward the manufacturing or assembly concepts intended to reduce the fabrication costs of the assembly times. Finally, a complete validation of the solution is only accomplished through prototype construction and physical validation.

Author Contributions: Conceptualization: A.F.G.B. and I.J.S.-A.; investigation, A.F.G.B. and I.J.S.-A.; methodology, R.D.S.G.C.; resources, C.P. and D.B.; supervision, C.P. and D.B.; validation, R.D.S.G.C. and F.J.G.S.; writing—original draft, A.F.G.B. and I.J.S.-A.; writing—review and editing, R.D.S.G.C., F.J.G.S., C.P. and D.B. All authors have read and agreed to the published version of the manuscript.

Funding: This research received no external funding.

Data Availability Statement: Not applicable.

Conflicts of Interest: The authors declare no conflict of interest.

References

1. Vaz, C.R.; Rauen, T.R.S.; Lezana, Á.G.R. Sustainability and Innovation in the Automotive Sector: A Structured Content Analysis. *Sustainability* **2017**, *9*, 880. [CrossRef]
2. Chen, C.-T.; Lien, W.-Y.; Chen, C.-T.; Wu, Y.-C. Implementation of an Upper-Limb Exoskeleton Robot Driven by Pneumatic Muscle Actuators for Rehabilitation. *Actuators* **2020**, *9*, 106. [CrossRef]
3. Sousa, V.F.C.; Silva, F.J.G.d.; Campilho, R.D.S.G.; Pinto, A.G.; Ferreira, L.P.; Martins, N. Developing a novel fully automated concept to produce bowden cables for the automotive industry. *Machines* **2022**, *10*, 290. [CrossRef]
4. Bowden, E.M. Mechanism for Transmitting Motion or Power. US Patent no. 609570, 23 August 1898.
5. Grosu, S.; Rodriguez-Guerrero, C.; Grosu, V.; Vanderborght, B.; Lefeber, D. Evaluation and analysis of push-pull cable actuation system used for powered orthoses. *Front. Robot. AI* **2018**, *5*, 105. [CrossRef]
6. Figueiredo, D.; Silva, F.J.G.; Campilho, R.D.S.G.; Silva, A.; Pimentel, C.; Matias, J.C.O. A new concept of automated manufacturing process for wire rope terminals. *Procedia Manuf.* **2020**, *51*, 431–437. [CrossRef]
7. Vieira, A.L.N.; Campilho, R.D.S.G.; Silva, F.J.G.; Faria, N.M.S.; Ferreira, L.P. Design of a thermoplastic micro over injection machine for the automotive component industry. *Procedia Manuf.* **2021**, *55*, 56–63. [CrossRef]
8. Chang, X.-D.; Peng, Y.-X.; Cheng, D.-Q.; Zhu, Z.-C.; Wang, D.-G.; Lu, H.; Tang, W.; Chen, G.-A. Influence of different corrosive environments on friction and wear characteristics of lubricated wire ropes in a multi-layer winding system. *Eng. Fail. Anal.* **2022**, *131*, 105901. [CrossRef]
9. *ISO 2408*; 2017-Steel Wire Ropes—Requirements. International Standards Organization: Geneva, Switzerland, 2017.
10. Oberg, E.; Jones, F.D.; Horton, H.L.; Ryffel, H.H. *Machinery's Handbook*, 26th ed.; Industrial Press, INC: New York, NY, USA, 2000.
11. GmbH, C.S. *Carl Stahl TECNOCABLES Catalog*, 7th ed.; Carl Stahl TECNOCABLES GmbH: Süßen, Germany; Available online: https://www.carlstahl-technocables.com/fileadmin/Resources/Public/Downloads/catalogue_cable_holder_and_suspension_systems.pdf (accessed on 1 August 2022).
12. Naveen, A.M.; Vishwanatha, H.M. Pneumatic automation of the assembly of spherical bearing and pin header into the gearbox housing. *Mater. Today Proc.* **2022**, in press. [CrossRef]
13. Moreira, B.M.D.N.; Gouveia, R.M.; Silva, F.J.G.; Campilho, R.D.S.G. A novel concept of production and assembly processes integration. *Procedia Manuf.* **2017**, *11*, 1385–1395. [CrossRef]
14. Cheng, L.; Tang, Q.; Zhang, L.; Meng, K. Mathematical model and enhanced cooperative co-evolutionary algorithm for scheduling energy-efficient manufacturing cell. *J. Clean. Prod.* **2021**, *326*, 129248. [CrossRef]
15. Zhang, X.; Li, Y.; Ran, Y.; Zhang, G. Stochastic models for performance analysis of multistate flexible manufacturing cells. *J. Manuf. Syst.* **2020**, *55*, 94–108. [CrossRef]
16. Martins, N.; Silva, F.J.G.; Campilho, R.D.S.G.; Ferreira, L.P. A novel concept of Bowden cables flexible and full-automated manufacturing process improving quality and productivity. *Procedia Manuf.* **2020**, *51*, 438–445. [CrossRef]
17. Ribeiro, R.; Silva, F.J.G.; Pinto, A.G.; Campilho, R.D.S.G.; Pinto, H.A. Designing a novel system for the introduction of lubricant in control cables for the automotive industry. *Procedia Manuf.* **2019**, *38*, 715–725. [CrossRef]
18. Li, B.; Zhang, S.; Du, J.; Sun, Y. State-of-the-art in cutting performance and surface integrity considering tool edge micro-geometry in metal cutting process. *J. Manuf. Processes* **2022**, *77*, 380–411. [CrossRef]
19. Llanto, J.M.; Tolouei-Rad, M.; Vafadar, A.; Aamir, M. Recent Progress Trend on Abrasive Waterjet Cutting of Metallic Materials: A Review. *Appl. Sci.* **2021**, *11*, 3344. [CrossRef]
20. Wang, Z.; Xie, T.; Ning, X.; Liu, Y.; Wang, J. Thermal degradation kinetics study of polyvinyl chloride (PVC) sheath for new and aged cables. *Waste Manag.* **2019**, *99*, 146–153. [CrossRef]
21. Krar, S.F.; Gill, A.R.; Smid, P. *Technology of Machine Tools*, 7th ed.; McGraw-Hill: New York, NY, USA, 2011.
22. Mohamed Akeel, A.; Kumar, R.; Chandrasekhar, P.; Panda, A.; Kumar Sahoo, A. Hard to cut metal alloys machining: Aspects of cooling strategies, cutting tools and simulations. *Mater. Today Proc.* **2022**, in press. [CrossRef]
23. Lei, M.K.; Miao, W.L.; Zhu, X.P.; Zhu, B.; Guo, D.M. High-performance manufacturing enabling integrated design and processing of products: A case study of metal cutting. *CIRP J. Manuf.* **2021**, *35*, 178–192. [CrossRef]
24. Kalpakjian, S.; Schmid, S.R. *Manufacturing Engineering and Technology*; Prentice Hall: Hoboken, NJ, USA, 2010.
25. Bhowmik, S.; Naik, R. Selection of Abrasive Materials for Manufacturing Grinding Wheels. *Mater. Today Proc.* **2018**, *5*, 2860–2864. [CrossRef]
26. Chen, Z.; Xiao, B.; Wang, B. Optimum and arrangement technology of abrasive topography for brazed diamond grinding disc. *Int. J. Refract. Met. Hard Mater.* **2021**, *95*, 105455. [CrossRef]
27. Mitaľová, Z.; Botko, F.; Vandžura, R.; Litecká, J.; Mitaľ, D.; Simkulet, V. Machining of wood plastic composite using AWJ technology with controlled output quality. *Machines* **2022**, *10*, 566. [CrossRef]
28. Madić, M.; Petrović, G.; Petković, D.; Antucheviciene, J.; Marinković, D. Application of a robust decision-making rule for comprehensive assessment of laser cutting conditions and performance. *Machines* **2022**, *10*, 153. [CrossRef]
29. Copertaro, E.; Perotti, F.; Castellini, P.; Chiariotti, P.; Martarelli, M.; Annoni, M. Focusing tube operational vibration as a means for monitoring the abrasive waterjet cutting capability. *J. Manuf. Processes* **2020**, *59*, 1–10. [CrossRef]
30. Naresh; Khatak, P. Laser cutting technique: A literature review. *Mater. Today Proc.* **2022**, *56*, 2484–2489. [CrossRef]
31. Krajcarz, D. Comparison Metal Water Jet Cutting with Laser and Plasma Cutting. *Procedia Eng.* **2014**, *69*, 838–843. [CrossRef]

32. Vaishnavi, V.K.; Kuechler, W. Introduction to Design Science Research in Information and Communication Technology. In *Design Science Research Methods and Patterns*; CRC Press: Boca Raton, FL, USA, 2015; pp. 52–91.
33. Abdullah, O.; Abbood, W.; Khalid, H. Development of Automated Liquid Filling System Based on the Interactive Design Approach. *FME Trans.* **2020**, *48*, 938–945. [CrossRef]
34. Ashby, M.F. *Materials Selection in Mechanical Design*; Butterworth-Heinemann: Oxford, UK, 2016.
35. Nunes, D.M.; Campilho, R.; Silva, F.J.G. Design of a transfer system for the automotive industry. *Proc. Inst. Mech. Eng. Part E J. Process Mech. Eng.* **2022**, *236*, 5. [CrossRef]
36. Norton. Flat Cutting off Wheel Non-Reinforced Cut-off-Norton NRCO-180x1x31.75-57A60RB25, Product Datasheet. Available online: https://www.nortonabrasives.com/en-gb/ (accessed on 1 August 2022).

machines

MDPI

Article

2D and 3D Wires Formability for Car Seats: A Novel Full-Automatic Equipment Concept towards High Productivity and Flexibility

Manuel Gaspar [1], Francisco J. G. Silva [1,2,*], Arnaldo G. Pinto [1] and Raul D. S. G. Campilho [1,2]

[1] ISEP, Polytechnic of Porto, Rua Dr. António Bernardino de Almeida, 4249-015 Porto, Portugal
[2] Associate Laboratory for Energy, Transports and Aerospace (LAETA-INEGI), Rua Dr. Roberto Frias 400, 4200-465 Porto, Portugal
* Correspondence: fgs@isep.ipp.pt; Tel.: +351-228340500; Fax: +351-228321159

Abstract: The automotive industry demands high quality at very low prices. To this end, it is necessary to constantly innovate, making processes increasingly competitive, while continuing to ensure high levels of quality. Model diversification has forced the automotive industry to make its manufacturing processes more flexible, without losing competitiveness. This has been the case for car seats, where the quantities to be produced per batch are significantly lowering due to the diversity of existing models. The objective of this work was to increase the production rate of bent wires used in car seat cushions and increase the flexibility of changing wire types in production. After benchmarking the existing solutions so far, it was verified that none are capable of complying with the required production rate, while also offering the desired flexibility. Thus, it is necessary to start with a new concept of conformation of the wires used in these seat cushions. The new concept developed and integrated some of the previously known solutions, developing other systems capable of providing the desired response in terms of productivity and flexibility. To this end, new mechanical solutions and automated systems were developed, which, together with other existing ones, made it possible to design equipment that complies with all the necessary requirements. The developed concept is innovative and can be employed to other types of products in which it can be applied. The new concept developed yields a production rate of 950 parts/hour (initial goal: 800 parts/hour), features a setup time of around 30 min, ensuring the desired flexibility, and the tool costs about 90% less than traditional tools. The payback period is around 5 months, given that the equipment cost was EUR 122.000 in terms of construction and assembly, and generated a gain of EUR 280.000 in the first year of service.

Keywords: full-automated solutions; competitiveness; high productivity; high flexibility; bent wires; cushions; car seats; automotive industry; automation; novel mechanical concepts

Citation: Gaspar, M.; Silva, F.J.G.; Pinto, A.G.; Campilho, R.D.S.G. 2D and 3D Wires Formability for Car Seats: A Novel Full-Automatic Equipment Concept towards High Productivity and Flexibility. *Machines* 2023, *11*, 410. https://doi.org/10.3390/machines11030410

Academic Editor: Dimitris Mourtzis

Received: 16 February 2023
Revised: 18 March 2023
Accepted: 20 March 2023
Published: 21 March 2023

1. Introduction

The automotive industry (AI) is one of the most innovative, competitive and economically important sectors worldwide [1]. In the last decade, it has experience constant growth, producing around 91.8 million vehicles during the year 2019 [2]. In the decade between 2010–2019, it showed constant growth, stabilizing in 2020 and showing a downward trend in the last two years [3]. This drop is related to the lack of definition in terms of vehicle motorization, as well as the pandemic situation experienced between 2020 and 2022 and the Russia–Ukraine conflict in 2022–2023, situations that induced a sudden rise in inflation, with a particular focus on fossil fuels and energy. Moreover, environmental concerns also affect this sector [4]. These data are particularly important to put into perspective the coming years, which are approaching as challenging ones [5]. However, AI has always shown the ability to reinvent itself, and this is, effectively, a transition situation [6].

The most competitive companies have two fundamental characteristics in their production processes, namely flexibility and agility, which allow them to be prepared for constant changes that the market is currently undergoing [7–10]. Industrial flexibility is the ability of a company to deal with variations in the components or products within its manufacturing processes, such as changes in raw materials, product size and weight, geometry and changes in product complexity [11–13]. Flexibility, in addition to allowing the system to absorb product variations, also possesses the ability to minimize interruptions in production changeovers [14–16]. These systems can produce different types of parts or, when they become obsolete, be easily converted, covering new needs [17–19]. Industrial agility reflects the ability of companies to react quickly to market needs or to new levels of quality and innovation presented by competitors, maintaining their capacity to follow new market trends without losing competitiveness [19].

Studies on wire bending mainly focus on phenomena related to fatigue [20–29]. As in other types of mechanical analyses, the simulation was also applied to wires to predict their behavior in certain loading cases, both in the form of wire and steel cables [30–32]. There are still studies on the manipulation of steel bent wires, both in the manipulation of wires to feed plastic injection equipment [33] and in the organization of wires in wire-bending equipment [34]. Moreover, the damping of the oscillation effect in the wire-bending process has also been studied [35]. However, research on wire and steel cable bending is very scarce, making this subject more attractive in terms of research due to the market's need to solve numerous problems related to this area of the industry. In addition, the automotive industry uses numerous products that base their working principle on steel cables or bent wires, such as Bowden cables and comfort systems related to the support of motor vehicle seats [36,37].

Indeed, cars are is made up of several components that complement each other. One of these components is the seat, which is one of the key elements of the safety and comfort of occupants. This makes it possible to ensure that the occupants remain solid with the structure of the car in the event of accidents or sudden movements. In addition to all the safety requirements that a seat must guarantee, it must ensure the comfort of its occupants, for both short and long trips, reducing the maximum the vibrations caused by the track, as well as by the various components that make up the seat.

Figure 1 illustrates the main components that make up a car seat:

1. Covers/lips;
2. Foam;
3. Headrest;
4. Suspension structure: cushion and suspension mat;
5. Plastic structure;
6. Interface structure between vehicle and mechanisms.

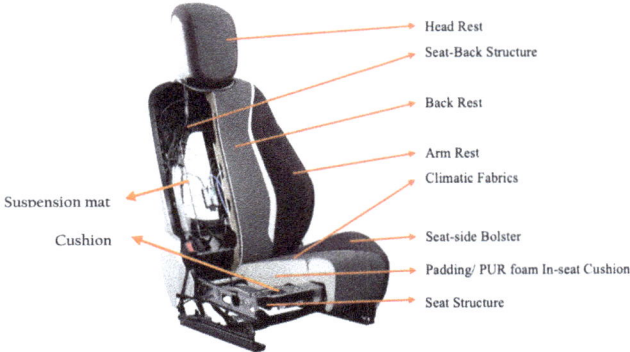

Figure 1. Components of a car seat (adapted from [38]).

In the cushion/suspension mat manufacturing industry, a significant loss of competitiveness was identified in the wire-bending process, as it had very flexible equipment with little productivity and equipment with high productivity, in which the implementation of new projects is quite expensive, i.e., they had poor flexibility. There exists equipment on the market that is flexible and has good productivity but has some limitations in terms of the manufacturing of wires for car seat comfort systems, namely in the minimum dimension between folds in the central part of the cushions/suspension mats. This equipment usually has two bending tools controlled by CNC and a wire rotation system positioned between the tools [39–41]. These rotation systems usually limit the size of the wire in the central part. This identified the literature gap in the development of automatic wire-bending systems for the automotive industry and justifies the development of this work. It then became necessary to think of a new system from scratch, while taking advantage of some of the existing concepts. To this end, the subsystems presenting characteristics that fit the new wire-bending philosophy to be implemented were firstly identified. Thus, the strengths and weaknesses of the equipment on the market and in the industry were dissected through a benchmark that included the systems that fit the development to be carried out, as well as the typical characteristics of cushions/suspension mats. After selecting these subsystems, they were integrated into an innovative concept that yields levels of flexibility and productivity not achieved so far.

A new concept was then developed following design science research (DSR) [42–45], which integrated some existing subsystems with others developed from scratch, forming a set that responds to the intended requirements. DSR is a methodology used to develop new products from existing ones and can also be applied in other types of situations. Teixeira et al. [46] refers to DRS as a useful methodology due to its "technological background and its focus on developing models and methods that address complex and ill-defined problems". Siedhoff [47] developed a new sequence of nominal processes for DSR based on the work of Devitt and Robbins [48] to be implemented after general problem identification, combining design thinking with pre-existing DSR phases, including research activities and the cycle principle. The process was divided into exploration (problem clarification/definition and solution establishment) and prescriptive investigation, which, as shown by Peffers et al. [49] and Lepenioti et al. [50], are solution recommendations that lead to optimal decision making ahead of time. For the development of the new concept of wire bending used in the cushion manufacturing industry, the DSR methodology used by Siedhoff [46], as well as by Tojal et al. [51], was realized as the best option. It was intended to design an evolution of the previous equipment, creating a new concept while incorporating some knowledge previously acquired.

This work also wants to show that it is possible to develop equipment with high flexibility and high productivity following the right methodology because the product is properly known.

2. Benchmarking: Concepts Existing on the Market

The market presents several solutions to bending metallic wire, ranging from the most generic equipment, which customers have to adapt themselves to the features of the equipment and create products based on their limitations, to the most personalized equipment that seeks to satisfy some of the most particular needs of customers, being much more flexible. Because this flexibility is sufficient for almost all applications, these devices do not always have the necessary productivity. The continuous use of this type of equipment and deeper knowledge of the products and techniques needed to make them translate into knowledge that can be used to develop even more flexible and productive equipment. Thus, benchmarking is then carried out on the existing equipment on the market with a qualitative evaluation of the same, showing the capabilities and limitations.

The wire-bending machines shown in Figure 2 are the most versatile and intuitive machines on the market (Robomac concept). They consist of the following features: (a) a motorized unwinder; (b) a two-plane roller straightener; (c) a drag system with four drive wheels; (d) an encoder; and (e) a cutting system and two folding heads. The wire is pulled by the drag system, straightening the wire on the roller straightener, thus arriving perfectly straight at the bending tools. The machine is only capable of performing a single bend at each step; thus, the subproduct remains suspended under its weight during the last bends. These machines have a very intuitive and accessible interface for the user, allowing a 3D perspective of the wire (Figure 3) and adjustment based on the dimensions of the wire, rather than the positional reference of the bending tools.

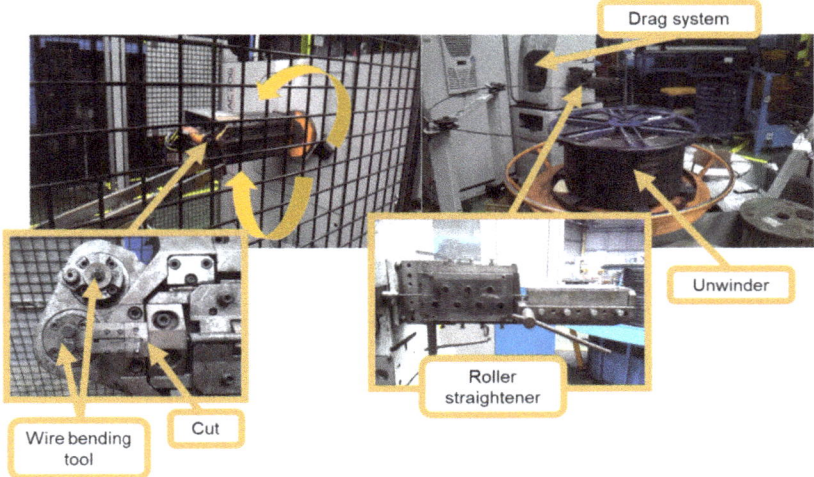

Figure 2. Common Robomac CNC wire-bending machine concept.

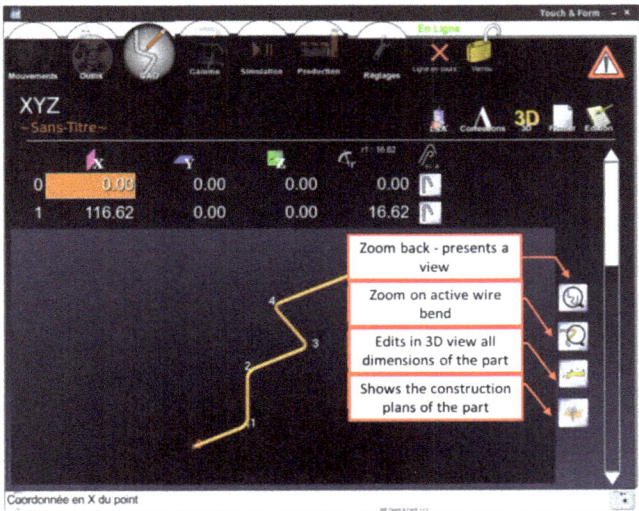

Figure 3. 3D view of the wire in the equipment console.

Wire-bending machines equipped with two CNC heads (Latour brand type) (Figure 4) maintain some of the sequence described in the previous example: unwinder, straightener and drag system. However, after dragging, the machine has a folding device that makes the hook and pre-folding (folds that most cushions need to have). Immediately afterwards, the cutting device is located, which is only actuated after the wire is in the CNC bending heads. Finally, the wire is transported by a transfer system to a stamping tool, where the wire is finished.

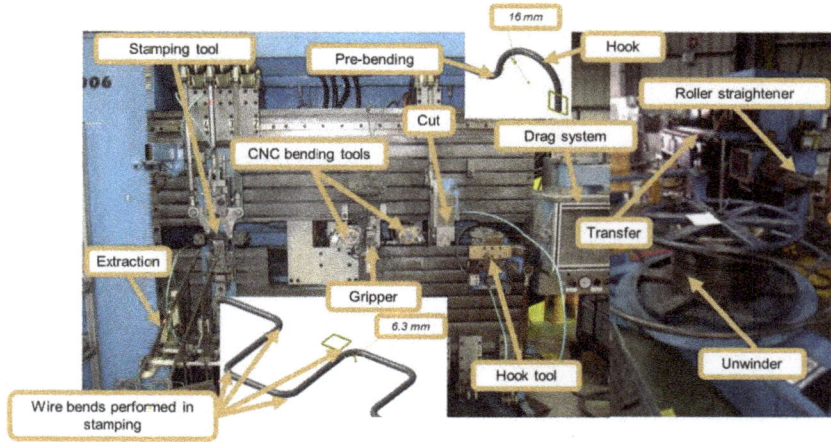

Figure 4. CNC double-head wire-bending machines.

There are also specific machines for bending wires, mainly for U-shaped bending (Inovmaq brand, for example), as shown in Figure 5. In this machine, the unwinder is not motorized, unlike the previous ones. After the unwinder, there is a two-plane roll straightener, a drag system with an encoder, a roll straightener, and a cutting device. After cutting, the wire is held by one of eight grippers that are on a rotating plate. The rotating plate moves the wire to the CNC bending stations, where the wire is formed. There are normally four CNC wire-bending stations, with the first station responsible for making the hooks and pre-bends and the last station responsible for making the two central bends.

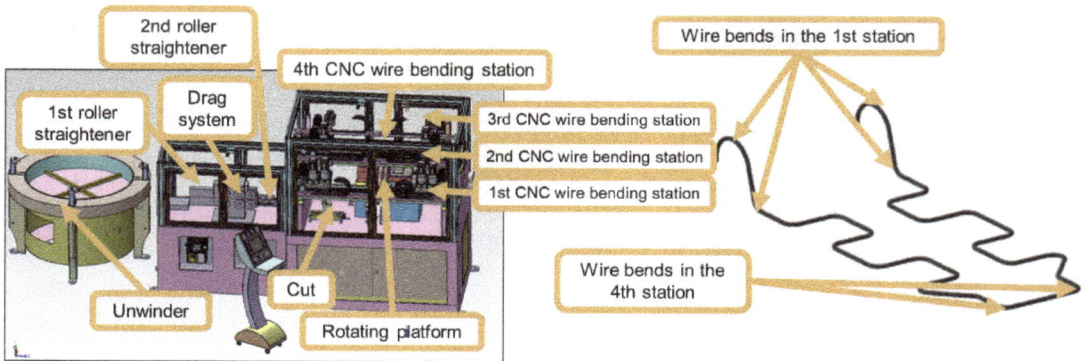

Figure 5. Specific wire-bending machine (U-shaped wires).

There are also specific wire-bending machines equipped with a transfer system (TEC, for example) shown in Figure 6, in which the wire placed in the buffer is supplied to the transfer clamps. On average, each rig has twenty clamps. The transfer, in its work cycle, places the wire at the bending stations and, on average, each machine has twenty of these stations. The bending stations have the particularity of being actuated by pneumatic cylinders, the adjustment of the bends being carried out by mechanical end-of-course stops.

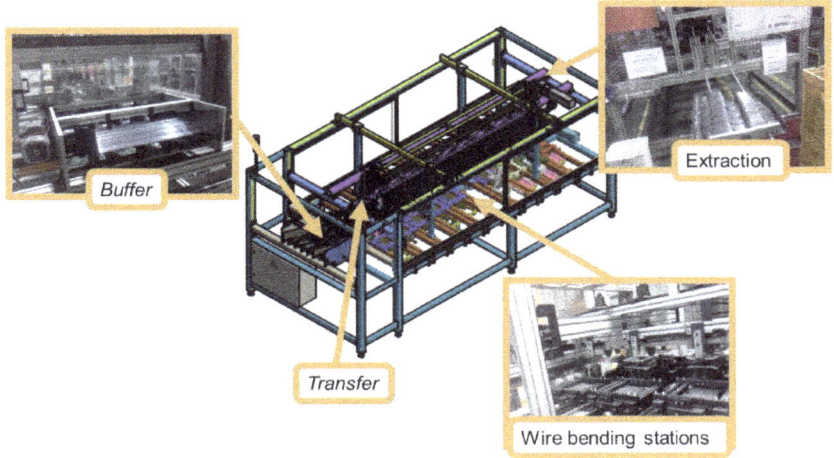

Figure 6. Specific wire-bending machine with transfer.

In the DH4012 VGP (BLM Group, Cantù, Italy) type bending equipment shown in Figure 7, the wire is used in a coil. The equipment consists of a roller straightener, a drag system, a CNC cutting system, two CNC bending heads and a rotating central gripper.

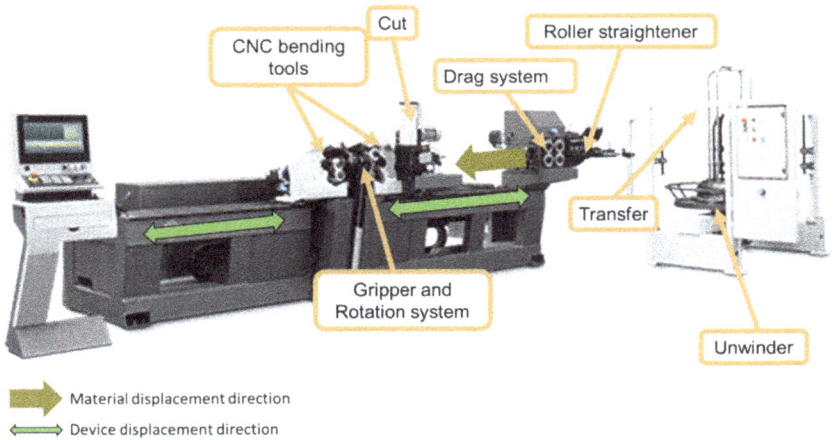

Figure 7. Wire-bending equipment DH4012 VGP (BLM Group).

The rotary gripper makes it possible to perform the different wire planes that need to be produced. Figure 8 details the configuration of the gripper with the bending heads. As the gripper is quite wide, it limits the central dimensions of the wire. This one has upward and downward vertical movement. When the gripper descends, it allows the bending heads to move closer together and shortens the central dimensions. However, in order to keep the wire fixed, these have clamps that are activated when the gripper is lowered.

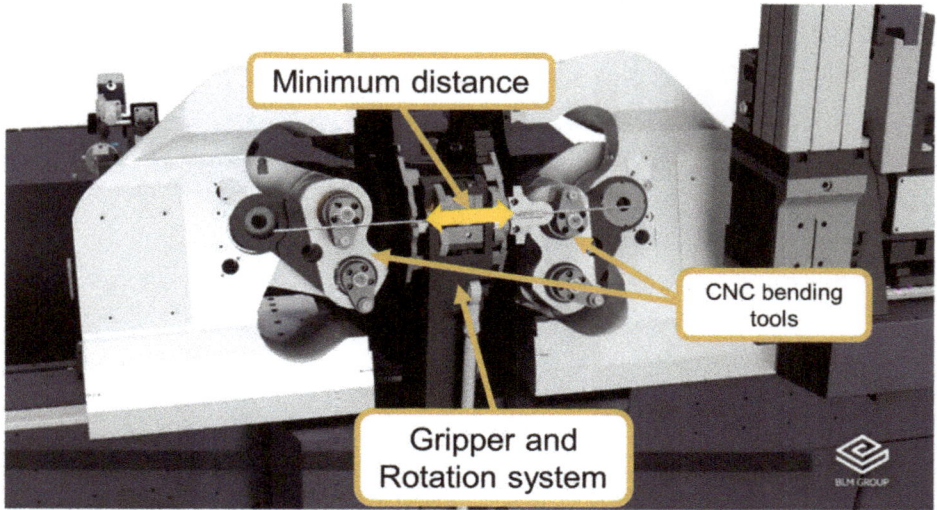

Figure 8. Central part of the equipment DH4012 VGP (BLM Group).

In the TWF-C07AD bending equipment shown in Figure 9, the concept is similar to the previous one, but wire fixation is promoted by a clamp provided with rotation movement. The system has two rotation axes, both with support at both ends of the machine. One has the center of the bending tools as a reference, while the other supports the gripper at the end in order to be able to supply and extract the wire.

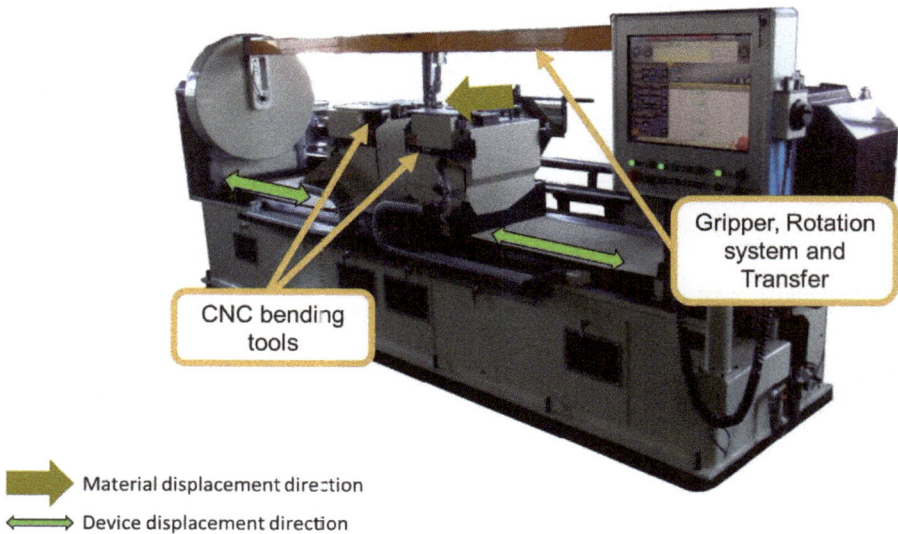

Figure 9. Wire-bending equipment TWF-C07 A D.

The BT5.2 bending equipment shown in Figure 10 presents a different approach. In this equipment, the bending heads are fixed, which is done in two different bending axes. The manipulators can move in X to change the folding position. In addition to this movement, they have two more rotation axes, one on the bending tool axis and another on the bottom to move between bending, feeding and extraction stations.

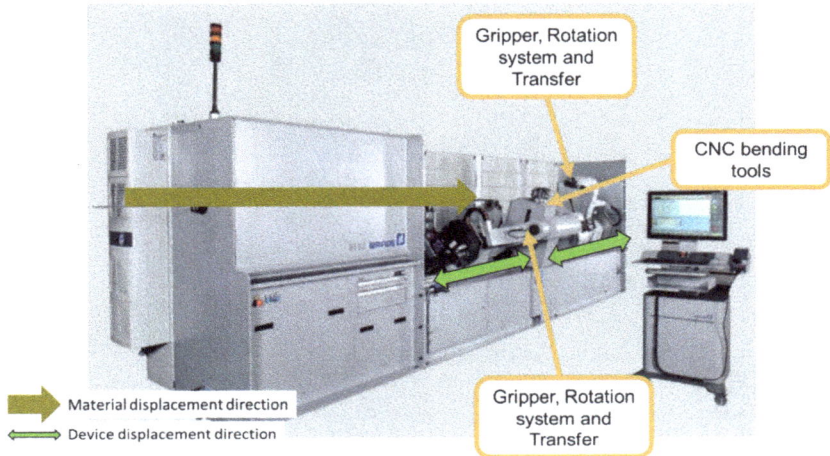

Figure 10. Wire-bending equipment BT5.2.

The BT5.2 equipment has a rotary straightener, whose approach is also different in relation to the previous ones, as can be seen in Figure 11. As it has two bending stations on different axes, the wire is forced to be fixed alternately by each one of the handlers. The manipulators have clamping systems, as well as folding stations, to ensure better transition between stations.

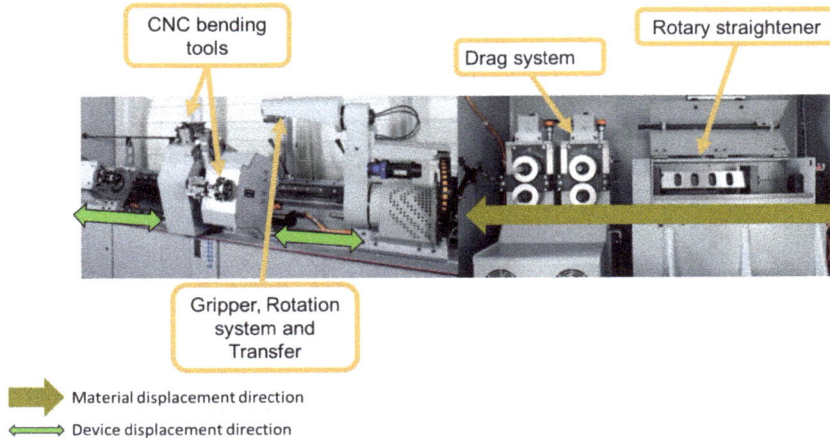

Figure 11. Details of the BT5.2 equipment.

The Gemini 7 equipment shown in Figure 12 has a central clamping system, and the bending heads rotate around the bending axis and move along the X axis, making the equipment more versatile.

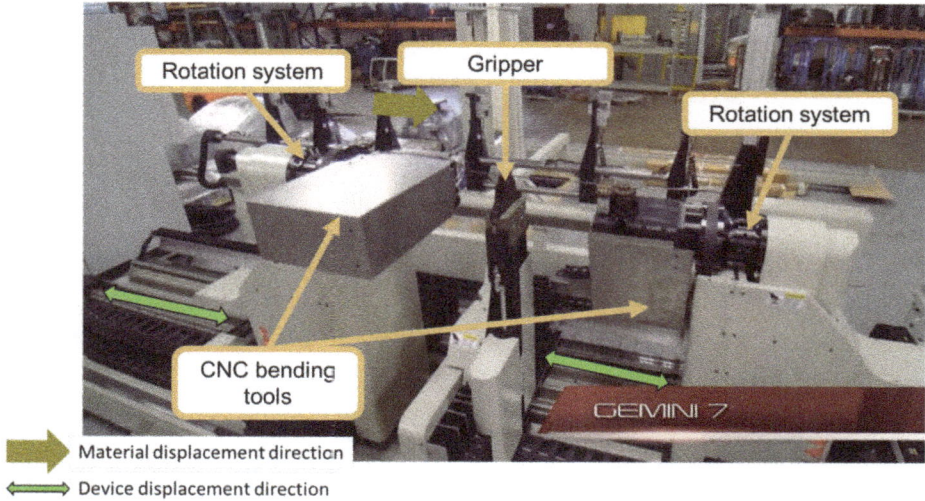

Figure 12. Wire-bending equipment Gemini 7.

The positioning and extraction of the wire are performed through two upper transfers, as can be seen in Figure 13. It has independent movement, thus not limiting the movement of the bending heads.

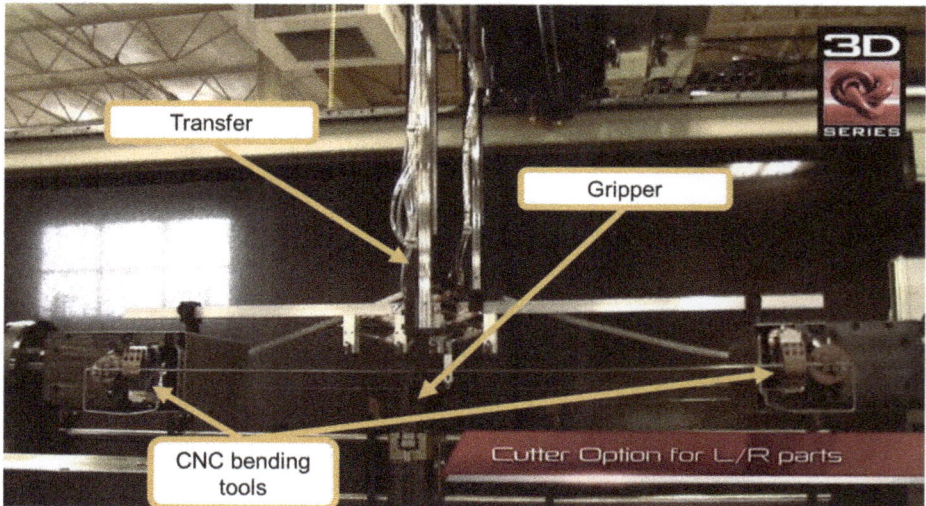

Figure 13. Upper transfers of Gemini 7.

The 3D-CN-8-M bending equipment shown in Figure 14 presents a different concept, in which the bending head is fixed and the manipulator is responsible for the displacement in X. However, it presents the particularity of having only a rotation axis in the manipulator and a very compact clamping system.

Figure 14. Wire-bending equipment 3D-CN-8-M [36].

Table 1 intends to summarize the main characteristics of the different concepts previously described in the perspective of the top management of any company. Thus, the main characteristics evaluated are flexibility, productivity and cost. This table does not intend to serve as a reference to further purchase decisions; thus, it is qualitative. Moreover, only some specific technical aspects are important to take them into consideration during the development of a new concept, taking the best ideas and trying to find solutions to overcome other ideas that do not work as well as expected regarding the requirements initially proposed.

Table 1. Some of the main features of the analyzed equipment.

Equipment	Flexibility	Productivity	Cost
Robomac concept	Very high	Low/Average	Average
Latour concept	High	Low	Expensive
Inovmaq concept	High	Low	Expensive
TEC concept	High	Low	Expensive
DH4012 VGP (BLM Group)	Average	Average	Expensive
TWF-C07AD concept	Average	Average	Average
BT5.2 concept	Average	Average	Expensive
Gemini 7 concept	High	Average	Average
3D-CN-8-M concept	Average	Average	Average

Through a summary description of the main characteristics of the equipment already on the market, it can be inferred that they have very particular advantages. However, they also have some limitations that are restrictive to combining a very high productive capacity and sufficient flexibility, in the same equipment, for any company to be able to easily change the type of wire produced without significantly interfering with its productivity, which would have obvious implications for the competitiveness of the manufactured products. This justifies the need to create a new equipment concept capable of combining flexibility with high productivity.

3. Methodology: Drawing the Formability Concept

The suspension system of a vehicle seat (Figure 15) has foam and flexible structures in wire following a spring shape, which have the capacity to filter vibrations. The level of vibration isolation varies depending on the type of vehicle. In sports cars, drivers seek to have a more direct connection with the car, with the aim of feeling the sensations that the car transmits better. As a result, more rigid seats are designed, that is, with a lower damping rate. Luxury cars have exactly the opposite requirements in that, in addition to elementary suspension systems, they have additional systems that allow adaptation to different occupants [52].

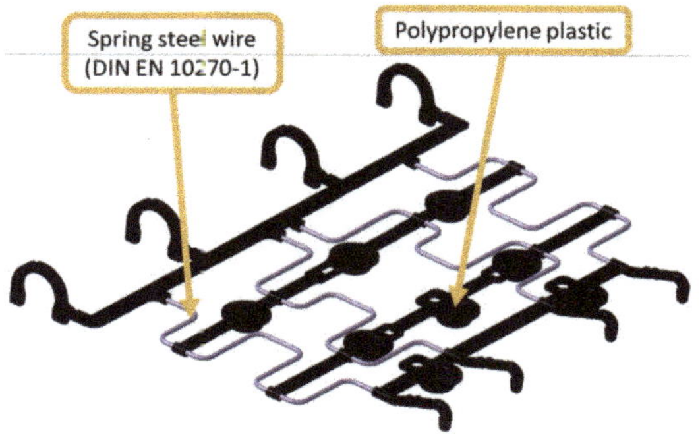

Figure 15. Car seat suspension system: cushion (own image).

Taking into account the components of the suspension system of a vehicle seat, in Figure 2, the marked component is the one that corresponds to the scope of this work. This is the cushion (also pointed out in Figure 1), which is conceived from the over-injection of plastic in previously formed wires. The plastic used is usually polypropylene and the wire is made from spring steel.

In the process of folding the cushions, as well as the suspension mats whose structure is very similar, equipment existing until now had a significant gap as the most flexible ones have low productivity, not being indicated for projects with a large production volume. On the other hand, there is also equipment that shows high productivity but is not flexible. Being more suitable for large-volume projects, that is, they follow the principle of fixed automation. For this reason, the modification of high-productivity equipment for new projects entails high costs, which customers are increasingly unwilling to pay, affecting the competitiveness of companies in the field.

In order to make the equipment more flexible, and after the wire class has been established, it is necessary to identify and characterize the wires produced. To this end, a survey about the following features of the wires needs to be performed:

- Wire diameter;
- Number of folds;
- Number of different radii per wire;
- Number of bending planes per wire;
- Wire length;
- Number of hooks;
- Minimum radius able to avoid crack induction on the wires.

A typical wire used in cushions is shown in Figure 16, which can be used as an example for the development of the concept. In this cushion, four wire references are used, in which the wire diameter is 3.4 mm, the wire class is SH (high tensile strength, following the EN 10270-1 standard), it has fourteen bends, two different radii, three planes of folding, two hooks and is 583 mm in length. The references are two symmetrical pairs.

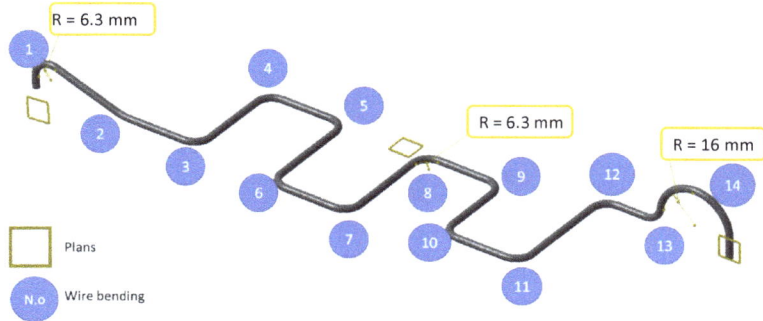

Figure 16. Typical cushion wire.

In Figure 16, folds 1 and 14 are commonly referred to as hooks, and at a minimum, each pad has a hook. Based on a company producing this type of components for seats of motor vehicles, a typification study of its main characteristics was carried out to guide the design of the new equipment. Table 2 shows a summary from which it can be concluded that the wire diameter varies between 3 and 4 mm, the number of bending planes varies between two and five, the number of hooks varies between one and two, the number of radii used in each wire cushion range varies between two and four, the length of the wire fluctuates between 400 mm and 1120 mm, the number of bends varies from thirteen to twenty-five and the wire grade is SH or SM.

Table 2. Characteristics of the wires used in cushions production.

Reference	Diameter	Wire Class	No. of Wire Bends	No. of Wire Rays	No. of Wire Plans	Length (mm)	No. of Hooks
1	3	SH	25	2	2	1115.8	2
2	4	SH	12	2	3	563.2	1
3	3.5	SM	12	2	5	465.4	1
4	3.5	SM	11	2	2	432.3	1

SH: high tensile strength (following the EN 10270-1 standard). SM: medium tensile strength (following the EN 10270-1 standard).

4. The New Concept

Thus, a new equipment concept was conceived and designed using Catia V6 software (Dassault Systèmes), which is schematically represented in Figure 17. In this new concept, it was understood that the wire was pulled by the drag system, being subsequently taken to the straight position through a two-plane roller straightener. After dragging, the wire is cut by the cutting device, which has movement capacity to perform various cut lengths, thus meeting the intended flexibility requirement. The wire piece is fixed in a system of clamps, ready to be picked up by the manipulator. The manipulator, in addition to transporting the wire between stations, has a rotation system, which yields several planes in the wire-bending process. The manipulator will take the wire to the cutting system, moving it to the CNC bending stations and to the final bending stations. For increased cadence, the center post has a hook fold and pre-fold station, similar to what Latour equipment uses. After the CNC folding stations, it has a folding station that makes the last two central folds. It is similar to the last station usually used by Latour and Inovmaq (Figure 17).

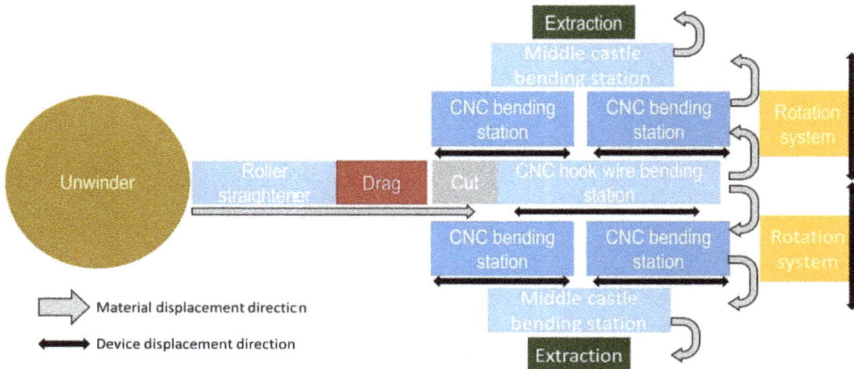

Figure 17. New concept of high-productivity/high-flexibility wires bending machine.

What makes this concept more innovative are the stations used to make the hooks and the final folds, as shown in Figure 18. The station used to make the hook, in addition to allowing the overlapping of operations as allowed by the Latour concept, also increases flexibility as its position is given by a servomotor, together with the cutting device.

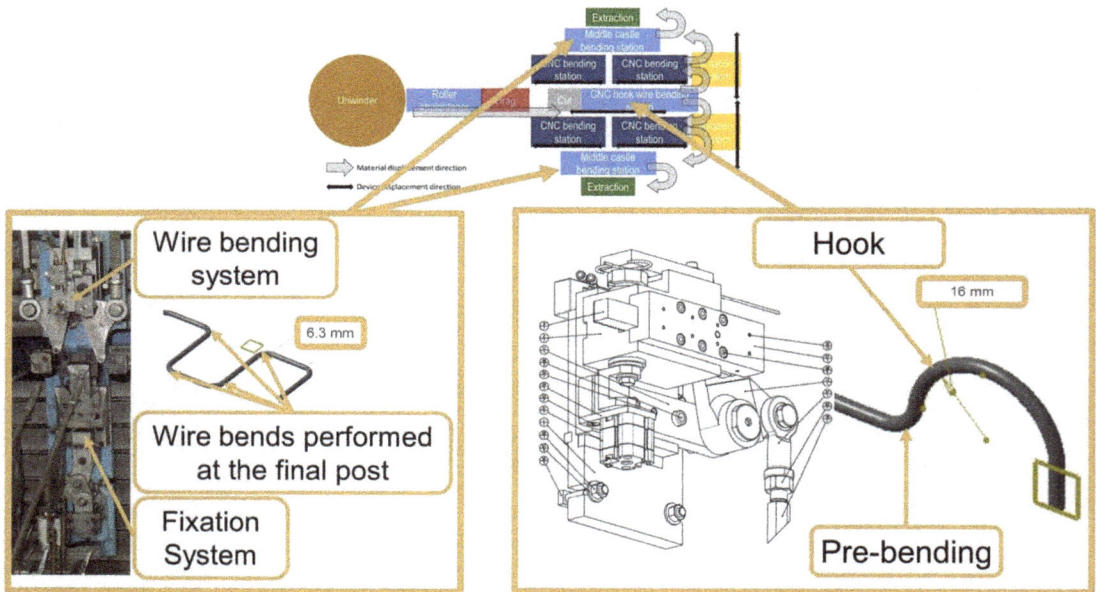

Figure 18. Wire-bending station added to the concept.

Like most of the concepts already presented, this equipment has an unwinder, roll straightener and drag, as can be seen in Figure 19. In this concept, the equipment can be divided into three large groups of subsets, as it is possible to observe in Figure 20. These groups are the following:

- Central post;
- Left post;
- Right post.

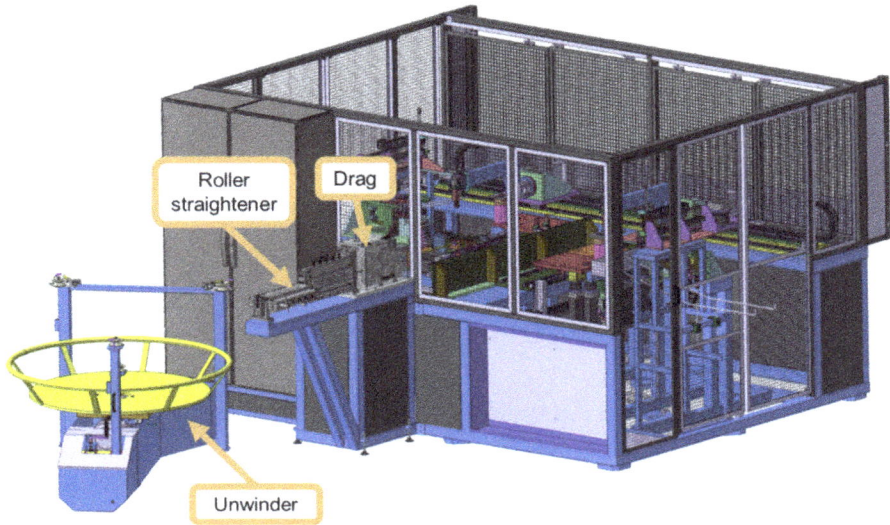

Figure 19. An overview of the entire concept developed as it can be installed on the shopfloor.

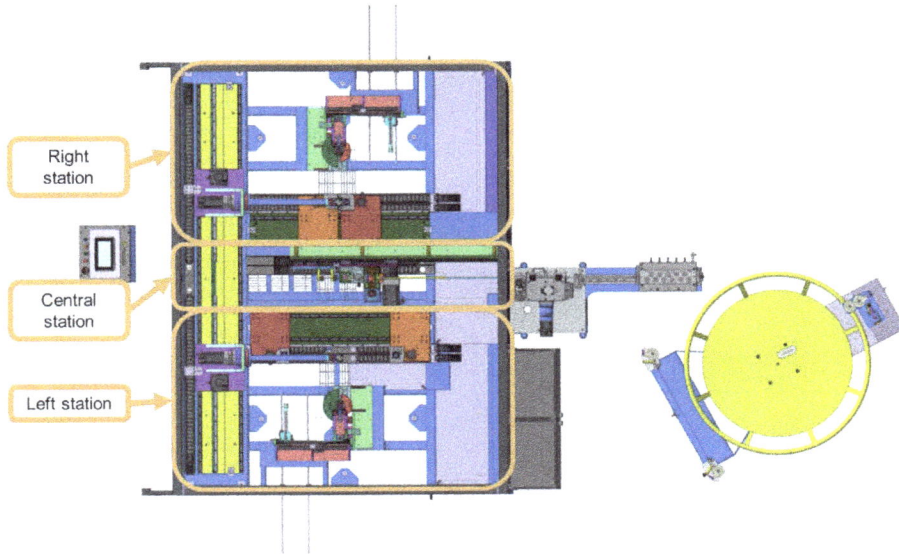

Figure 20. Wire-bending equipment, new concept.

At the central station, which can be seen in Figure 21, the raw material is fed into coils, thus avoiding the use of cutting equipment and the existence of intermediate stocks, as in the TEC concept (Figure 8). At the central station, the coil is placed in the unwinder where it is unwound, maintaining constant tension. The wire is pulled by the drag system, forcing it to unwind and straighten. Wire alignment is carried out when passing through the two-plane roller straightener. After dragging, the wire passes through the cutting system and reaches the hook tool, where hook bending and pre-bending takes place. With these operations carried out, more wire is fed until it reaches its cutting position. Once in the cutting position, the support claws are activated, and then the cut is performed. Finally, the device that performs the Hook + Pre-bend + Cut set withdraws, and the wire is available for the subsequent operations. The devices that make up the central station are as follows:

- Device that performs the Hook + Pre-fold + Cut;
- Support claws.

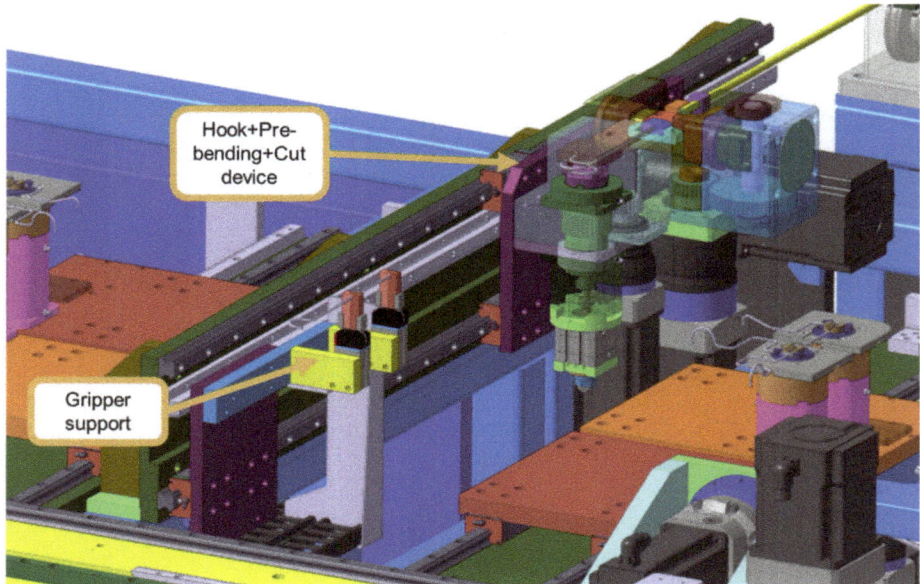

Figure 21. Central station.

The wire available at the central station is picked up by the left or right handler (Figure 22). Afterwards, it moves to the CNC bending devices. Here, several bends are performed, always with the wire fixed to the manipulator. With the bends completed, it moves to the middle fixation device. The wire is fixed in this device and the central bends of the wire are performed. Finally, the extraction system removes the shaped wire and drop-off it onto the equipment's exit ramps.

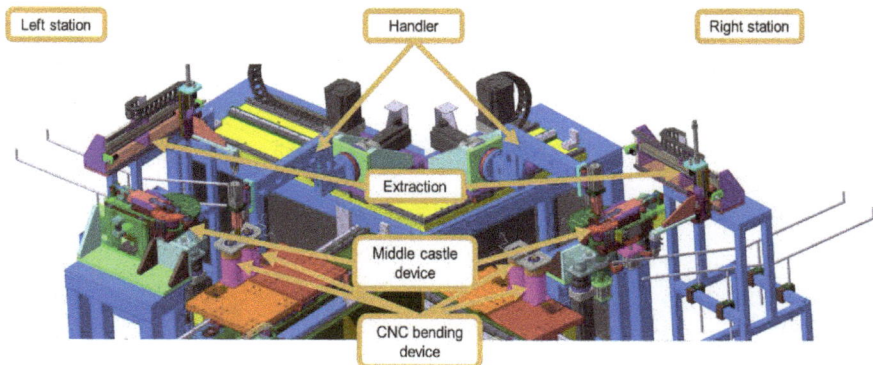

Figure 22. Left and right station.

The devices that make up the left and right stations are as follows:

- CNC bending device;
- Middle fixing device;
- Handler;
- Extraction.

The equipment is capable of bending wire from 3 to 4 mm, with a maximum length of 1100 mm. The wire is dispensed on both sides of the equipment, as shown in Figure 23.

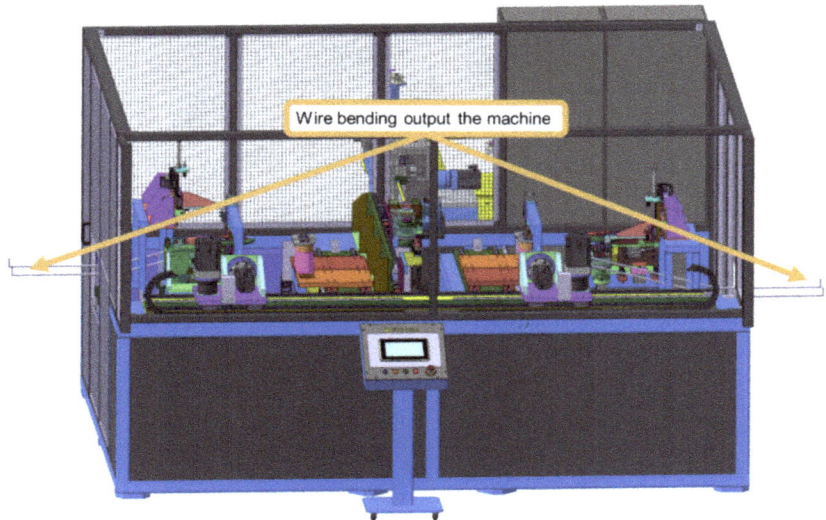

Figure 23. Wire-bending output machine.

In terms of productivity, and considering only five of several references produced by the company used as target for the implementation of this concept and considering the concept now developed face to face with the TEC and the Robomac concepts previously described, a comparative study is presented in Table 3, yielding an understanding of the differences.

In addition to productivity gains, this concept yields the implementation of new wire references, changing only the tool kit, as can be seen from the different references already implemented. Therefore, this concept is proven to be much more flexible and productive than any of the concepts already existing on the market. The flexibility of this concept is also verified in the ability to produce two symmetrical references simultaneously, or even different references on each side of the equipment, a situation that is also innovative, as there is currently no equipment that uses this concept and is capable of allowing different programming for each side of the machine. This point is extremely important for the manufacture of bent wires for car seats, as it is extremely common for each component on one side to have a corresponding one on the other side (symmetrical). However, in certain situations, there are differences in detail (no symmetry), which can be easily managed by this new equipment, producing equivalent parts, although with specific formats. This possibility significantly facilitates the management of stocks and setups, as production is flexible, allowing the perfect adjustment of the quantity to be produced to the quantity ordered by the customer and, when it comes to symmetrical components, the quantity obtained from one side of the equipment is exactly equal to that obtained from the other (imposed by the cadence of the equipment), which makes it possible to complete the amount of shaped wires necessary so that the cushions or suspension mats can be produced immediately (the next product where the bent wires are applied), with a shorter lead time. Thus, the advantages presented by this innovative concept are clear, both in terms of productivity and flexibility, as well as the advantages in terms of internal logistics that it provides.

Table 3. Productivity analysis regarding the different considered concepts.

Type of Wire	Output Robomac	Output TEC	Output of the New Concept
	270	800	900
	260	780	880
	250	765	875
	240	Never produced	860
	220	Never produced	850

The cost of this new concept is 3% lower than that of the TEC concept because the last one uses a high number of mechanical parts. Taking into account the Robomac concept, it is 19% higher. However, the productivity gain is around 230%, largely offsetting any disadvantage in the initial investment. With regard to the necessary investment in tools, there are significant differences in terms of their cost. The cost of the tools for the new equipment is about the double that of the Robomac tools cost, which is compensated by increased productivity. On the other hand, the tools of the new concept are 87% lower than the cost of new tools for the TEC concept. It is easy to understand that, although some differences exist, the increased productivity provided by the new concept overlaps the small costs of the tools and allows a flexibility level completely incomparably higher.

5. Discussion

This work was developed with the aim of solving the need for concrete of companies that produce components for car seats, whose products have a relatively low added value, and where any increase in productivity without deteriorating the quality is extremely welcome. Analyzing existing equipment on the market (benchmarking), it was possible to detect an opportunity to develop a new concept that would significantly increase productivity, without jeopardizing flexibility. The concept aims to take advantage of the dead time that the wire feeding system has in relation to bending systems, which are necessarily slower. A concept was then conceived that aimed to achieve double production, doubling the bending systems around a common wire feeding system.

Given that vehicle customization led to an increase in the model variants available to customers, setup time has become a problem that needs to be resolved quickly as tool changes and process fine-tuning heavily penalizes productivity [13]. It also became necessary to create an agile system for changing tools, which were not excessively expensive, but which fully complied with what was intended, i.e., to switch production from one model to another in an extremely expeditious manner [14–16].

In addition to this, it was possible to establish two ways in the same equipment, programmable in a different way, which allow two different references or mirrored references, to be carried out simultaneously. In terms of bending-wire operations with this complexity, this is a novel concept.

In addition to extremely significant gains in terms of productivity and flexibility, it was also found that, in terms of internal logistics, there are significant advantages. In fact, two symmetrical models can be produced simultaneously, producing exactly the same amount of both products at the same time, which allows these wires to move simultaneously to the next production step (manufacturing cushions or suspension mats) simultaneously, which allows a much more organized production management. In case there is a need to produce different products, lead time is also greatly reduced, as it is not necessary to finish the production of a model to start the production of another. Thus, industrial management is much more agile, allowing different combinations of products as necessary to respond to demand.

Thus, it can be said that this work is in line with some work that has already been carried out in this area [7,33,34,52], where an attempt was made to increase productivity, also significantly improving the logistics around these processes that normally have low added value. Management techniques, in many cases, solve serious problems in terms of lack of competitiveness in products of this nature [16,36,37], but sometimes it is even necessary to promote necessary changes in terms of the equipment involved. Automation plays an extremely relevant role in this regard, mainly flexible automation, but it is necessary to select the best set of devices so that the desired results can be obtained [4]. In addition, a high critical state of mind is necessary, so that truly innovative solutions can be obtained, disruptive with what has been conducted so far, in order to achieve new standards of productivity and flexibility [4,19].

Figure 24 intends to illustrate the implementation of the previously described concept, which only became viable through a full demonstration of the advantages that it would entail for an increase in flexibility and productivity. The equipment is based on 1400 different parts and devices, comprising electrical actuators, pneumatic actuators, sensors, electrical components, standard parts and several mechanical parts specifically designed to this equipment, especially conceived to be easily replaced through planned maintenance operations.

Figure 24. Image of the real equipment after assembling.

6. Conclusions

Given the characteristics of the wires and the volumes of the projects, budgetary restrictions and production objectives are usually imposed. An innovative solution for wire-bending equipment is developed, requiring a low budget and complying with the requirements and constraints initially pointed out. The concept consists of equipment provided with three stations, a central one where the wire is unwound from the coil, straightened, shaped (double bends are made) and cut. In the lateral stations, left and right, the wire, after being picked up by the manipulator at the central station, is shaped in CNC bending devices with the wire always fixed in the manipulator. At the end, it is placed by the manipulator at the final bending station to finish shaping the wire. After forming, the wire is extracted onto material exit guides.

The implementation of this new concept yielded an output in productivity of around 230% in relation to flexible equipment previously existing, and a gain of 12.5% in relation to high-cadence equipment on the market. In addition to productivity gains, this concept yields the implementation of new wire references by changing only the tool kit. The flexibility of this concept is also verified in the ability to produce two symmetrical references simultaneously, or even different references (albeit similar). This point is extremely important in the management of stocks and setups.

The cost of this new concept is 3% lower than the TEC concept and 19% higher, relative to the Robomac concept, but with the previously described productivity gain.

The cost of new tools is double that of the Robomac concept, which is 87% cheaper than the cost of new tools for the TEC concept. It is thus verified that this concept is much more competitive in the implementation of new projects. The new concept was also thought to be easier to repair. Knowing the main drawbacks of the other equipment, some solutions were developed, taking into account forecasted problems. Moreover, because all development is made "in house" mode, the domain of all devices and mechanical parts are easier to deal with. A maintenance guide is provided to the maintenance department with all care needed in the different zones of the equipment, identifying and solving the most common problems.

Table 4 intends to summarize the main advantages and limitations of the developed concept.

Table 4. Advantages and limitations of the new concept concepts.

	Advantages	Limitations
New concept	Lower budget needed Feeding system feds two bending stations Production time optimized Higher production than flexible equipment (+230%) Higher production than high pace equipment on the market (+12.5%) Possibility of producing simultaneously different product references Setup: tool kit change Tool kit cheaper than TEC concept Easier maintenance	Cost 19% higher than Robomac concept Tool kit costs the double of the Robomac concept

A video of the equipment working in its initial phase can be observed in Supplementary Materials S1 attached to this work.

This work followed the design search research methodology, taking solutions already implemented in the market as a starting point and promoting an evolution of knowledge from these initial ideas, reaching a level of knowledge that can be shared and expanded to other business areas, equally successfully. In fact, the knowledge produced can be used in other industries linked to the same sector, or even to different sectors, yielding the development of new concepts with greater efficiency. Indeed, the DSR methodology structured the thought and achieved the desired results, thus verifying that it is extremely useful in the study of upgrading production systems.

Supplementary Materials: The following supporting information can be downloaded at: https://www.mdpi.com/article/10.3390/machines11030410/s1, Video S1: Equipment working.

Author Contributions: M.G.: Conceptualization, main research and data collection; F.J.G.S.: Work orientation, Investigation, Methodology, Supervision, Writing—Reviewing; Investigation, Formal analysis; A.G.P.: Co-Supervision, Formal analysis, Visualization; R.D.S.G.C.: Formal analysis, Visualization, Writing—Reviewing. All authors have read and agreed to the published version of the manuscript.

Funding: This research received no external funding.

Data Availability Statement: Not applicable.

Acknowledgments: The authors would like to thank Mário Cardoso from FicoCables, Lda. due to its permanent strong support. F.J.G. Silva, A.G. Pinto and R.D.S.G. Campilho also thank INEGI—Institute of Science and Innovation in Mechanical and Industrial Engineering, Porto, Portugal, due to their continuous support on applied research.

Conflicts of Interest: The authors declare no conflict of interest.

References

1. Munten, P.; Vanhamme, J.; Maon, F.; Swaen, V.; Lindgreen, A. Addressing tensions in coopetition for sustainable innovation: Insights from the automotive industry. *J. Bus. Res.* **2021**, *136*, 10–20. [CrossRef]
2. OICA. "After 2019's Halt to 10 Years of Industry Growth, the World Auto Industry Faces a Huge 2020 Crisis, but will Once Again Show Its Resilience". Available online: http://www.oica.net/category/media-center/ (accessed on 20 November 2019).
3. OICA, Production Statistics. Available online: https://www.oica.net/category/production-statistics/2019-statistics/ (accessed on 5 November 2019).
4. Lenort, R.; Wicher, P.; Zapletal, F. On influencing factors for Sustainable Development goal prioritisation in the automotive industry. *J. Clean. Prod.* **2023**, *387*, 135718. [CrossRef]
5. Mohamad, M.; Songthaveephol, V. Clash of titans: The challenges of socio-technical transitions in the electrical vehicle technologies—The case study of Thai automotive industry. *Technol. Forecast. Soc. Chang.* **2020**, *153*, 119772. [CrossRef]
6. Palea, V.; Santhià, C. The financial impact of carbon risk and mitigation strategies: Insights from the automotive industry. *J. Clean. Prod.* **2022**, *344*, 131001. [CrossRef]
7. Araújo, W.F.S.; Silva, F.J.G.; Campilho, R.D.S.G.; Matos, J.A. Manufacturing cushions and suspension mats for vehicle seats: A novel cell concept. *Int. J. Adv. Manuf. Technol.* **2017**, *90*, 1539–1545. [CrossRef]
8. Zhang, X.; Ming, X.; Bao, Y. A flexible smart manufacturing system in mass personalization manufacturing model based on multi-module-platform, multi-virtual-unit, and multi-production-line. *Comput. Ind. Eng.* **2022**, *171*, 108379. [CrossRef]
9. Da Rold, A.; Furiato, M.; Zaki, A.M.A.; Carnevale, M.; Giberti, H. Deep learning-based robotic sorter for flexible production. *Procedia Comput. Sci.* **2023**, *217*, 1579–1588. [CrossRef]
10. Bhatta, K.; Li, C.; Chang, Q. Production Loss Analysis in Mobile Multi-skilled Robot Operated Flexible Serial Production Systems. *Manuf. Lett.* **2022**, *33*, 835–842. [CrossRef] [PubMed]
11. Costa, R.J.S.; Silva, F.J.G.; Campilho, R.D.S.G. A novel concept of agile assembly machine for sets applied in the automotive industry. *Int. J. Adv. Manuf. Technol.* **2017**, *91*, 4043–4054. [CrossRef]
12. Moreira, B.M.D.N.; Gouveia, R.M.; Silva, F.J.G.; Campilho, R.D.S.G. A novel concept of production and assembly processes integration. *Procedia Manuf.* **2017**, *11*, 1385–1395. [CrossRef]
13. Sarkar, M.; Park, K.S. Reduction of makespan through flexible production and remanufacturing to maintain the multi-stage automated complex production system. *Comput. Ind. Eng.* **2023**, *177*, 108993. [CrossRef]
14. Silva, F.J.G.; Campilho, R.D.S.G.; Sousa, V.F.C.; Coelho, L.F.P.; Ferreira, L.P.; Pereira, M.T.; Matos, J. A New Concept of Jig Rotary Holder System for 3-Axis CNC Milling Machine Operated by the Main Machine Control. *J. Test. Eval.* **2022**, *50*, 20210723. [CrossRef]
15. Sousa, E.; Silva, F.J.G.; Ferreira, L.P.; Pereira, M.T.; Gouveia, R.; Silva, R.P. Applying SMED methodology in cork stoppers production. *Procedia Manuf.* **2018**, *17*, 611–622. [CrossRef]
16. Vieira, A.M.; Silva, F.J.G.; Campilho, R.D.S.G.; Ferreira, L.P.; Sá, J.C.; Pereira, T. SMED methodology applied to the deep drawing process in the automotive industry. *Procedia Manuf.* **2020**, *51*, 1416–1422. [CrossRef]
17. Ikome, J.M.; Laseinde, O.T.; Katumba, M.G.K. The Future of the Automotive Manufacturing Industry in Developing Nations: A Case Study of its Sustainability Based on South Africa's Paradigm. *Procedia Comput. Sci.* **2022**, *200*, 1165–1173. [CrossRef]
18. Anzolin, G.; Andreoni, A.; Zanfei, A. What is driving robotisation in the automotive value chain? Empirical evidence on the role of FDIs and domestic capabilities in technology adoption. *Technovation* **2022**, *115*, 102476. [CrossRef]
19. Silva, F.J.G.; Gouveia, R.M. *Cleaner Production—Toward a Better Future*; Springer Nature: Cham, Switzerland, 2020; ISBN 978-3-030-23165-1.

20. Guo, X.; Liu, X.; Long, G.; Zhao, Y.; Yuan, Y. Data-driven prediction of the fatigue performance of corroded high-strength steel wires. *Eng. Fail. Anal.* **2023**, *146*. 107108. [CrossRef]
21. Groover, M.P. *Automation, Production Systems, and Computer-Integrated Manufacturing (Fourth Edition)*; Pearson High Education: New York, NY, USA, 2015; ISBN 978-0-13-349961-2.
22. Lavvaf, H.; Lewandowski, J.R.; Lewandowski, J.J. Flex bending fatigue testing of wires, foils, and ribbons. *Mater. Sci. Eng. A* **2014**, *601*, 123–130. [CrossRef]
23. Gupta, S.; Pelton, A.R.; Weaver, J.D.; Gong, X.-Y.; Nagaraja, S. High compressive pre-strains reduce the bending fatigue life of nitinol wire. *J. Mech. Behav. Biomed. Mater.* **2015**, *44*, 96–108. [CrossRef]
24. Zhang, D.; Feng, C.; Chen, K.; Wang, D.; Ni, X. Efect of broken wire on bending fatigue characteristics of wire ropes. *Int. J. Fatigue* **2017**, *103*, 456–465. [CrossRef]
25. Chen, Y.; Chen, J.; Zhang, Y.; He, Y.; Xu, J.; Xiang, J. Effect of internal defect on the low-cycle bending fatigue behavior of a single-strand wire rope. *Constr. Build. Mater.* **2022**, *350*, 128874. [CrossRef]
26. Wang, G.; Ma, Y.; Guo, Z.; Bian, H.; Wang, L.; Zhang, J. Fatigue life assessment of high-strength steel wires: Beach marks test and numerical investigation. *Constr. Build. Mater.* **2022**, *323*, 126534. [CrossRef]
27. Chen, C.; Jie, Z.; Wang, K. Fatigue life evaluation of high-strength steel wires with multiple corrosion pits based on the TCD. *J. Constr. Steel Res.* **2021**, *186*, 106913. [CrossRef]
28. Weaver, J.D.; Sena, G.M.; Falk, W.M.; Sivan, S. On the influence of test speed and environment in the fatigue life of small diameter nitinol and stainless-steel wire. *Int. J. Fatigue* **2022**, *155*, 106619. [CrossRef] [PubMed]
29. Xue, S.; Shen, R.; Chen, W.; Miao, R. Corrosion fatigue failure analysis and service life prediction of high strength steel wire. *Eng. Fail. Anal.* **2020**, *110*, 104440. [CrossRef]
30. Pal, U.; Mukhopadhyay, G.; Sharma, A.; Bhattacharya, S. Failure analysis of wire rope of ladle crane in steel making shop. *Int. J. Fatigue* **2018**, *116*, 149–155. [CrossRef]
31. Cao, X.; Wu, W. The establishment of a mechanics model of multi-strand wire rope subjected to bending load with finite element simulation and experimental verification. *Int. J. Mech. Sci.* **2018**, *142*, 289–303. [CrossRef]
32. Bonneric, M.; Aubin, V.; Durville, D. Finite element simulation of a steel cable—Rubber composite under bending loading: Infuence of rubber penetration on the stress distribution in wires. *Int. J. Solids Struct.* **2018**, *160*, 158–167. [CrossRef]
33. Silva, F.J.G.; Soares, M.R.; Ferreira, L.P.; Alves, A.C.; Brito, M.; Campilho, R.D.S.G.; Sousa, V.F.C. A Novel automated system for the handling of car seat wires on plastic over-injection molding machines. *Machines* **2021**, *9*, 141. [CrossRef]
34. Magalhães, A.J.A.; Silva, F.J.G.; Campilho, R.D.S.G. A novel concept of bent wires sorting operation between workstations in the production of automotive parts. *J. Braz. Soc. Mech. Sci. Eng.* **2019**, *41*, 25. [CrossRef]
35. Baraldo, A.; Bascetta, L.; Caprotti, F.; Ferretti, G.; Ponti, A.; Sakcak, B. Damping oscillations in a wire bending process. *Mechatronics* **2022**, *86*, 102846. [CrossRef]
36. Rosa, C.; Silva, F.J.G.; Ferreira, L.P.; Campilho, R. SMED methodology: The reduction of setup times for Steel Wire-Rope assembly lines in the automotive industry. *Procedia Manuf.* **2017**, *13*, 1034–1042. [CrossRef]
37. Rosa, C.; Silva, F.J.G.; Ferreira, L.P.; Pereira, T.; Gouveia, R. Establishing standard methodologies to improve the production rate of assembly lines used for low added value products. *Procedia Manuf.* **2018**, *17*, 555–562. [CrossRef]
38. Saggu, G. 3D Silicone Whipping Additive Manufacturing (SWAM): Technology, Applications, and Research Needs. Master's Thesis of Applied Science in Chemical Engineering, University of Waterloo, Waterloo, ONT, Canada, 2022. Available online: https://uwspace.uwaterloo.ca/bitstream/handle/10012/17932/Saggu_Gurkamal.pdf?sequence=3 (accessed on 13 February 2023).
39. Numalliance. Available online: https://www.rumalliance.com/en/machines/robomac-tf-en/ (accessed on 5 January 2022).
40. Wafios. Available online: https://www.wafios.com/en/downloads/wire-bending-machines/ (accessed on 5 January 2022).
41. Ashfield-Springs. Available online: https://www.ashfield-springs.com/robomac/ (accessed on 5 January 2022).
42. Webber, R. Design-science research. In *Research Methods*, 2nd ed.; Williamson, K., Johanson, G., Eds.; Chandos Publishing, Elsevier: Amsterdam, The Netherlands, 2018. [CrossRef]
43. Abdullah, O.I.; Abbood, W.T.; Hussein, H.K. Development of automated liquid filling system based on the interactive design approach. *FME Trans.* **2020**, *48*, 838–945. [CrossRef]
44. Tamada, S.; Chandra, M.; Patra, P.; Mandol, S.; Bhattacharjee, D.; Dan, P.K. Modeling for design simplification and power-flow efficiency improvement in an automotive planetary gearbox: A case example. *FME Trans.* **2020**, *48*, 707–715. [CrossRef]
45. Van der Borgh, M.; Xu, J.; Sikkenk, M. Identifying, analyzing, and finding solutions to the sales lead black hole: A design science approach. *Ind. Mark. Manag.* **2020**, *88*, 136–151. [CrossRef]
46. Teixeira, J.G.; Patrício, L.; Huang, K.H.; Fisk, R.P.; Nóbrega, L.; Constantine, L. The MINDS Method: Integrating Management and Interaction Design Perspectives for Service Design. *J. Serv. Res.* **2017**, *20*, 240–258. [CrossRef]
47. Siedhoff, S. *Design Science Research, Seizing bus. Model Patterns and Disruptive Innovations*; Springer: Cham, Switzerland, 2019; pp. 29–43, ISBN 978-3658263355.
48. Devitt, F.; Robbins, P. Design, Thinking and Science. In *Design Science: Perspectives from Europe. EDSS 2012*; Helfert, M., Donnellan, B., Eds.; Communications in Computer and Information Science; Springer: Cham, Switzerland, 2013; Volume 388. [CrossRef]
49. Peffers, K.; Tuunanen, T.; Rothenberger, M.A.; Chatterjee, S. Positioning and presenting design science research for maximum impact. *J. Manag. Inf. Syst.* **2007**, *24*, 45–77. [CrossRef]

50. Lepenioti, K.; Bousdekis, A.; Apostolou, D.; Mentzas, G. Prescriptive analytics: Literature review and research challenges. *Int. J. Inf. Manag.* **2020**, *50*, 57–70. [CrossRef]
51. Tojal, M.C.; Silva, F.J.G.; Campilho, R.D.S.G.; Pinto, A.G.; Ferreira, L.P. Case-based product development of a high-pressure die casting injection subset using design science research. *FME Trans.* **2022**, *50*, 32–45. [CrossRef]
52. Silva, F.J.G.; Swertvaegher, G.; Campilho, R.D.S.G.; Ferreira, L.P.; Sá, J.C. Robotized solution for handling complex automotive parts in inspection and packing. *Procedia Manuf.* **2020**, *51*, 156–163. [CrossRef]

machines

MDPI

Article

A Method for Measurement of Workpiece form Deviations Based on Machine Vision

Wei Zhang [1,*] , Zongwang Han [1,2], Yang Li [1] , Hongyu Zheng [1] and Xiang Cheng [1]

[1] School of Mechanical Engineering, Shandong University of Technology, Zibo 255000, China
[2] School of Mechanical Engineering, University of Shanghai for Science and Technology, Shanghai 200093, China
* Correspondence: zw062003@163.com

Abstract: Machine vision has been studied for measurements of workpiece form deviations due to its ease of automation. However, the measurement accuracy limits its wide implementation in industrial applications. In this study, a method based on machine vision for measurement of straightness, roundness, and cylindricity of a workpiece is presented. A subsumed line search algorithm and an improved particle swarm optimization algorithm are proposed to evaluate the straightness and roundness deviations of the workpiece. Moreover, an image evaluation method of cylindricity deviation by the least-square fitting of the circle's center coordinates is investigated. An image acquisition system incorporating image correction and sub-pixel edge positioning technology is developed. The performance of the developed system is evaluated against the measurement results of the standard cylindricity measuring instrument. The differences in the measurement of straightness, roundness, and cylindricity are −4.69 μm, 3.87 μm, and 8.51 μm, respectively. The proposed method would provide a viable industrial solution for the measurement of workpiece form deviations.

Keywords: machine vision; form deviation; evaluation algorithm; image-based process; edge detection

check for updates

Citation: Zhang, W.; Han, Z.; Li, Y.; Zheng, H.; Cheng, X. A Method for Measurement of Workpiece form Deviations Based on Machine Vision. *Machines* **2022**, *10*, 718. https://doi.org/10.3390/machines10080718

Academic Editors: Raul D. S. G. Campilho and Francisco J. G. Silva

Received: 17 July 2022
Accepted: 5 August 2022
Published: 22 August 2022

Publisher's Note: MDPI stays neutral with regard to jurisdictional claims in published maps and institutional affiliations.

1. Introduction

Shafts are one of the most important machinery parts for a wide range of industrial applications. Geometric deviations of a shaft will affect its functional performance. Conventionally, manual inspection and measurement are usually conducted for the measurement of straightness and roundness, which has the disadvantage of larger errors and lower efficiency [1]. Machine vision technology has been employed to measure industrial parts of different sizes with high efficiency and accuracy, which is widely used in automatic measuring [2–4]. Many researchers have conducted research on measurements based on machine vision technology. Lu et al. [5] developed a straightness measurement system based on the combination of a laser and machine vision, where multiple groups of vision sensors were adopted to realize on-line detection of seamless steel pipe straightness. Cho et al. [6] explored a new method of support vector regression to detect roundness to improve the accuracy and speed of the fitting algorithm, and it was proved to be more robust to noises, including measurement deviations for the tested problems. Liu et al. [7] proposed a binocular-vision-based deviation detection system and an identification algorithm to achieve deviation detection with three-dimensional measurement capability and to simplify the complex error identification formulations of position-independent geometric deviations in the rotary axis. Xiao et al. [1] proposed an on-line dimensional accuracy measurement method by machine vision, where three surface sources were placed in the positions of left, middle, and right to ensure uniform illumination, which realized the real-time measurement of the straightness and roundness on the conical spun workpiece. Tan et al. [8] studied the measurement of shaft diameter with the structured light system composed of a laser linear light source and a camera. The test results show that when the shaft diameter was 36.162 mm, the speed was 1250 r/min, and the maximum average measurement deviation was 0.019 mm. Li. [9] developed a geometric measurement system for shaft parts by machine vision, and an improved single-pixel edge detection method was proposed by the Canny detection algorithm. The experimental results show that the repeatability deviation of the system was less than 0.01 mm. Luo et al. [10] proposed an improved differential evolution algorithm (IoCoDE) for the accurate evaluation of minimum zone axis straightness deviation. The results indicate that the evaluation accuracy of IoCoDE was better than linking ends, the least-square method (LSM), and other common evaluation algorithms, and it was basically around 1 s. Hao et al. [11] suggested a coded references and geometric-constraints-based method to solve the inconsistency problem of measurement range and accuracy for slender shafts. The systematic deviation of the experimental system was 0.01754 mm. Min et al. [12] measured the high-precision geometric deviation of thread with machine vision and optical enlargement. The geometric deviations of thread were calculated by using the thread cross-sectional image. The linear precision of this system was less than 10μm. Chai et al. [13] proposed a non-contact optical measurement scheme to measure the co-axiality of a composite gear shaft; the least-square circle (LSC) and the particle swarm optimization (PSO) were used, and the measurement deviation range was less than 0.065 mm.

In the literature mentioned above, the machine vision measurement is more precise and efficient as compared with the traditional measurements. However, there is a lack of standard vision detection instrument for form deviations measurement in industry. Therefore, the structure of vision measurement and related algorithms need to be further studied.

In this study, a new method is presented for measuring the form accuracy of a shaft workpiece. In order to obtain images of the shaft, an image acquisition system is designed. The image pre-processing method and form deviations evaluation algorithm are studied. By integrating the computer control and calculation algorithm, form deviations, such as straightness, roundness, and cylindricity deviation of the shaft part, are calculated automatically and efficiently.

2. Image Acquisition System and Camera Calibration

2.1. Composition of the Image Acquisition System

The composition of the form deviation measurement system is depicted in Figures 1 and 2. It is mainly composed of the X, Y, Z linear electrokinetic displacement platform, rotating electric platform, CMOS camera, optical lens, LED light source, three-jaw chuck, tailstock, motion control card, and computer. Information on the main hardware models of the system is provided in Table 1. The parameters of the motion stages are listed in Table 2.

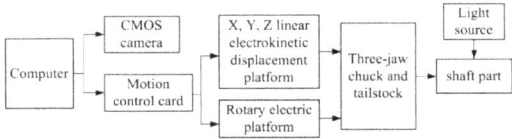

Figure 1. Schematic diagram of an image acquisition system.

Table 1. Main hardware of the system.

Computer	CMOS Camera	Lens	Light Source
ADLINK IPC-610	DAHENG MER-2000-19U3C	Computer V1228-MPY	KOMA JS-LT-180-32

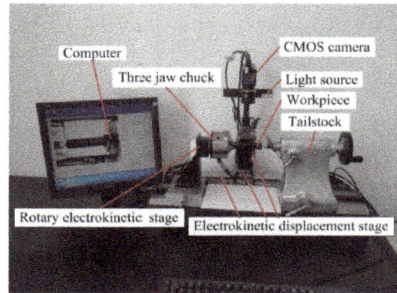

Figure 2. Form deviation detection system by machine vision.

Table 2. Specification of motion stages.

Motorized Stage		Parameters		
		Travel Range (mm)	Resolution (µm)	Repeatability Positioning (µm)
X	KXL06200-C2-F	200	0.2	±1
Y, Z	KXG06030-C	30	0.1	±1
Rotating	KS401-40	360°	0.003°	±0.05°

2.2. Camera Calibration

In the process of two-dimensional images, there is nonlinear deformation in different degrees, which is usually called geometric distortion. In addition, there are other factors, such as the instability of the camera imaging process and the quantization deviation caused by the low image resolution. Thus, there is a complex nonlinear relationship between the object points in the image and the corresponding points in the World Coordinate System. Because of the existence of these distortions, the calibration coefficients of different image zones even in one direction are different [14,15].

Figure 3 shows the relationship between the camera coordinate system and the World Coordinate System. Let p be a point in the field of camera view. The homogeneous coordinates of point p in the camera coordinate system and the World Coordinate System are $(X_C, Y_C, Z_C, 1)$ and $(X_W, Y_W, Z_W, 1)$, respectively. The homogeneous transformation relationship between the World Coordinate System and pixel image coordinate system is as follows:

$$Z_c \begin{bmatrix} u \\ v \\ 1 \end{bmatrix} = \begin{bmatrix} f_u & 0 & u_0 & 0 \\ 0 & f_v & v_0 & 0 \\ 0 & 0 & 1 & 0 \end{bmatrix} \begin{pmatrix} \mathbf{R} & \mathbf{t} \\ 0^T & 1 \end{pmatrix} \begin{bmatrix} X_W \\ Y_W \\ Z_W \\ 1 \end{bmatrix} = \mathbf{M}_1 \mathbf{M}_2 \begin{bmatrix} X_W \\ Y_W \\ Z_W \\ 1 \end{bmatrix} \tag{1}$$

where $\mathbf{R}$ is a 3 × 3 rotation matrix, $\mathbf{t}$ is a 3 × 1 translation matrix, $\mathbf{M}_1$ is the inner camera parameter matrix, $\mathbf{M}_2$ is the outer camera parameter matrix, f_u and f_v are the equivalent focal lengths in x and y directions, (u_0, v_0) represents the coordinate of the main point on the camera. The space point p in the camera coordinate system is set as $p(X_C, Y_C, Z_C)$. Suppose that the projected coordinates of point p normalized are $p(x_n, y_n)$, and the projected coordinates of point p after adding distortion are $p(x_d, y_d)$ [16]; then, the relationship is as follows:

$$x_d = x_n(1 + k_1(x_n^2 + y_n^2) + k_2(x_n^2 + y_n^2)^2) \tag{2}$$

$$y_d = y_n(1 + k_1(x_n^2 + y_n^2) + k_2(x_n^2 + y_n^2)^2) \tag{3}$$

where k_1, k_2 are radial distortion coefficients.

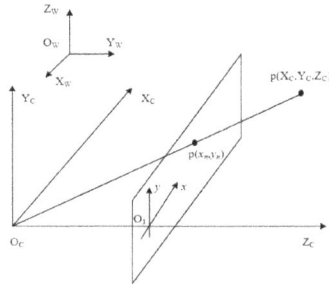

Figure 3. Relationship between the camera coordinate system and the World Coordinate System.

In order to make vision measurement more accurate, the ceramic calibration board with 1 μm possession precision is selected in this experiment. The sum of the calibration grid is 20 × 16, and the side lengths are 4 mm. The process for camera calibration includes, firstly, the collection of a calibration board image under the same condition on the form deviation measurement, such as focal length, object distance, lighting strength, and then, the collection of nine different orientation board images. The flow chart for camera calibration is shown in Figure 4.

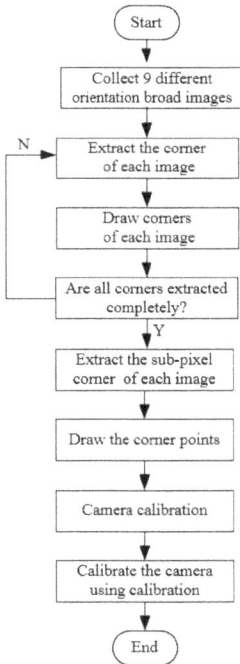

Figure 4. Flow chart of camera calibration.

3. Main Algorithms

3.1. Image Correction

Due to deviations of camera installation and equipment assembly, the acquired part image may have a small tilt angle. When evaluating form deviations, it is necessary to obtain the element line coordinates in 360 measuring images, whose coordinate results are affected by the small tilt angle. This increases the detection deviation and detection complexity. In order to improve the detection accuracy and efficiency, it is necessary to rectify the measured part images. We propose an image correction algorithm based on the centerline slope. Firstly, the distorted image is eliminated by camera calibration data, and then, the coordinates of the top and bottom edges on the workpiece image are obtained by the Canny algorithm [17]. The average data are computed as a centerline by the corresponding addition algorithm. The slope of the centerline is obtained by the least-square fitting, by which the rotation angle θ is obtained. Finally, the measured part image is revised by θ. Suppose that the point $P_0(x_0, y_0)$ rotated anticlockwise by θ is $P_0(x, y)$, the coordinate point matrix expression after rotation is as follows:

$$\begin{bmatrix} x \\ y \\ 1 \end{bmatrix} = \begin{bmatrix} \cos\theta & -\sin\theta & 0 \\ \sin\theta & \cos\theta & 0 \\ 0 & 0 & 1 \end{bmatrix} \begin{bmatrix} x_0 \\ y_0 \\ 1 \end{bmatrix} \tag{4}$$

3.2. Sub-Pixel Edge Detection Algorithm

The conventional sub-pixel edge detection methods include: sub-pixel edge detection by moment, sub-pixel edge detection by fitting, sub-pixel edge detection by interpolation, etc. [18,19]. The measured workpiece in this experiment is an axis, so it is assumed there will be no burrs, roughness, etc., in the edge area of the image. The top and bottom edges of the sampling image are in accordance with the image characteristics of a polynomial function. In order to realize the sub-pixel edge position, the parameters describing the edge features in the image can be obtained by establishing equations of polynomial parameters and using the principle of the least-square method [20,21].

The method used is to filter and denoise gray images and use the Otsu method to complete threshold segmentation [22]. The Canny algorithm is used to obtain the rough edges of the part image, selecting rough edges in order to turn them into a single row. An appropriate zoom is chosen, and then, the fine edges are accurately obtained by using the polynomial fitting algorithm. The edge point formula fitted by polynomial $y(x)$ can be expressed as Equation (5):

$$y(x) = a_0 + a_1 x + a_2 x^2 + \ldots + a_m x^m = \sum_{j=0}^{m} a_j x^j \tag{5}$$

By calculating the quadratic sum of the least squares and making partial derivatives of a_m equal to 0, the result is obtained.

$$F(a_0, a_1, \ldots, a_m) = \sum_{i=1}^{n} [y(x_i) - y_i]^2 \tag{6}$$

$$\frac{\partial F}{\partial a_j} = 2 \sum_{i=1}^{n} [y(x_i) - y_i] x_i^j = 0 \quad (j = 0, 1, \ldots, m) \tag{7}$$

By solving the above equations, the fitting polynomial coefficients can be determined.

3.3. Calculation of Straightness Deviation

3.3.1. Axis of workpiece fitting

By extracting the middle line on the top pixel edges and the corresponding bottom pixel edges of a workpiece image as the central axis, the calculation is rendered simple and easy. However, if the workpiece in the image has a tilted angle or straightness deviation, the upper and the corresponding lower edge pixel edges are asymmetric, which leads to a large straightness deviation of the central axis. In order to reduce this deviation, the radial local zone search method is used to determine the position of the shaft axis.

In order to determine the coordinates of the top and bottom edges, the top edge point (x_j, y_j) closest to the bottom edge point (x_i, y_i) is shown in Figure 5.

$$\sqrt{(y_i - y_j)^2 + (x_i - x_j)^2} = \min\{H\} \tag{8}$$

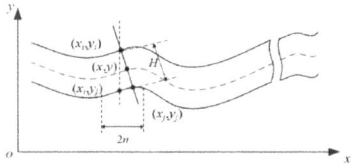

Figure 5. Graphical demonstration of calculating pixel coordinates of the central axis.

The geometric center coordinates of the workpiece axis are as follows.

$$\begin{cases} x = \frac{x_i + x_j}{2} \\ y = \frac{y_i + y_j}{2} \end{cases} \tag{9}$$

To decrease computing time, extract each n point on the left and right of the bottom edges corresponding relatively to the abscission of the top edge point, which is a total of $2n$ points. According to Equations (8) and (9), the geometric coordinates of a workpiece axis are obtained. Then, the axis of the part is estimated.

3.3.2. Straightness deviation algorithm

The straightness deviation according to ISO 1101 is the difference between the largest and smallest distances between the workpiece line and the reference line. According to ISO 1101, the form tolerance zone has the direction of the minimum [23,24]. The minimum zone method [25] is needed to search for the minimum value of the distance between two parallel lines containing the measured contours according to the minimum condition principle. So, an enveloped line searching algorithm is proposed to obtain the straightness deviation. The algorithm is discussed below.

The least-square method is used to obtain the baseline L_1, as shown in Figure 6, and the linear equation is set as

$$y = k_1 x + m \tag{10}$$

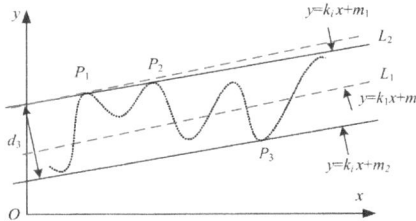

Figure 6. Graphical demonstration of enveloped line search algorithm.

Regarding baseline L_2 as the boundary, all sampling points are divided into two category points, i.e., high points and low points, from which the farthest point P_1 is found. Serving P_1 as the base point, a line L_2 is generated with the slope k_1. By changing the slope of line L_2, the sampling points are located below or on the line, and the critical point P_2 is determined.

$$k_i = k_1 + a \tag{11}$$

where a is the minimum value. Calculate the corresponding intercept based on k_i using the expression below.

$$m_i = -k_i x P_1 + y P_1 \tag{12}$$

In order to calculate P_2, let

$$W = k_i x + m_i - y \tag{13}$$

All coordinates of the measured points, except P_1, are substituted into Equation (13), and the sampling points are calculated when k_i varies. In theory, when W is zero, P_2 is obtained. However, in an actual situation, by setting small step size a, P_2 appears between two scanning lines. In this case, the following requirement should be met:

$$\min\{k_i x + m_i - y\} < 0 \tag{14}$$

The enveloped line $y = k_i x + m_1$ is confirmed by two points, P_1 and P_2, and then, the minimum enveloped line $y = k_i x + m_2$ is parallel to the line through the farthest point P_3. At this point, the three points, P_1, P_2, and P_3, are satisfied with the 'high-low-high' rule criterion of the minimum zone. The straightness deviation is calculated by the distance between the two enveloped lines.

$$d_3 = |m_1 - m_2| / \sqrt{k^2 + 1} \tag{15}$$

In the same way, search for the bottom line and three points of the 'low-high-low' criterion to satisfy the minimum zone. The distance of a pair of enveloped lines is also obtained after calculation, according to the similar steps above. The minimum distance value of the two results obtained will be used as the straightness deviation.

3.4. Calculation of Roundness Deviation

3.4.1. Three-Dimensional Reconstruction by Monocular Camera

The workpiece three-dimensional coordinate system $xoyz$ is seen in Figure 7. Let the initial position angle be $0°$; the point coordinates on the workpiece contour surface collected by the camera are (x_1, y_1, z_1), and the rotating β point A on the workpiece is moved to point B, whose coordinate is (x_2, y_2, z_2). According to the geometric relationship, as shown in Figures 7 and 8, the relationship is as follows:

$$y_1 = AO \cdot \cos 0° \tag{16}$$

$$z_1 = AO \cdot \sin 0° \tag{17}$$

$$y_2 = BO \cdot \cos \beta \tag{18}$$

$$z_2 = BO \cdot \sin \beta \tag{19}$$

$$x_1 = x_2 \tag{20}$$

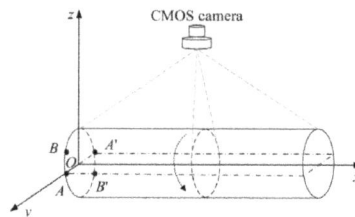

Figure 7. Position relation of camera and workpiece in the measuring coordinate.

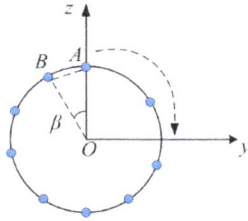

Figure 8. Graphical demonstration of the axis section acquisition point.

Among them, AO is distance between point A and the workpiece axis. The workpiece contour information is collected by the camera through the workpiece rotation, and the 3D contour model of the workpiece surface is established by the camera.

Due to the installation eccentricity of the workpiece, the circle center of a certain section workpiece is changed. In order to eliminate the eccentricity deviation, the half distance between the top and bottom edge is used as a radius value, that is, $AO = |y_1' - y_1|/2$, so the radius of the corresponding positions under other rotation angles can be obtained. The radius under every x position can be determined on the corresponding circle, so as to realize the three-dimensional reconstruction of the workpiece contour.

3.4.2. Roundness Deviation Algorithm

The roundness deviation according to ISO 1101 is the difference between the largest and smallest radial distance of the workpiece circumference from the reference circle [23,24]. There are four commonly used methods for evaluating roundness deviation, which are the minimum zone circle method, the minimum circumscribed circle method, the maximum inner circumscribed circle method, and the least-square circle method [26,27]. The minimum zone method is an evaluation method in accordance with the definition, but it is difficult to solve directly because it is a nonlinear problem, and it is complicated to calculate the collected data [28].

As shown in Figure 9, assuming that (x_i, y_i) are the measured coordinates on the actual contour of the workpiece, and (x_k, y_k) are the center coordinates of the minimum zone method to be solved, then the distance H_{ik} from the measured point to the center is

$$H_{ik} = \sqrt{(x_i - x_k)^2 + (y_i - y_k)^2} \tag{21}$$

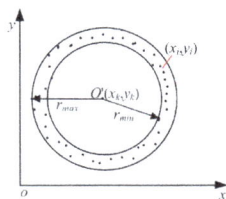

Figure 9. Schematic of roundness deviation evaluation.

The objective function F to be optimized is

$$F = \max(H_{ik}) - \min(H_{ik}) \tag{22}$$

What needs to be solved is how to determine the value of (x_k, y_k), so that F is the minimum. When $F = f$, f is the roundness deviation.

Particle swarm optimization (PSO) is an intelligent optimization algorithm proposed by Kennedy and Eberhart in 1995, inspired by the movement of flock birds [29]. In this paper, the PSO algorithm of synchronously changing learning factors is used to solve roundness. This algorithm has clear advantages in optimization accuracy and convergence speed.

The synchronous learning factor refers to setting the range of learning factors c_1 and c_2 as $[c_{min}, c_{max}]$, and the value formula of the learning factor in the t-th iteration is as follows:

$$c_1 = c_2 = c_{max} - \frac{c_{max} - c_{max}}{t_{max}} \cdot t \qquad (23)$$

The algorithm flow chart is shown in Figure 10 as follows.

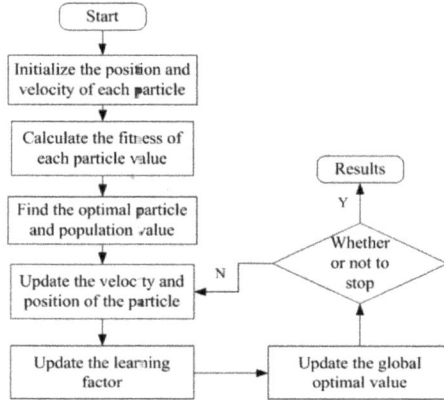

Figure 10. Improved PSO flow chart.

3.4.3. Cylindricity Deviation Algorithm

The cylindricity deviation according to ISO 1101 is the difference between the largest and smallest radial distances of the workpiece surface from the reference [23,24]. According to the form deviation evaluation principle, when the actual cylindrical surface is compared with the ideal surface, the minimum enveloped zone should be determined according to the actual cylindrical surface. When the actual surface is tightly contained by two identical co-axial cylindrical surfaces, between which the radius difference is the minimum, it is the minimum enveloped zone.

In this paper, a method of spatial cylindricity deviation detection is proposed based on the PSO and the least-square algorithm. Suppose that the circle center coordinate of each cross-section in Figure 11 is obtained by the improved particle swarm algorithm. According to the circle center coordinates, the space axis is fitted by the least-square algorithm. Assume that E_1 and E_2 are the points with the largest and smallest distances, respectively, from the measured contour points on the fitted straight line. The cylindricity deviation d_2 can be expressed as the difference between the maximum distance and the minimum distance from the measured contour points to the spatial axis.

$$d_2 = d_{max} - d_{min} \qquad (24)$$

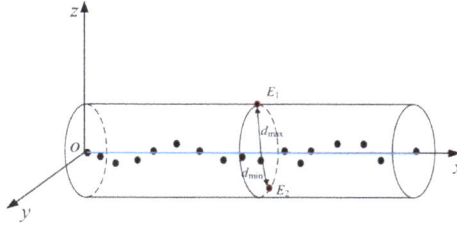

Figure 11. Schematic of the space axis fitting.

4. Experiment and Results Analysis

4.1. Calibration Results

Camera calibration is performed before image collection. A total of nine images are collected, as shown in Figure 12, for calibration. The corner detection result of the calibration plate is as shown in Figure 13.

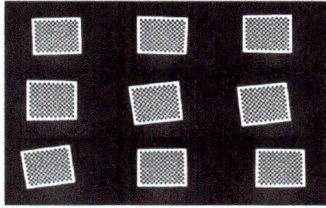

Figure 12. Calibration plate image.

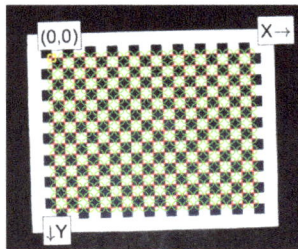

Figure 13. Corner detection of the calibration plate.

The camera is calibrated according to the flow chart in Figure 4, and the inner parameters and distortion matrix are obtained as follows:

$$[k_1 \ k_2] = [-0.0868 \ 0.0127]$$

$$M_1 = \begin{vmatrix} 5395.86 & 0 & 2749.23 \\ 0 & 5394.83 & 1809.22 \\ 0 & 0 & 1 \end{vmatrix}$$

Extract subpixel image corrected angular coordinates,. The pixel distance between the adjacent corner point is obtained as h. Internal and external calibration parameters are used to calculate the sub-pixel corner coordinates in the image. Assume the distance between the adjacent corner points is M, according to the pixel distance, the ratio of the proportion relation coefficient k is calculated out, which is shown in Equation (25). The values k in the horizontal direction and vertical direction are 30.32 μm and 30.44 μm, respectively.

$$k = \frac{M}{h} \tag{25}$$

After calibration, the measured workpiece is clamped by the three-jaw chuck, and the other end of the part is held by the tail. The X, Y-direction linear electrokinetic displacement stages are moved to the initial position of the camera, and the measured workpiece is moved to the focal plane of the camera lens by controlling the Z-direction linear electrokinetic displacement stage. When the workpiece is rotated by 1°, an image of the workpiece is acquired by the camera. A total of 360 images are obtained and stored in the computer. Before image pre-processing, the image is corrected by the methods described in Section 3.1. Figure 14 is the collected original workpiece image, and Figure 15 is the corrected workpiece image.

Figure 14. Original workpiece image.

Figure 15. Workpiece image after correction.

4.2. Image Pre-Processing

Since the measurement of the form deviation is only related to the edges, the complex background and noise are present in the collected images. It is necessary to carry out pre-processing, such as region of interest (ROI) extraction, filtering, and image enhancement, to eliminate additional interference in order to precisely measure the workpiece form. When collecting the image, the workpiece, three-jaw chuck, and tailstock part are photographed, so a rectangular area is used for ROI extraction. According to the requirements of form deviation and characteristics of the workpiece image, various filters are used to deal with the same image. After comparison, Gaussian filtering is more suitable for image pre-processing for our study. The processed images are shown in Figure 16. The original workpiece image is obtained by machine vision, as shown in Figure 16a. ROI extraction is performed by a rectangular area, as shown in Figure 16b. Threshold segmentation is completed by the Otsu method, as shown in Figure 16c. Rough edges of the part image are extracted by the Canny algorithm, as shown in Figure 16d. Parts of the upper-edge sub-pixel-position fine edges are accurately obtained by using the polynomial fitting algorithm, as shown in Figure 16e.

(**a**) Original image

(**b**) ROI image

(**c**) Threshold segmentation

(**d**) Edge contour extracted by Canny

(**e**) Part of the upper edge sub-pixel position

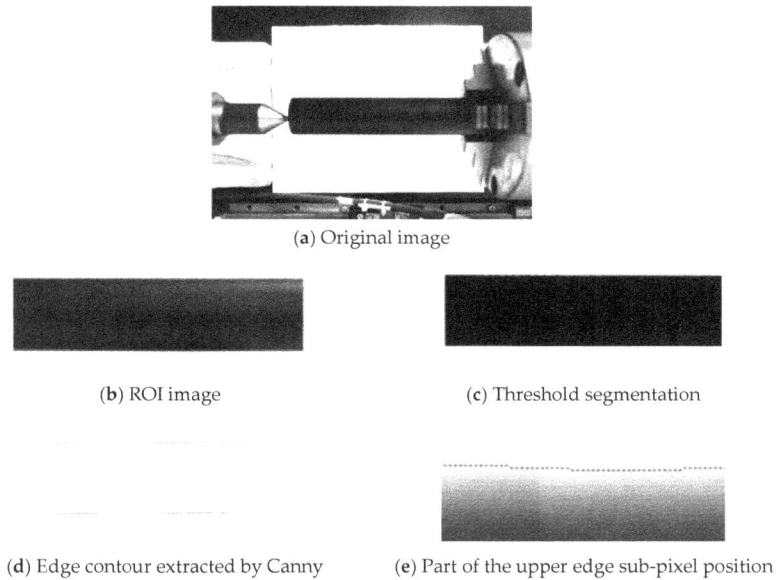

Figure 16. Images after image processing.

4.3. Straightness Deviation Results

According to the vision measurement algorithm of straightness deviation, several measuring experiments are completed on the φ 20 mm × 100 mm shaft workpiece. Results of the containment line search algorithm at the rotation angle of 0° are shown in Figure 17. All fitting axis points are successfully contained by the two containment lines. The straightness deviation is the distance between the two lines multiplied by the calibration coefficient k. The results of 360 straightness measurement are shown in Figure 18, where the maximum straightness deviation is 30.18 μm, the minimum value is 4.71 μm, and the average value is 11.12 μm, respectively.

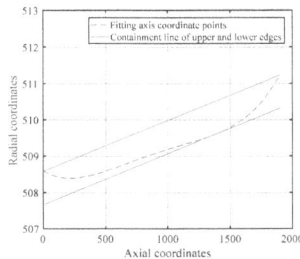

Figure 17. Graph of straightness deviation evaluation results.

In Figure 18, the variation in straightness deviation is somewhat large. There are two possible reasons. Firstly, it takes about 3 min to collect 360 images from different angles, during which the image quality may be affected by an unstable illumination. Secondly, the workpiece axis is not the center of rotation. These factors affect the measured results.

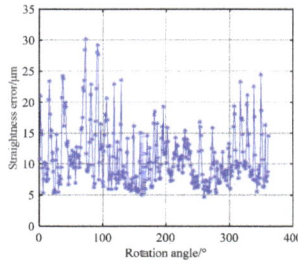

Figure 18. Graph of 360 straightness deviation results.

4.4. Roundness Deviation Results

Since the images of the entire axis are not detected in the roundness measurement, several cross-sections in the same distance are used to calculate roundness deviation, so that the ROI region required for roundness deviation measurement is smaller than that of straightness deviation. Seven equidistance cross-section positions are selected in group A with 2000 pixels and group B with 1100 pixels in the diameter direction, and the comparison results are shown in Table 3 by using methods described in Section 3.3. Figure 19 shows the graph of the improved particle optimization algorithm (PSO) calculation results for position 1 in group A (the first position).

Table 3. Roundness deviation results.

Section Position	Roundness Deviation of Group A/μm	Roundness Deviation of Group B/μm
1	15.89	14.53
2	14.90	14.91
3	13.83	13.77
4	13.24	13.68
5	12.82	13.25
6	13.05	12.86
7	12.36	13.71
Average value	13.73	13.82

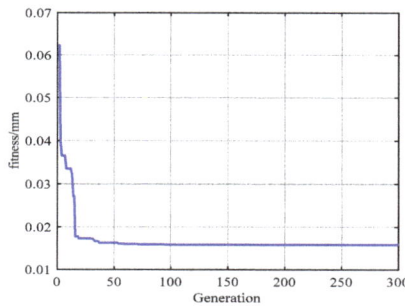

Figure 19. Graph of the improved PSO calculation results for position 1 in group A.

4.5. Cylindricity Deviation Results

The number of cross-sections usually used in the measuring cylinder should be no less than 5. In this study, by the rotation angle interval of 1°, 10 cylindricity sections and 20 cylindricity sections of the parts are acquired to calculate cylindricity deviation. This calculation process is the same as the roundness deviation calculation. Data fitting results are shown in Figure 20; the cylindricity deviations are 26.91 μm and 29.81 μm, respectively.

The cylindricity deviation of the workpiece changes slightly with the different number of sections under the same experimental conditions.

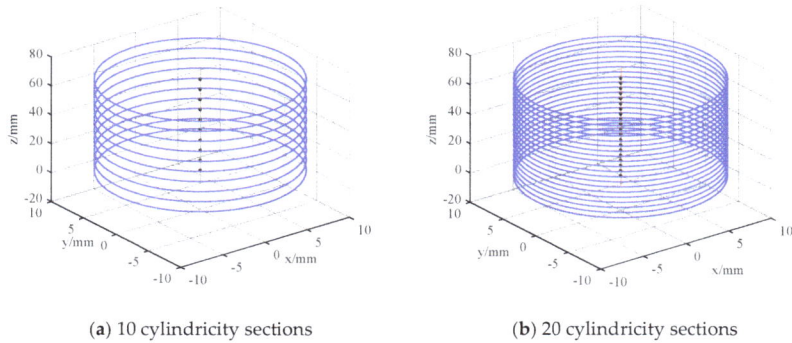

(**a**) 10 cylindricity sections   (**b**) 20 cylindricity sections

Figure 20. Graph fitting of workpiece axis and measured points.

4.6. Verification of Measurement Results

In order to judge whether the measurement proposed is correct, a commercial instrument, RD602 cylindricity measuring instrument, is used to measure the same workpiece, as shown in Figure 21. RD602 cylindricity measuring instrument has a high precision, with the deviation of less than 0.5 μm at full working motion, which is used to verify the accuracy of the on-line measurement system. The repeatability of this instrument is 0.2 μm. The comparison results of form deviation are shown in Table 4.

Figure 21. Workpiece form deviation measurement by cylindricity measuring instrument RD602.

Table 4. Comparison of deviation results.

	Average Value by RD602/μm	Average Value by the Designed Instrument/μm	Error/μm
Straightness	15.80	11.11	−4.69
Roundness	9.95	13.82	3.87
Cylindricity	21.30	29.81	8.51

Through some experiments, the standard deviations of straightness, roundness, and cylindricity measured by this instrument are 0.52, 0.31, and 1.56, respectively. From the comparison results shown in Table 4, it can be concluded that the straightness, roundness, and cylindricity deviations of the form deviation measurement system are −4.69 μm, 3.87 μm, and 8.51 μm, respectively. The various form deviation values were measured accurately, which can reach the measurement accuracy of commonly used three-coordinate measuring machine and provides a reference for testing form deviation in production. When calculating the straightness deviation, in order to use less time, it is not necessary to calculate the $\bar{c}60$ data values. Instead, the rotation angle can be appropriately increased to reduce the calculation. When calculating cylindricity, about 20 sections can be selected, which can be time saving and efficient, satisfying the accuracy requirements.

5. Conclusions

In this paper, a new workpiece form deviation measurement method based on machine vision is presented for the measurement of straightness, roundness, and cylindricity of a workpiece. An image acquisition system for obtaining images of shaft workpiece is developed. Edge detection technology and sub-pixel edge positioning technology are used to extract the edge information. A subsumed line search algorithm and an improved particle swarm optimization algorithm are proposed to evaluate the straightness and roundness deviations of the workpiece. Moreover, a method of spatial cylindricity deviation solution based on an improved synchronous PSO algorithm is proposed. The results of straightness, roundness, and cylindricity deviations of the workpiece are obtained by the above algorithms. Their standard deviations are 0.52, 0.31, and 1.56, respectively, implying consistency in the measurement. From the contrast experiments, the differences of straightness, roundness, and cylindricity deviations of the form deviation measurement system are -4.69 μm, 3.87 μm, and 8.51 μm, respectively, which are comparable to the traditional measurement methods. Therefore, the proposed method meets the precision requirements and is applicable for non-contact measurement, which has the advantages in measuring vulnerably scratched workpieces and in quickly obtaining form outlines of the workpieces. The proposed method would provide a viable industrial solution for the measurement of form deviations.

Author Contributions: Conceptualization, W.Z. and Z.H.; Data curation, Z.H. and Y.L.; Formal analysis, H.Z. and X.C.; Funding acquisition, H.Z.; Investigation, Y.L.; Methodology, W.Z.; Project administration, H.Z.; Resources, Z.H.; Validation, W.Z. and X.C.; Writing—original draft, W.Z. and Z.H. All authors have read and agreed to the published version of the manuscript.

Funding: The paper is financially supported by the National Key Research and Development Program of China (2022YFE0199100) and the Natural Science Foundation of Shandong Province (ZR2020ME164, ZR2016FL15).

Institutional Review Board Statement: Not applicable.

Informed Consent Statement: Not applicable.

Data Availability Statement: Not applicable.

Conflicts of Interest: The authors declare no conflict of interest.

References

1. Xiao, G.; Li, Y.; Xia, Q.; Cheng, X.; Chen, W. Research on the on-line dimensional accuracy measurement method of conical spun workpieces based on machine vision technology. *Measurement* **2019**, *148*, 106881. [CrossRef]
2. Derganc, J.; Likar, B.; Pernuš, F. A machine vision system for measuring the eccentricity of bearings. *Comput. Ind.* **2003**, *50*, 103–111. [CrossRef]
3. Kakaley, D.E.; Altieri, R.E.; Buckner, G.D. Non-contacting measurement of torque and axial translation in high-speed rotating shafts. *Mech. Syst. Signal Process.* **2020**, *138*, 106520. [CrossRef]
4. Dong, C.Z.; Ye, X.W.; Jin, T. Identification of structural dynamic characteristics based on machine vision technology. *Measurement* **2018**, *126*, 405–416. [CrossRef]
5. Lu, R.S.; Li, Y.F.; Yu, Q. On-line measurement of the straightness of seamless steel pipes using machine vision technique. *Sens. Actuators A Phys.* **2001**, *94*, 95–101. [CrossRef]
6. Cho, S.; Kim, J.-Y.; Asfour, S.S. Machine learning-based algorithm for circularity analysis. *Int. J. Inf. Decis. Sci.* **2014**, *6*, 70–86. [CrossRef]
7. Liu, W.; Li, X.; Jia, Z.; Li, H.; Ma, X.; Yan, H.; Ma, J. Binocular-vision-based error detection system and identification method for PIGEs of rotary axis in five-axis machine tool. *Precis. Eng.* **2018**, *51*, 208–222. [CrossRef]
8. Tan, Q.; Kou, Y.; Miao, J.; Liu, S.; Chai, B. A Model of Diameter Measurement Based on the Machine Vision. *Symmetry* **2021**, *13*, 187. [CrossRef]

9. Li, B. Research on geometric dimension measurement system of shaft parts based on machine vision. *EURASIP J. Image Video Process.* **2018**, *2018*, 1–9. [CrossRef]
10. Luo, J.; Liu, Z.; Zhang, P.; Liu, X.; Liu, Z. A method for axis straightness error evaluation based on improved differential evolution algorithm. *Int. J. Adv. Manuf. Technol.* **2020**, *110*, 413–425. [CrossRef]
11. Hao, F.; Shi, J.; Meng, C.; Gao, H.; Zhu, S. Measuring straightness errors of slender shafts based on coded references and geometric constraints. *J. Eng.* **2020**, *2020*, 221–227. [CrossRef]
12. Min, J. Measurement method of screw thread geometric error based on machine vision. *Meas. Control.* **2018**, *51*, 304–310. [CrossRef]
13. Chai, Z.; Lu, Y.; Li, X.; Cai, G.; Tan, J.; Ye, Z. Non-contact measurement method of coaxiality for the compound gear shaft composed of bevel gear and spline. *Measurement* **2021**, *168*, 108453. [CrossRef]
14. Zhengyou, Z. Camera calibration with one-dimensional objects. *IEEE Trans. Pattern Anal. Mach. Intell.* **2004**, *26*, 892–899. [CrossRef]
15. Lv, Y.; Feng, J.; Li, Z.; Liu, W.; Cao, J. A new robust 2D camera calibration method using RANSAC. *Optik* **2015**, *126*, 4910–4915. [CrossRef]
16. Bu, L.; Huo, H.; Liu, X.; Bu, F. Concentric circle grids for camera calibration with considering lens distortion. *Opt. Lasers Eng.* **2021**, *140*, 106527. [CrossRef]
17. Bao, P.; Lei, Z.; Xiaolin, W. Canny edge detection enhancement by scale multiplication. *IEEE Trans. Pattern Anal. Mach. Intell.* **2005**, *27*, 1485–1490. [CrossRef] [PubMed]
18. Ye, J.; Fu, G.; Poudel, U.P. High-accuracy edge detection with Blurred Edge Model. *Image Vis. Comput.* **2005**, *23*, 453–467. [CrossRef]
19. Xie, X.; Ge, S.; Xie, M.; Hu, F.; Jiang, N. An improved industrial sub-pixel edge detection algorithm based on coarse and precise location. *J. Ambient. Intell. Humaniz. Comput.* **2019**, *11*, 2061–2070. [CrossRef]
20. Sun, Q.; Hou, Y.; Tan, Q. A subpixel edge detection method based on an arctangent edge model. *Optik* **2016**, *127*, 5702–5710. [CrossRef]
21. Li, C.-M.; Xu, G.-S. Sub-pixel Edge Detection Based on Polynomial Fitting for Line-Matrix CCD Image. In Proceedings of the 2009 Second International Conference on Information and Computing Science, Manchester, UK, 21–29 May 2009; pp. 262–264.
22. Xu, X.; Xu, S.; Jin, L.; Song, E. Characteristic analysis of Otsu threshold and its applications. *Pattern Recognit. Lett.* **2011**, *32*, 956–961. [CrossRef]
23. *ISO 1101*; Geometrical Product Specification (GPS)—Geometrical Tolerancing—Tolerances of Form, Orientation, Location and Run-Out. ISO: Geneva, Switzerland, 2017.
24. Henzold, G. *Geometrical Dimensioning and Tolerancing for Design, Manufacturing and Inspection*, 2nd ed.; Butterworth-Heinemann: Oxford, UK; Elsevier: Burlington, VT, USA, 2006.
25. Pratheesh Kumar, M.R.; Prasanna Kumaar, P.; Kameshwaranath, R.; Thasarathan, R. Roundness error measurement using teaching learning based optimization algorithm and comparison with particle swarm optimization algorithm. *Int. J. Data Netw. Sci.* **2018**, *2*, 63–70. [CrossRef]
26. Srinivasu, D.S.; Venkaiah, N. Minimum zone evaluation of roundness using hybrid global search approach. *Int. J. Adv. Manuf. Technol.* **2017**, *92*, 2743–2754. [CrossRef]
27. Rossi, A.; Antonetti, M.; Barloscio, M.; Lanzetta, M. Fast genetic algorithm for roundness evaluation by the minimum zone tolerance (MZT) method. *Measurement* **2011**, *44*, 1243–1252. [CrossRef]
28. Pathak, V.K.; Singh, A.K. Effective Form Error Assessment Using Improved Particle Swarm Optimization. *Mapan* **2017**, *32*, 279–292. [CrossRef]
29. Zhang, Y.; Wang, S.; Ji, G. A Comprehensive Survey on Particle Swarm Optimization Algorithm and Its Applications. *Math. Probl. Eng.* **2015**, *2015*, 1–38. [CrossRef]

machines

MDPI

Article

An Improved Automation System for Destructive and Visual Measurements of Cross-Sectional Geometric Parameters of Microdrills

Wen-Tung Chang * and Yu-Yun Lu

Department of Mechanical and Mechatronic Engineering, National Taiwan Ocean University, Keelung 20224, Taiwan
* Correspondence: wtchang@mail.ntou.edu.tw

Abstract: Microdrills are specific cutting tools widely used to drill microholes and microvias. For certain microdrill manufacturers, a conventional sampling inspection procedure is still manually operated for carrying out the destructive and visual measurements of two essential cross-sectional geometric parameters (CSGPs), called the cross-sectional web thickness (CSWT) and the cross-sectional outer diameter (CSOD), of their straight (ST) and undercut (UC) type microdrill products. In order to comprehensively automate the conventional sampling inspection procedure, a destructive and visual measuring system improved from an existing vision-aided automation system, for both the hardware and the automated measuring process (AMP), is presented in this paper. The major improvement of the hardware is characterized by a machine vision module consisting of several conventional machine vision components in combination with an innovative and lower cost optical subset formed by a set of plano-concave achromatic (PCA) lenses and a reflection mirror, so that the essential functions of visually positioning the drilltip and visually measuring the CSGPs can both be achieved via the use of merely one machine vision module. The major improvement of the AMP is characterized by the establishment of specific image processing operations for an auto-focusing (AF) sub-process based on two-dimensional discrete Fourier transform (2D-DFT), for a web thickness measuring (WTM) sub-process based on an iterative least-square (LS) circle-fitting approach, and for an outer diameter measuring (ODM) sub-process based on integrated applications of an iterative LS circle-fitting approach and an LS line-fitting-based group-dividing approach, respectively. Experiments for measuring the CSGPs of microdrill samples were conducted to evaluate the actual effectiveness of the developed system. It showed that the developed system could achieve good repeatability and accuracy for the measurements of the CSWTs and CSODs of both ST and UC type microdrills. Therefore, the developed system could effectively and comprehensively automate the conventional sampling inspection procedure.

Keywords: microdrill; cross-sectional geometric parameter (CSGP); cross-sectional web thickness (CSWT); cross-sectional outer diameter (CSOD); automation system; image processing operations; destructive and visual measurement

check for updates

Citation: Chang, W.-T.; Lu, Y.-Y. An Improved Automation System for Destructive and Visual Measurements of Cross-Sectional Geometric Parameters of Microdrills. *Machines* 2023, 11, 581. https://doi.org/10.3390/machines11060581

Academic Editors: Raul D. S. G. Campilho and Francisco J. G. Silva

Received: 12 April 2023
Revised: 13 May 2023
Accepted: 14 May 2023
Published: 23 May 2023

1. Introduction

Microdrilling processes, based on mechanical and laser microdrilling techniques, have been widely adopted by printed circuit board (PCB) industries for the mass production of microholes and microvias in single- and multi-layer PCBs [1–3]. Mechanical microdrilling is based on the fundamentals of conventional cutting processes [4–16] and the use of specific cutting tools called microdrills to drill microholes and microvias. As compared with laser microdrilling, higher hole quality and lower equipment cost are the advantages of mechanical microdrilling, although such a conventional cutting process may also lead to slightly lower positioning accuracy and drilling efficiency. Therefore, microdrills are necessary and high-demand consumables for PCB industries, and superior design, rapid and precision manufacture, and reliable inspection techniques must be involved for their mass production.

Nowadays, commercially available microdrill products can be classified into the straight (ST) and undercut (UC) type microdrills, as shown by the illustrations depicted in Figure 1. Each type of microdrill can be functionally divided into two main portions: the drill body and the shank. (Note that for the purpose of clarity of illustration, the drill body of each microdrill depicted in Figure 1 is exaggerated.) The drill body, usually with a nominal diameter under 0.5 mm, can be further functionally divided into two sub-portions: the drill point and the helical flutes. The drill point, with the geometry of a so-called four-facet chisel point [2,4,5], is designed to produce cutting action inside the PCB (i.e., the workpiece). The helical flutes are designed to provide screw-shaped passageways for chip removal. The main difference between the ST and UC type microdrills is that the drill body of an ST type one has a diametral continuity along its axial direction, but that of a UC type one has a diametral discontinuity occurring at a certain axial location. The drill body of a UC type microdrill, besides its front portion adjacent to the drill point, is further ground to yield a smaller outer diameter, called an undercut diameter. The design of an undercut diameter is to largely reduce the extent of friction and heat generation caused by the surfaces of the drill body rubbing against the hole walls, which is especially suitable for producing microholes and microvias, with diameters under 0.25 mm, in multi-layer PCBs.

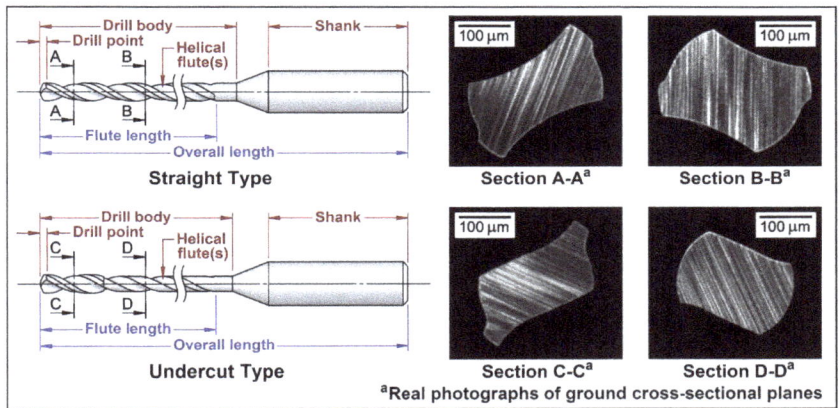

Figure 1. Illustrations and ground cross-sections of ST and UC type microdrills.

Furthermore, as shown in the real photographs of the ground cross-sections of the drill bodies shown in Figure 1, the exterior contours of the cross-sections of an ST type microdrill are theoretically formed by six piecewise curves (i.e., Sections A-A and B-B). However, those of a UC type microdrill are theoretically formed by six and four piecewise curves at the front portion (i.e., Section C-C) and the undercut portion (i.e., Section D-D) of the drill body, respectively. For the cross-section of the ST type microdrill depicted in Figure 2, two important cross-sectional geometric parameters (CSGPs) called the cross-sectional web thickness (CSWT, denoted by w) and the cross-sectional outer diameter (CSOD, denoted by D) are highlighted and must be considered. The centroid of the cross-section is located at point O_a, where the rotational axis of the microdrill passes through. Two concentric circles C_t and C_b centered at point O_a and of diameters D_t and D_b, respectively, are indicated in Figure 2. The circle C_t is mathematically a circle common tangent to the pair of cross-sectional flute contours (the concave contours denoted by Φ_F) at points P_{t1} and P_{t2}, respectively, and the CSWT w is theoretically equal to its diameter D_t and the distance $P_{t1}P_{t2}$. The circle C_b is mathematically the minimum bounding circle of the entire cross-section that overlaps the pair of cross-sectional margin contours (denoted by Φ_M), and the CSOD D is theoretically equal to its diameter D_b. For the cross-section of a

UC type microdrill, whether formed by six or four piecewise curves (and characterized by a pair of nonconcave flute contours), its CSWT and CSOD can both be defined in the same manner. The trade-off of the magnitude of CSWT will influence both the chip removal ability and rigidity of the drill body, while the accuracy and stability of the CSOD will influence the hole quality. For microdrill manufacturers, the cross-sectional geometry measurements (for both the CSWTs and CSODs of the drill body), as well as the drill point defect inspections, are essential quality control tasks that can benefit the mass production of their microdrill products.

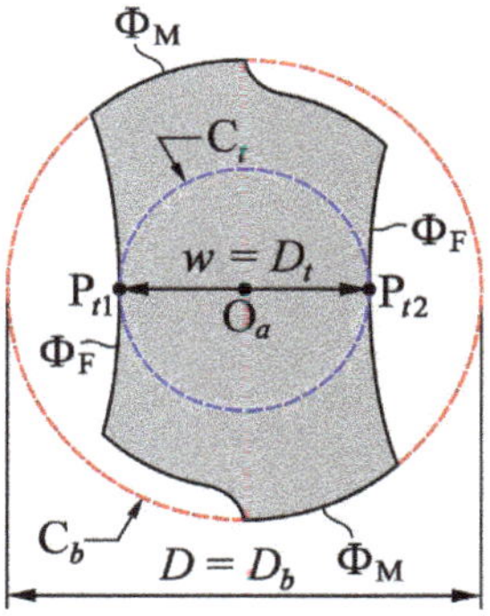

Figure 2. Cross-section and geometric parameters of the drill body of an ST type microdrill.

In the last two decades, optical-based nondestructive methods and systems for dealing with the defect inspections and geometry measurements of certain types of microdrills have been developed by a variety of researchers [17–26]. In order to inspect some types of drill point defects and flank wear, visual measuring methods based on dedicated image processing procedures in combination with specific machine vision devices have been developed and applied [17–21]. Tien et al. [17,18] have proposed two specific image processing algorithms for detecting some types of drill point defects from captured drill point images. Huang et al. [19] have developed an automated visual inspection system and also an image processing procedure for automatically capturing the drill point images and identifying the drill point defects. Su et al. [20] have proposed a machine vision-based approach for carrying out the flank wear measurement. Duan et al. [21] have proposed a specific image processing procedure for optically measuring the flank wear. As for the geometry measurements, some industrial measuring devices with excellent precision (repeatability) and accuracy, such as laser micro-gauges (LMGs), laser confocal displacement meters (LCDMs), and optical micrometers, have been applied to develop some nondestructive measuring systems [22–25]. Huang et al. [22] have adopted an LMG to measure the CSODs and runout amounts of microdrills. For the measurement of the CSWT, a laser-based inspection system and method have been developed by Chuang et al. [23] via applying an LMG and a conventional LCDM. Chang et al. [24] have applied an optical micrometer and a surface-scanning LCDM for measuring the CSWT with runout compensation. Chang and Wu [25] have applied a rotatable optical micrometer-based design for measuring the CSWT with an innovative optical-based method. The methods developed by Chuang et al. [23] and by Chang et al. [24], respectively, are suitable for measuring the CSWTs of ST type microdrills, while that developed by Chang and Wu [25] can be applicable to both ST and UC type ones. However, some of the above-mentioned nondestructive geometry measuring systems may not be easily universalized because of the higher hardware cost and/or lower availability (due to the import/export control on strategic commodities) of some of their key measuring devices. Besides the measurement of CSGPs, Jaini et al. [26] have developed a specific computer vision algorithm for measuring the axial straightness errors and radial runout amounts of microdrills. Without the use of optical-based devices,

Beruvides et al. [27] have studied the runout detection in microdrilling via online measured force signals and a neural network-based model.

In addition, a conventional procedure for carrying out the cross-sectional geometry measurements (for the purpose of sampling inspection) is still employed by certain microdrill manufacturers. In the conventional sampling inspection procedure, the drill body of a microdrill sample is ground off by a microdrill grinder to yield certain ground cross-sections (similar to those shown in Figure 1), while their corresponding CSWTs (and also CSODs) are measured by a measuring microscope. Such a destructive measuring procedure is completely manually operated by experienced inspectors and may not have good efficiency and accuracy. In order to automate and improve the manually operated procedure, Chang et al. [28] have proposed a vision-aided automation system, characterized by the integration of a grinding module and two specific machine vision modules, for carrying out the destructive web thickness measurement via a dedicated automation process. As compared with the nondestructive measuring systems [23–25], such a vision-aided automation system could have benefits of lower hardware cost and higher availability of key devices, although the arrangement of its two machine vision modules could be disadvantageous to achieve a compact design. It should also be noted that the automated measuring process (AMP) developed in Chang et al.'s work [28] is suitable for measuring the CSWTs of ST type microdrills, but not the CSWTs of UC type ones nor the CSODs of ST or UC type ones. In this process, two approaches called the shortest-distance (SD) and the minimum common-tangent-circle (MCTC) approaches are developed to calculate the CSWTs of ST type microdrills. However, both the SD and MCTC approaches cannot be applied to calculate the CSWTs of UC type microdrills, since they theoretically cannot be applied to cross-sections characterized by a pair of nonconcave flute contours. In addition, the approach for measuring the CSODs of ST or UC type microdrills was not developed in their study. Hence, further improvements on hardware design and AMP are still required to develop a more comprehensive system for carrying out the destructive and visual measurements of the CSWTs and CSODs of both ST and UC type microdrills.

According to the literature review, it can be understood that the destructive and visual measuring system and method for the essential CSGPs of microdrills still need to be improved for comprehensively automating the conventional sampling inspection procedure. To this end, an automation system improved from the one proposed by Chang et al. [28], for both the hardware and the AMP, is presented in this study for carrying out the destructive and visual measurements of the CSWTs and CSODs of both ST and UC type microdrills. Experiments for measuring the CSWTs and CSODs of certain microdrill samples were conducted to evaluate the actual effectiveness of the developed system.

2. Hardware Design and Construction of the Improved Automation System

In order to automate the conventional sampling inspection procedure, the hardware of the automation system proposed by Chang et al. [28] primarily consists of a dual-axis motion module, a grinding module, and two machine vision modules. Its dual-axis motion module is applied to move (or feed) and position a microdrill sample to be measured, and its grinding module is applied to grind off the drill body of that sample. One of its machine vision modules is applied to visually position the drilltip of that sample (relative to the grinding wheel of the grinding module) before the grinding off operation, and the other machine vision module is applied to visually measure the CSWTs of ground cross-sections of that sample. The essential functions of such an automation system must be equally realized by the improved one developed in this study.

The hardware of an improved automation system was innovatively developed according to the essential functions and was then constructed. In concept, the improved hardware primarily consists of a dual-axis motion module, a grinding module, and merely one machine vision module that can be innovatively applied to position the drilltip and to measure the CSGPs. As merely one innovative machine vision module is required, a more compact and lower cost hardware design, as compared with that of the previous system [28], can be achieved. The setup of the constructed hardware of the improved automation system is shown in Figures 3 and 4, and its hardware modules and system integration are described in the following sections.

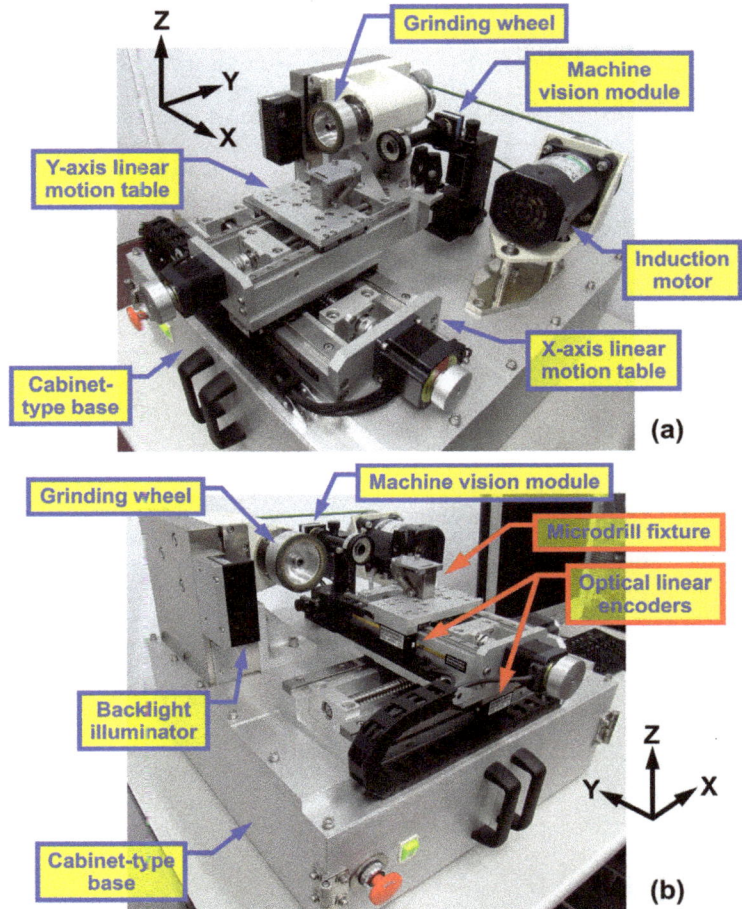

Figure 3. Entire views of the constructed hardware of the improved automation system: (**a**) the first perspective view; (**b**) the second perspective view.

2.1. Hardware Modules

As indicated in Figure 3, the foundation of the constructed hardware was a cabinet-type base with high rigidity, and a dual-axis motion module, a grinding module, and a machine vision module were all mounted on the top of the cabinet-type base. The dual-axis motion module primarily consisted of an X-axis linear motion table (LMT) serially combined with a Y-axis LMT, while a microdrill fixture was installed on the moving stage of the Y-axis LMT. The driving mechanism of the X-axis LMT consisted of a Misumi BSX1505 C3 class ballscrew unit (with a nominal lead of 5 mm) coupled to an Oriental Motor (OM) VEXTA RK566BAE 5-phase stepping motor, while that of the Y-axis LMT consisted of Misumi BSX1202 C3 class ballscrew unit (with a nominal lead of 2 mm) coupled to an OM VEXTA PK545BW 5-phase stepping motor. In addition, two sets of Renishaw RGH41X/RGS40S optical linear encoders (with a fine resolution of 1 μm) were installed in the X- and Y-axis LMTs, respectively, as their positioning sensors. According to the orthogonal traveling directions of X- and Y-axis LMTs, the X-, Y-, and Z-directions of the entire machine could thus be defined, as indicated in Figures 3 and 4a. As indicated in Figure 4, the microdrill fixture, designed for holding the shank of a microdrill sample, was characterized by a simple and reliable three-point clamping formed by a V-grooved bracket, a hinged pressing plate, and an extension spring used for a pre-loading purpose. The central axis (i.e., the rotational axis) of a held microdrill sample would be parallel to the Y-direction. As a result, the setup of the dual-axis motion module could enable a held microdrill sample to be precisely moved (or fed) and positioned along the X- and/or Y-directions.

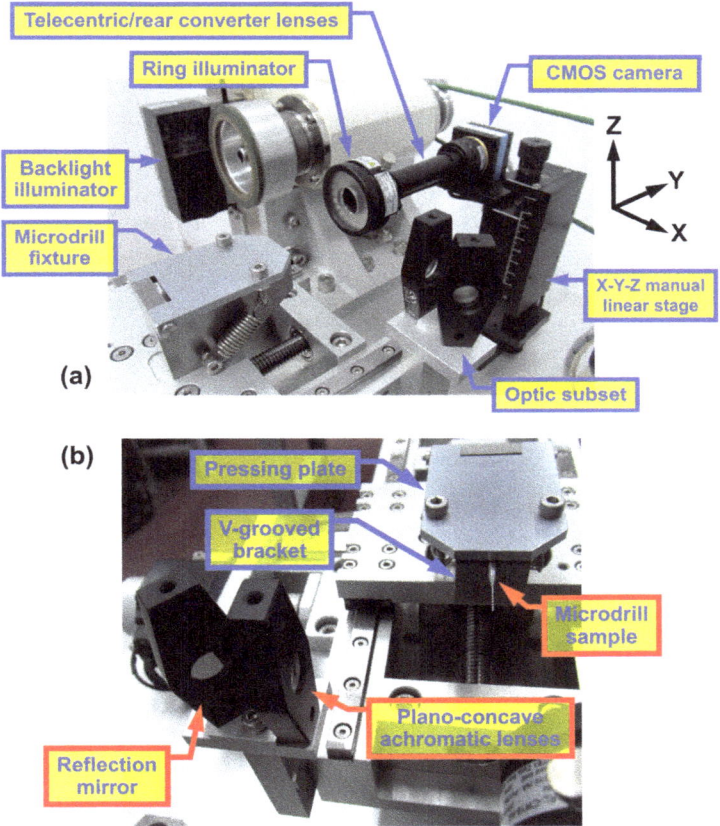

Figure 4. Closeup views of the constructed hardware of the improved automation system: (**a**) the setup of the machine vision module; (**b**) the details of the microdrill fixture and the optical subset.

As shown in Figure 3a, the grinding module primarily consisted of a customized grinder unit driven by an OM 5IK90A-BW2 induction motor. The grinder unit primarily consisted of a GMN Paul Müller Industrie TSA-50x160-6004 grinding spindle (with its radial and axial runouts within 3 and 1 µm, respectively) with an installed Asahi Diamond Industry SD1000-600P100B3.0 grinding wheel. The end face of the installed grinding wheel is specially formed by an inner annular portion (the ocher-colored portion with a code of SD600P100B3.0) and an outer annular portion (the dark-green-colored portion with a code of SD1000P100B3.0); the inner and outer annular portions were applied for roughly grinding off the drill body and finishing the ground cross-sections, respectively, via the surface grinding process. The rotational axes of the grinding unit and the induction motor were both parallel to the Y-direction. The induction motor was set to rotate at a constant rotary speed of 3200 rpm, and a belt-pulley mechanism with a velocity ratio of 2 was used to connect the induction motor and the grinding spindle. As a result, the setup of the grinding module could enable the grinding wheel to be driven at a constant rotary speed of 6400 rpm for the grinding off operation.

As shown in Figure 4a, the machine vision module primarily consisted of an Imaging Source DMK72AUC02 complementary metal-oxide-semiconductor (CMOS) camera, a Moritex MML8-ST65S telecentric lens (with a nominal magnification of 8) coupled with a Moritex ML-Z2X rear converter lens (with a nominal magnification of 2), a Moritex MDRL-CW16-NS diffuse ring illuminator, a Moritex MCBP-CW3430 collimated backlight illuminator, and a specially designed optical subset. The rear converter lens, the telecentric lens, and the ring illuminator were sequentially and coaxially mounted on the CMOS camera to form a camera-lens subassembly, while the CMOS camera itself was installed on an X-Y-Z manual linear stage for fine positioning and focusing purposes. The camera-lens subassembly could achieve an overall magnification of 16 due to the combined effects of the telecentric and rear converter lenses, and its optical axis was set parallel to the rotational axis of the grinding unit (i.e., along the Y-direction). The backlight illuminator was fixed to one side of the supporting structure (block) of the grinder unit, so that the grinding wheel could be placed between it and the camera-lens subassembly and its collimated (parallel) light beam could be projected along the X-direction (toward the grinding wheel and the camera-lens subassembly). As shown in Figure 4b, the optical subset, consisting of a set of Newport BAC21AR.14 plano-concave achromatic (PCA) lenses (a set of achromatic doublets) and a Newport 05D20ER.2 broadband silver-coated reflection mirror, was fixed to one side of the base of the Y-axis LMT and could be precisely moved and positioned by the X-axis LMT. That is, the optical subset and the microdrill fixture (and the held microdrill sample) could have no relative movement along the X-direction.

Figure 5a,b show the closeup view photographs of the machine vision module under its two actual usage conditions, while their corresponding working principles are illustrated in Figure 6a,b, respectively. As shown in Figures 5a and 6a, when the machine vision module is applied to visually position the drilltip of an original microdrill sample (relative to the end face of the grinding wheel) before the grinding off operation, the microdrill sample is moved to a prescribed position where its drilltip (and its central axis) would align with the outer boundary edge of the grinding wheel end-face (GWEF), i.e., the edge point G, but also keep a very small distance along the Y-direction. Meanwhile, the optical subset is synchronously moved to the front of the telecentric lens, where the X-directional distance between point G and the primary principle point L of the set of the PCA lenses would equal the effective focal length f of the PCA lenses (whose nominal value is 50 mm). That is, the ideal condition of GL = f = 50 mm would exist. When an object is placed at a Y-Z plane where the effective focal point of the PCA lenses (i.e., point G at that ideal condition) is located, a half-size virtual image of that object would be exactly formed at another Y-Z plane where point V is located, for which the ideal condition of VL = $f/2$ = 25 mm would exist. When the parallel light beam is projected from the backlight illuminator, it would be reflected by the reflection mirror (set at an inclination angle of 45°) and would then reach the telecentric lens and finally be sensed by the CMOS camera. A ray passing through points G, V, and L would reach a reference point R on the reflection mirror and would then be perpendicularly reflected to reach a reference point T on the telecentric lens. The ideal condition of this optical design is that the sum of the three distances VL, LR (denoted by d_{LR}), and RT (denoted by d_{RT}) would equal the working distance of the telecentric lens (denoted by d_w with a nominal value of 64.9 mm), that is, the ideal condition of VL + LR + RT = $f/2 + d_{LR} + d_{RT} = d_w$ = 64.9 mm or LR + RT = $d_{LR} + d_{RT}$ = 39.9 mm would exist. Therefore, when the ring illuminator is not switched on, a half-size virtual image of the dark shadows of the drilltip of the microdrill sample and of part of the grinding wheel would be sensed and captured by the CMOS camera. Such a virtual image of dark shadows (called a side-viewed image hereinafter) would be captured under a resultant magnification of 8, because of the optical effect of the PCA lenses, and would further be used to position the drilltip (i.e., to measure the axial distance between the drilltip and the GWEF). Due to a lower magnification, the area of the resultant field of view (FOV) of this optical design would be four times that of the camera-lens subassembly itself; such a larger FOV should be more appropriate for the application of positioning the drilltip. In other words, a larger FOV means that a larger axial distance between the drilltip and the grinding wheel could be maintained when capturing their side-viewed images, so that the interference between the two objects could be more easily avoided.

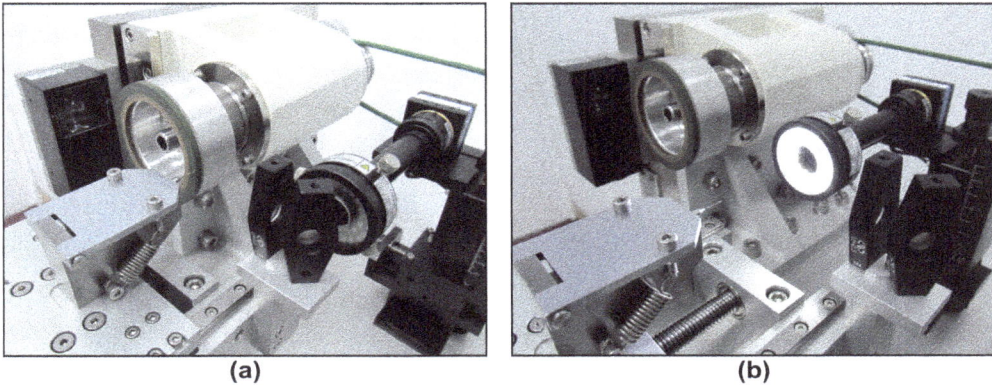

(a)   **(b)**

Figure 5. Closeup views of the machine vision module under its actual usage conditions: (**a**) the application for positioning the drilltip; (**b**) the application for measuring the CSGPs.

As shown in Figures 5b and 6b, when the machine vision module is applied to visually measure the CSGPs of a ground microdrill sample, the microdrill sample would be moved to another prescribed position where its ground cross-section would be located at the front of the telecentric lens and be focused properly; for which, the axial (Y-directional) distance between the ground cross-section and the telecentric lens would ideally equal the nominal working distance of the telecentric lens (i.e., $d_w = 64.9$ mm). Meanwhile, the optical subset would be synchronously moved to keep a proper distance from the front of the telecentric lens. When the ring illuminator is switched on, part of the diffused light beams projected from it would reach the ground cross-section and would then be reflected into the telecentric lens and finally be sensed by the CMOS camera. Therefore, a bright surface image of the ground cross-section (called an axial-viewed image hereinafter) would be sensed and captured by the CMOS camera under a magnification of 16. Such a high magnification would lead to a tiny FOV and a fine resolution, which should be appropriate for the application of measuring the CSGPs. In addition, the parallel light beam projected from the backlight illuminator would not influence the measurement, since it would not illuminate the ground cross-section and not be reflected into the telecentric lens by the optical subset. Thereby, the backlight illuminator would not need to be switched off under this usage condition. As a result, the setup of the machine vision module, characterized by several conventional machine vision components in combination with the innovative and lower cost optical subset, could enable the drilltip to be visually positioned and the CSGPs to be visually measured via captured side- and axial-viewed images, respectively.

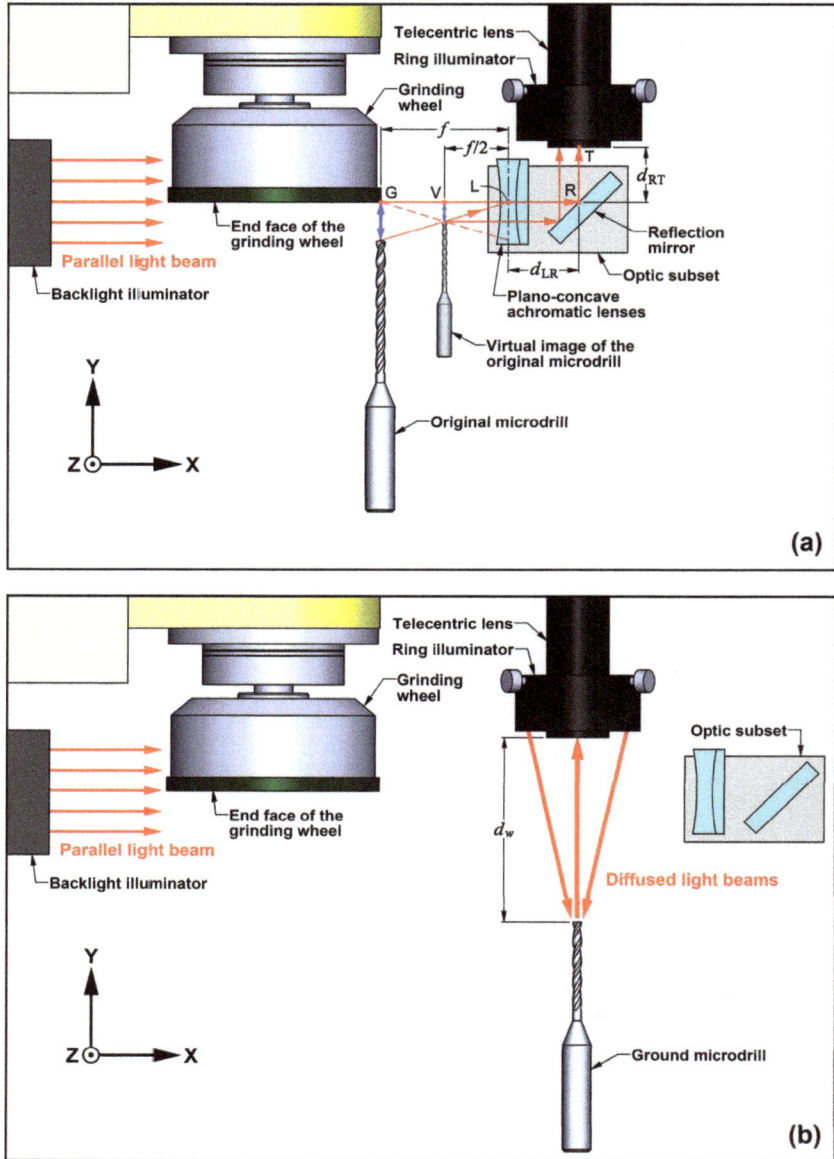

Figure 6. Illustrations of the working principles of the machine vision module: (**a**) the application for positioning the drilltip; (**b**) the application for measuring the CSGPs.

2.2. System Integration

The entire functional block diagram of the developed system is depicted in Figure 7. For accomplishing the system integration, a host personal computer (PC) was prepared to control, manipulate, and monitor the hardware modules of the entire machine. The functional connections between the host PC and the hardware modules were divided into three subsets: the motion control subset, the switching subset, and the image acquisition subset. The motion control subset was applied to manipulate the stepping motors of the dual-axis motion module via a National Instruments (NI) PCI-7340 motion control card installed on the motherboard of the host PC. The motion control card was electrically connected to the drivers of the two stepping motors. The stepping motor of the X-axis LMT was driven via an OM VEXTA RKD514L-A stepping motor driver, and that of the Y-axis LMT via an OM VEXTA RKD507-A stepping motor driver, while their angular resolutions were set to 5000 and 2000 step/rev, respectively. Thereby, the X- and Y-axis LMTs could both achieve a fine positioning resolution of 1 μm/step according to the nominal leads of their ballscrew units. The signals of the installed optical linear encoders were fed back to the motion control card for sensing the reference positions of both LMTs as well as for executing closed-loop control, and the extreme positions of both LMTs were detected by Renishaw A-9531-0251 limit switches (magnets) installed in them.

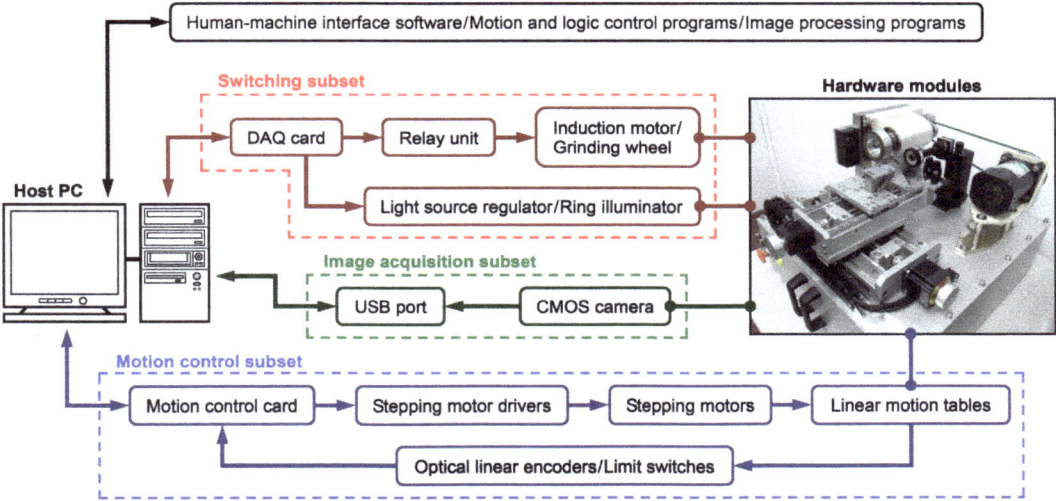

Figure 7. Functional block diagram of the improved automation system.

The switching subset was applied to switch on/off the induction motor and the ring illuminator via an NI PCI-6221 data acquisition (DAQ) card installed on the motherboard of the host PC. The DAQ card was electrically connected to a relay unit attached to the induction motor and the light source regulator of the ring illuminator. (Two customized Fadracer Technology VR-ML1505 type light source regulators were used to power and adjust the ring and backlight illuminators, respectively; that of the backlight illuminator did not need to be connected to and manipulated by the switching subset.) Two independent analog voltage signals were generated and outputted by the DAQ card for triggering/untriggering the relay unit and the light source regulator, respectively, thereby the induction motor and the ring illuminator were accordingly switched on/off.

The image acquisition subset was applied to manipulate the CMOS camera for capturing the required digital images and then storing them in the host PC, for which the CMOS camera was electrically connected to the host PC via a universal serial bus (USB) port. Each captured digital image (a side- or axial-viewed image) consisted of a 2592 × 1944 array of pixels with 8-bit grayscale values ranging from 0 (black) to 255 (white). For a captured side-viewed image, its conversion factor corresponding to the physical dimension was found to be 0.323 μm/pixel by means of spatial calibration for computer vision [29,30]; that for a captured axial-viewed image was found to be 0.165 μm/pixel. Thereby, real-world FOV areas of the captured side- or axial-viewed images were approximately 837 × 628 or 428 × 321 μm^2, respectively. Additionally, by means of the sub-pixel localization algorithm [29,30], theoretically estimated resolutions of 1/25 pixel for the side- and axial-viewed images (i.e., 0.0129 and 0.0066 μm, respectively) could be achieved. Two captured side-viewed images are shown in Figure 8a,b, in which dark shadows of part of the grinding wheel and of the drilltip of an original or ground microdrill sample with a bright background can be clearly observed. Two captured axial-viewed images are shown in Figure 8c,d, in which ground cross-sections of a UC or ST type microdrill sample (the bright objects) with a dark background can be clearly observed.

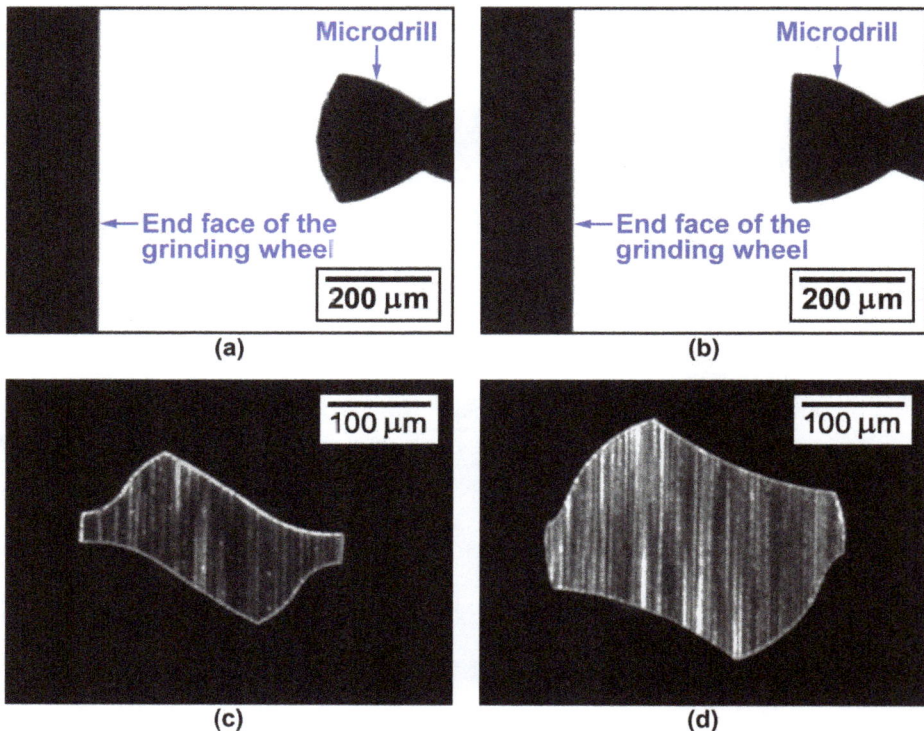

Figure 8. Digital images captured by the machine vision module: (**a**) a side-viewed image of an original microdrill sample; (**b**) another side-viewed image of a ground microdrill sample; (**c**) an axial-viewed image of a ground cross-section; (**d**) an axial-viewed image of another ground cross-section.

After the three subsets had been successfully set up, the required human-machine interface software and key programs (for motion control, logic control, and image processing) were developed, integrated, and tested in the NI LabVIEW environment, which enabled the constructed hardware system to be applied to execute an AMP that is introduced in Section 3.

3. The Automated Measuring Process (AMP)

In order to establish an AMP that is suitable for measuring the CSWTs and CSODs of both ST and UC type microdrills with the use of the improved automation system, the AMP proposed by Chang et al. [28] is extended and improved in this work. As a result, a more comprehensive AMP with specific image processing operations, which primarily consists of five essential sub-processes, is established, and its main flowchart is shown in Figure 9. The main steps in this flowchart are described as follows:

Step 1. Start the AMP, and let the moving stages of both the X- and Y-axis LMTs return to their prescribed home positions, respectively. Subsequently, go to the next step.

Step 2. Put a microdrill sample into the microdrill fixture to hold its shank (which should be manually operated). Subsequently, go to the next step.

Step 3. Input N_c specified axial positions of the drill body of the microdrill sample to be measured, for which an axial position, denoted by $l_c = l_{c(i)}$ for a counting index $i = 1, 2, \ldots, N_c$, is defined as the axial distance from the drilltip to a cross-section to be ground off and measured. Subsequently, let the counting index $i = 1$, and go to the next step.

Step 4. Move the microdrill sample to the first prescribed position, where its drilltip (or a certain ground cross-section) would keep a small distance from the GWEF, as illustrated in Figure 6a. Subsequently, go to the next step.

Step 5. Execute a drilltip positioning (DP) sub-process (described in Section 3.1), so that the drilltip (or a certain ground cross-section) of the microdrill sample would be precisely positioned to keep a small axial distance d_p from the GWEF (via the cooperation of the Y-axis LMT and the machine vision module). Subsequently, go to the next step.

Step 6. Switch on the induction motor for activating the grinding wheel. Subsequently, go to the next step.

Step 7. Execute a grind-off (GO) sub-process (described in Section 3.2), so that the drill body of the microdrill sample would be ground off to the ith specified axial position (via the cooperation of the dual-axis motion and grinding modules). Subsequently, go to the next step.

Step 8. Switch off the induction motor to stop the grinding wheel. Subsequently, move the ground microdrill sample to the second prescribed position where its ground cross-section would be located in front of the telecentric lens at a distance close to the nominal working distance, as illustrated in Figure 6b, and go to the next step.

Step 9. Switch on the ring illuminator, and execute an auto-focusing (AF) sub-process (described in Section 3.3), so that the ground cross-section of the microdrill sample would be clearly focused and its axial-viewed image would be captured (via the cooperation of the Y-axis LMT and the machine vision module). Subsequently, go to the next step.

Step 10. Execute a web thickness measuring (WTM) sub-process (described in Section 3.4), so that the CSWT would be visually measured via the axial-viewed image captured in Step 9. Subsequently, go to the next step.

Step 11. Execute an outer diameter measuring (ODM) sub-process (described in Section 3.5), so that the CSOD would be visually measured via the axial-viewed image captured in Step 9. Subsequently, go to the next step.

Step 12. Switch off the ring illuminator. Subsequently, let the counting index $i = i + 1$, and check if $i > N_c$. If so, go to the next step. Else, go to Step 4.

Step 13. Output all the measured values (and related data) of the CSWTs and CSODs. Subsequently, go to the next step.

Step 14. Once again let the moving stages of both the X- and Y-axis LMTs return to their prescribed home positions. Subsequently, remove the destructivelly measured microdrill sample from the microdrill fixture (which should be manually operated), and stop the AMP.

As can be seen, the DP, GO, AF, WTM, and ODM sub-processes must be included in the established AMP. Although the DP, GO, and WTM sub-processes are adopted in Chang et al.'s AMP [28], the AF and ODM sub-processes are newly added ones, while the WTM sub-process presented in this work is an improved one with better applicability.

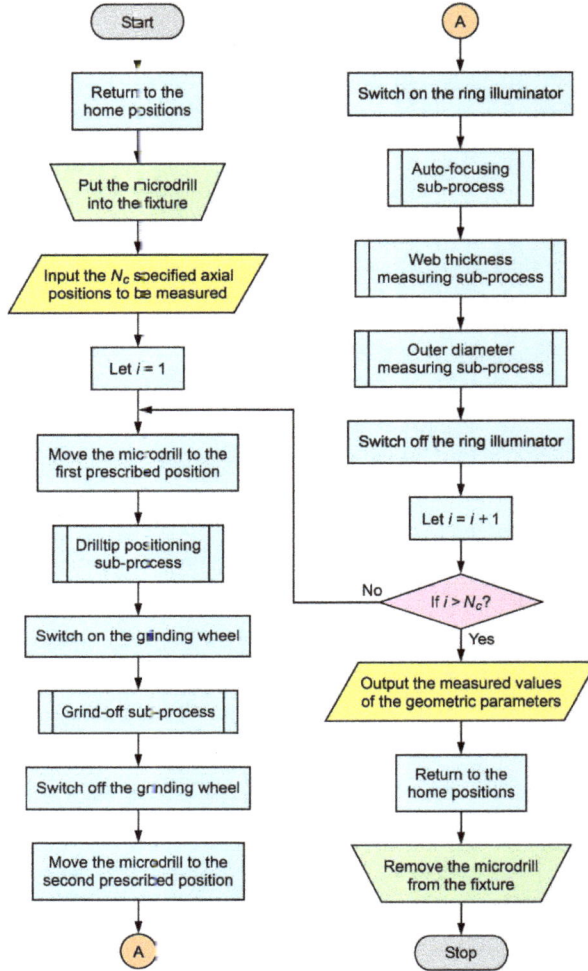

Figure 9. Main flowchart of the established AMP.

3.1. Drilltip Positioning (DP) Sub-Process

This sub-process is basically identical to that adopted by Chang et al. [28]. When the microdrill sample has been moved to the first prescribed position (during Step 4), it is then moved a small axial distance (whose value was set to 50 μm ($\approx$154.80 pixels in a side-viewed image)) toward the GWEF (via the Y-axis LMT) and a side-viewed image is captured and stored as a temporary file. The two captured images shown in Figure 8a,b correspond to the counting index $i = 1$ and $i > 1$, respectively.

Subsequently, edge detection for the captured side-viewed image is performed, as shown in Figure 10a or Figure 10b; a rectangular region of interest (ROI), numbered ROI #1, is assigned to appropriately cover the shadows of the two objects appearing in the FOV. The concept of one-dimensional (1D) edge detection [25,28–34] (described in Appendix A) is applied for performing a bilateral edge detection. To this end, multiple horizontal grayscale profiles (with an equal vertical interval of 1 pixel) are scanned along the axial direction (i.e., the Y-direction) with their search directions being bilaterally outward from the middle of ROI #1. According to the detected locations where sudden changes of grayscale values occur, two groups of edge points (i.e., the left and right edges indicated in the figures) with a Y-directional sub-pixel resolution can thus be obtained. Furthermore, as shown in Figure 10c or Figure 10d, the rightmost edge point among the left group (denoted by G) and the leftmost edge point among the right group (denoted by D) are extracted to calculate their horizontal distance, d_p. The visually measured distance d_p is regarded as the positioning distance between the drilltip (or a certain ground cross-section) and the GWEF at that moment.

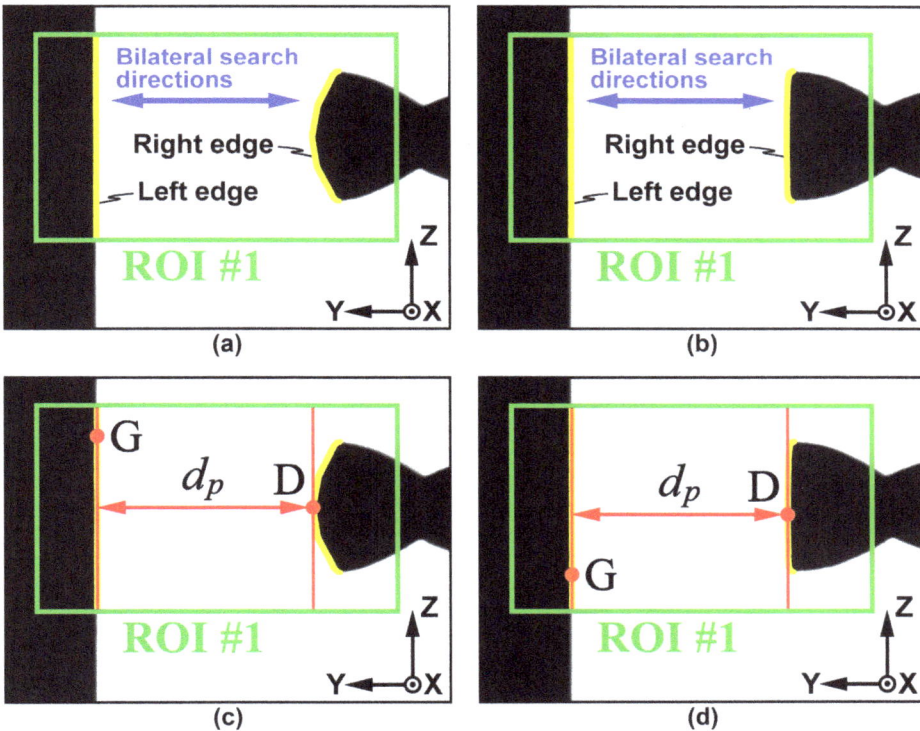

Figure 10. Illustrative examples of the image processing procedure for the DP sub-process: (**a**) edge detection for a side-viewed image; (**b**) edge detection for another side-viewed image; (**c**) measuring the positioning distance between the drilltip and the GWEF; (**d**) measuring the positioning distance between a certain ground cross-section and the GWEF.

The criterion for stopping this sub-process is d_p being less than or equal to a prescribed positioning distance (the condition of $d_p \leq 500$ μm ($\approx$1547.98 pixels in a side-viewed image) was given). If not, the cycle of moving the microdrill sample forward a small axial distance, capturing and storing a side-viewed image, and performing the bilateral edge detection is repeated until the given criterion can be achieved. The actual positioning distance d_p obtained in the final cycle is stored for further usage (in the GO sub-process).

3.2. Grind-Off (GO) Sub-Process

This sub-process is similar to that adopted by Chang et al. [28]. Two situations corresponding to the counting index $i = 1$ (for grinding off an original microdrill) and $i > 1$ (for regrinding off a ground microdrill) both exist in this sub-process and their schematic diagrams are depicted in Figure 11a,b, respectively.

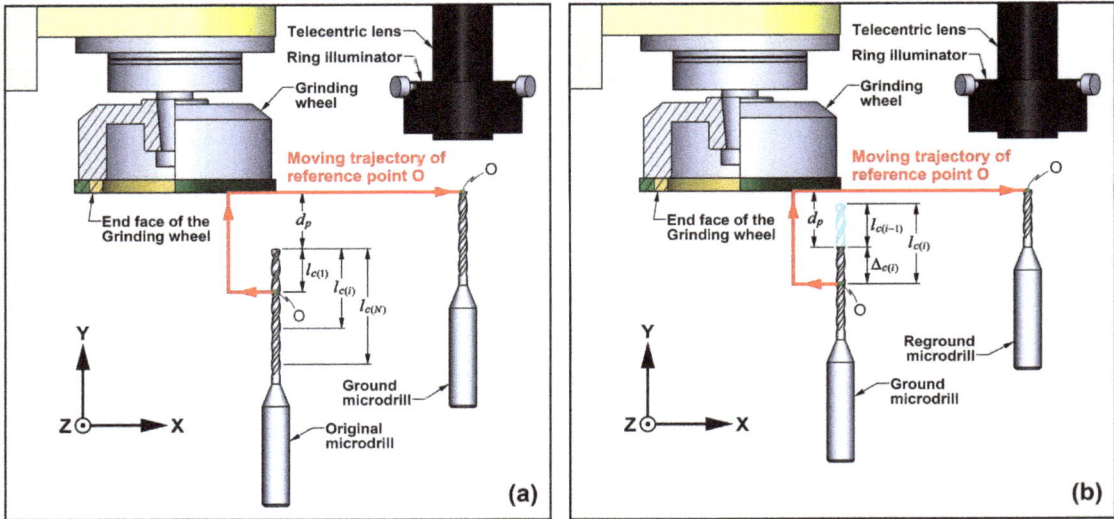

Figure 11. Schematic diagrams of the GO sub-process: (**a**) grinding off an original microdrill; (**b**) regrinding off a ground microdrill.

For the situation of $i = 1$, the axial distance between the cross-section of the original microdrill sample to be ground off and the GWEF is $(d_p + l_{c(1)})$, as shown in Figure 11a; the actual positioning distance d_p has been obtained in the DP sub-process, and $l_{c(1)}$ is the specified axial distance from the drilltip to the reference point O of the 1st specified axial position. The microdrill sample is moved a prescribed transverse distance along the negative X-direction (via the X-axis LMT) and is then moved an axial distance of $(d_p + l_{c(1)})$ along the positive Y-direction (via the Y-axis LMT), so that the drill body to be ground off can be moved to a specific location with a small transverse distance (of several millimeters) from the inner annular portion of the grinding wheel, where the reference point O can be aligned with the GWEF. Subsequently, the microdrill sample is fed transversely along the positive X-direction (via the X-axis LMT) to perform the grinding off operation, that is, to roughly grind off the drill body and finish the ground cross-section.

For the situation of $i > 1$, the axial distance between the cross-section of the ground microdrill sample to be reground off and the GWEF is $(d_p + \Delta_{c(i)})$, as shown in Figure 11b; the actual positioning distance d_p has been obtained in the DP sub-process, and $\Delta_{c(i)}$ is an increment amount calculated by

$$\Delta_{c(i)} = l_{c(i)} - l_{c(i-1)} \tag{1}$$

which is the axial distance from the previously ground cross-section to the reference point O of the ith specified axial position. Similar to the situation of $i = 1$, the microdrill sample is moved a prescribed transverse distance along the negtive X-direction (via the X-axis LMT) and is then moved an axial distance of $(d_p + \Delta_{c(1)})$ along the positive Y-direction (via the Y-axis LMT), so that the remaining drill body to be reground off can also be moved to a specific location with a small transverse distance (of several millimeters) from the inner annular portion of the grinding wheel, where the reference point O can be aligned with the GWEF. Subsequently, the microdrill sample is fed transversely along the positive X-direction (via the X-axis LMT) to perform the grinding off operation, i.e., to roughly regrind off the remaining drill body and finish the reground cross-section.

The feed rate for performing the grinding off operations was set to 0.1 mm/s. When the microdrill sample has been fed to a specific position leaving a small transverse distance (of several millimeters) from the outer boundary of the GWEF, this sub-process is stopped. The ground microdrill sample is then transversely moved to the front of the telecentric lens (during Step 8), as shown by the ends of the moving trajectories of point O shown in Figure 11.

3.3. Auto-Focusing (AF) Sub-Process

This sub-process is based on the application of two-dimensional discrete Fourier transform (2D-DFT) [29,31–33]. When the microdrill sample has been moved to the second prescribed position (during Step 8), an axial-viewed image of its ground cross-section is then captured and stored as a temporary file. For a digital grayscale image $g(x, y)$ consisting of an $N \times M$ matrix of grayscale values, its 2D-DFT can be computed by [29,31–33]

$$E_F(u, v) = \frac{1}{MN} \sum_{x=0}^{M-1} \sum_{y=0}^{N-1} g(x, y) e^{-j2\pi(\frac{ux}{M} + \frac{vy}{N})} \tag{2}$$

in which u and v are the 2D frequency variables and $j = \sqrt{-1}$. For the captured axial-viewed image (with $N = 1944$ and $M = 2592$), its Fourier spectrum can thus be computed and depicted.

In Figure 12a, four axial-viewed images captured from an unfocused condition to a clearly focused condition are presented, while their corresponding centered (shifted) Fourier spectra are depicted in Figure 12b. (Note that the four Fourier spectra are all depicted from $u = 0$ to $u = 0.35M$ and from $v = 0$ to $v = 0.35N$, and the origin (i.e., $u = 0$ and $v = 0$) is located at the center of each 2D spectral plot.) As can be observed, higher spectrum intensities (with brighter colors) are concentrated in the low-frequency domains of the 2D spectral plots, especially the clearly focused one. Therefore, a weighted sum of squares of the spectrum intensities in the low-frequency domain (from $u = 0$ to $u = 0.25M$ and from $v = 0$ to $v = 0.25N$), denoted by S_{wl}, can be obtained by

$$S_{wl} = \sum_{u=0}^{(M/4)-1} \sum_{v=0}^{(N/4)-1} (u^2 + v^2) |E_F(u, v)|^2 \tag{3}$$

which can be used to evaluate the focusing condition. By specifying a focusing search range according to the reference position of the Y-axis LMT being increased from $Y = Y_{f(b)}$ to $Y = Y_{f(f)}$, the value of the weighted sum of squares, S_{wl}, is a function of the Y-axial position, i.e., $S_{wl} = S_{wl}(Y)$. Thereby, a relative focusing index, Γ, can defined as

$$\Gamma(Y) = \frac{S_{wl}(Y)}{S_{wl(\max)}} \quad \text{for } Y_{f(b)} \leq Y \leq Y_{f(f)} \tag{4}$$

in which $S_{wl(\max)}$ is the global maximum value of $S_{wl}(Y)$ within the specified focusing search range. Figure 13 shows an illustrative example of the variation curve of the relative focusing index Γ with respect to the Y-axial relative position (within a range of 880 μm), for which 89 axial-viewed images were captured sequentially with a positional increment of 10 μm. As can be observed, when the condition of $\Gamma(Y) = 1$ (i.e., $S_{wl}(Y) = S_{wl(\max)}$) is achieved, the axial-viewed image captured at that Y-axial position can be regarded as an optimally focused one. Nevertheless, obtaining the variation curve of $\Gamma(Y)$ for each ground cross-section is quite time-consuming.

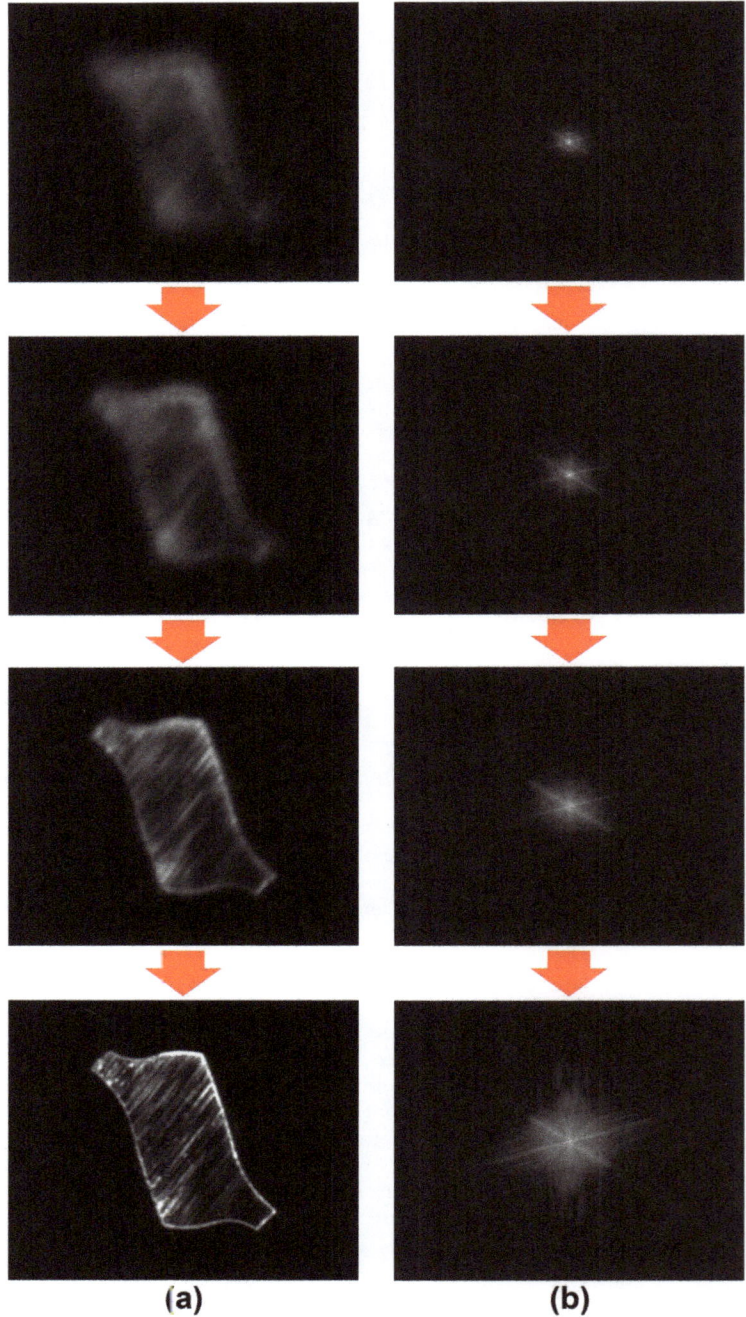

Figure 12. Illustration of the AF sub-process: (**a**) axial-viewed images captured from an unfocused condition to a clearly focused condition; (**b**) the corresponding centered Fourier spectra.

In order to efficiently search for the optimal focusing position within the specified focusing search range, the bisection method [35] is applied in this sub-process. To this end, three axial-viewed images corresponding to the backmost, middle, and frontmost Y-axial positions (within the search range) are captured and stored, and their yielded values of S_{wl} are calculated and ordered (denoted by $S_{wl(\mathrm{BS,min})}$, $S_{wl(\mathrm{BS,mid})}$, and $S_{wl(\mathrm{BS,max})}$ in ascending order). For evaluating the focusing condition, a relative difference ratio, r_f, can be defined as

$$r_f = \frac{S_{wl(\mathrm{BS,max})} - S_{wl(\mathrm{BS,mid})}}{S_{wl(\mathrm{BS,max})}} \tag{5}$$

The criterion for stopping this sub-process is r_f being less than or equal to a prescribed small value (the condition of $r_f \leq 0.05$ was given). If not, the search range is updated (according to the range between the two Y-axial positions where $S_{wl(\mathrm{BS,mid})}$ and $S_{wl(\mathrm{BS,max})}$ occur, respectively), and the cycle of capturing and storing the three axial-viewed images within the updated search range and then calculating and ordering their yielded values of S_{wl} is repeated until the given criterion can be achieved. Among the three axial-viewed images captured in the final cycle, the one yielding a value of $S_{wl(\mathrm{BS,max})}$ can meet the condition of $\Gamma(Y) \approx 1$ (i.e., $S_{wl(\mathrm{BS,max})} \approx S_{wl(\mathrm{max})}$) and is stored as an optimally focused one for further usage (in the WTM and ODM sub-processes).

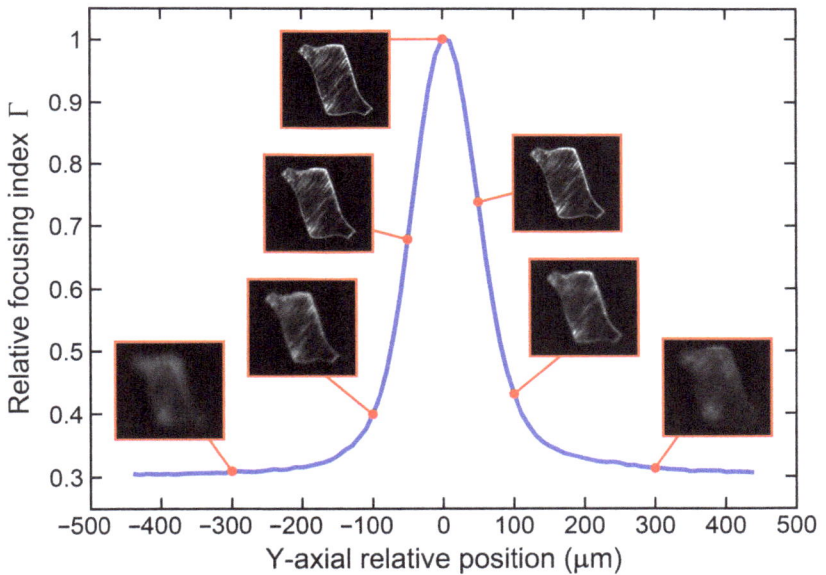

Figure 13. Illustrative example of the variation curve of the relative focusing index Γ with respect to the Y-axial relative position.

3.4. Web Thickness Measuring (WTM) Sub-Process

This sub-process is improved from that adopted by Chang et al. [28]. An illustrative example of this sub-process is presented in Figure 14. As the sequential image processing operations shown in Figure 14a–k are basically identical to those used by Chang et al. [28], their concepts are described briefly. The sequential operations shown in Figure 14l–o regard a specific approach called the best-fit circle (BFC) approach, whose fundamental has been preliminarily studied by the authors [34] and is explained in this section.

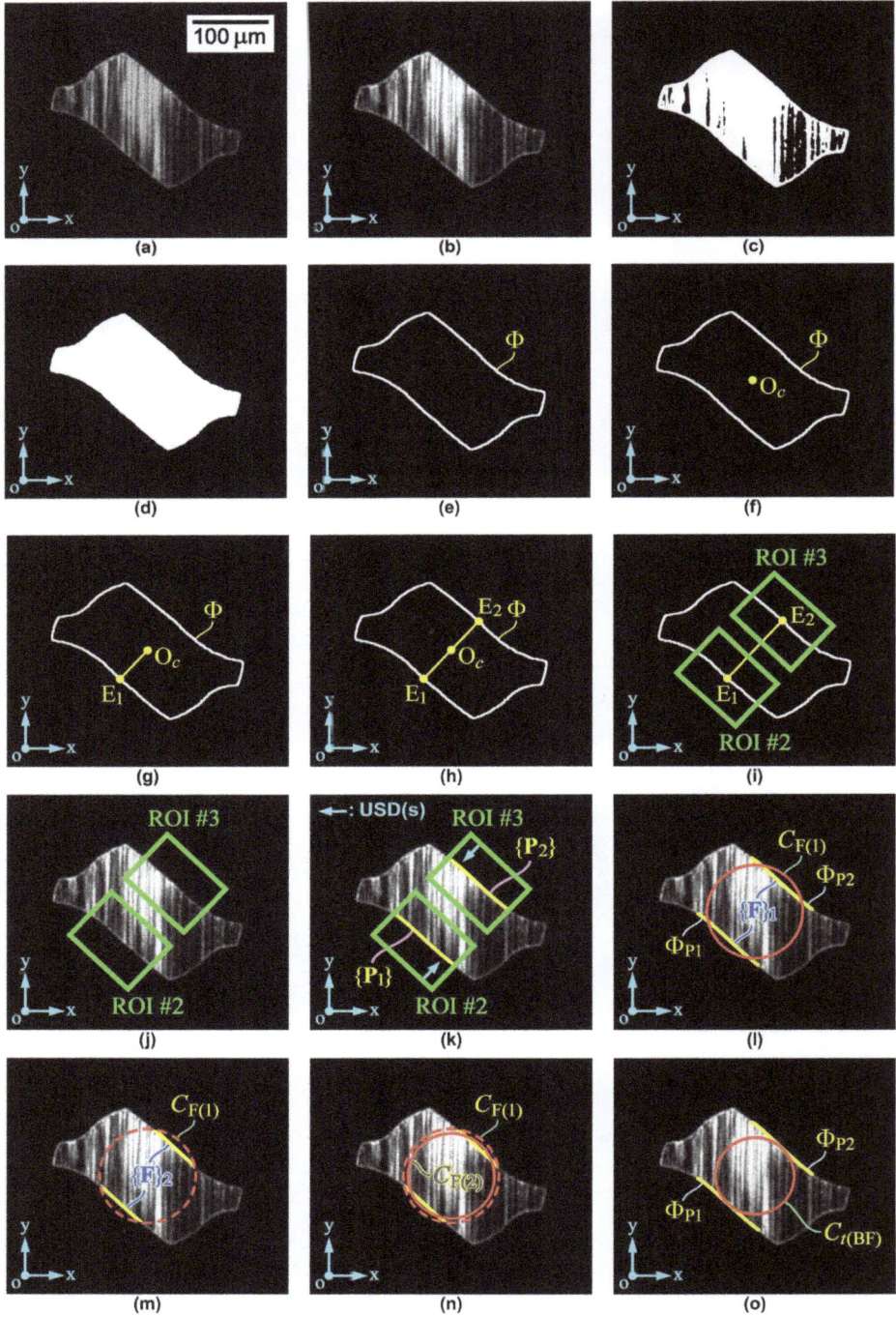

Figure 14. Illustration of the sequential operations of the WTM sub-process.

Figure 14a shows an original axial-viewed image obtained at the end of Step 9, and a global coordinate system o-xy is consistently set on the image. This original grayscale image is enhanced via a grayscale transformation function [28] for adjusting its brightness, contrast, and gamma values, so that the ground cross-section can be highlighted relative to the background. The enhanced grayscale image, shown in Figure 14b, is stored as a temporary file and is transformed into a binary image (consisting of pixels with 1-bit binary values of zeros (black) and ones (white)) via a thresholding approach [29,30]. The transformed binary image, shown in Figure 14c, is dealt with by specific morphological operations [29–31] to fill holes (black pixels) within the cross-section and to eliminate thin protrusions and isolated particles (white pixels) around the cross-section, as shown by the modified binary image shown in Figure 14d. The exterior contour of the cross-section (formed by a group of white pixel points) in this modified binary image (denoted by Φ) is extracted via a boundary extraction operation [31], as shown in Figure 14e, and the centroid (i.e., the average position of all pixel points) of contour Φ (denoted by O_c) is calculated, as indicated in Figure 14f. As shown in Figure 14g, a certain pixel point E_1 on contour Φ is extracted as a reference point by finding a pixel point having the shortest Euclidean distance among those between all pixel points on contour Φ and centroid O_c. Thereby, another reference point E_2 can be determined by finding a point satisfying the condition of $\mathbf{O}_c\mathbf{E}_1 + \mathbf{O}_c\mathbf{E}_2 = \mathbf{0}$ (see Figure 14h), which may not exactly coincide with (but is quite close to) contour Φ. Points E_1 and E_2 are regarded as two reference points for the pair of flute contours of the cross-section (as shown by contour Φ_F of Figure 2). Subsequently, two rectangular ROIs centered at points E_1 and E_2 with their orientations dependent on vector $\pm\mathbf{E}_1\mathbf{E}_2$, numbered ROI #2 and ROI #3, respectively, are assigned to appropriately cover parts of the pair of flute contours, as shown by the operation shown in Figure 14i. In addition, the length of each ROI (vertical to vector $\pm\mathbf{E}_1\mathbf{E}_2$) can be 80 to 95 percent of the Euclidean distance between points E_1 and E_2, and the width of each ROI (parallel to vector $\pm\mathbf{E}_1\mathbf{E}_2$) is approximately 80 percent of that Euclidean distance. The determined ROI #2 and ROI #3 are then assigned to the enhanced grayscale image (see Figure 14j) for performing unidirectional edge detections within the two ROIs, as shown by the operation shown in Figure 14k. The inward direction parallel to vector $\pm\mathbf{E}_1\mathbf{E}_2$ is defined as the unitary search direction (USD) of each ROI, and the concept of 1D edge detection [25,28–34] (described in Appendix A) is applied for the unidirectional edge detection. To this end, multiple USD-parallel grayscale profiles (with an equal interval of 1 pixel vertical to the USD) are scanned along the USD of each ROI. Two groups (sets) of detected edge points on the pair of flute contours (denoted by $\{P_1\}$ and $\{P_2\}$, respectively) with a sub-pixel resolution (along their USDs) can be obtained. Furthermore, two planar curves fitted from $\{P_1\}$ and $\{P_2\}$, respectively, can be obtained, as shown by the two fitted flute contours Φ_{P1} and Φ_{P2} shown in Figure 14l, while their parametric vector equations can be expressed as

$$\Phi_{P1} = \Phi_{P1}(w_1) = \begin{Bmatrix} x_{\Phi_{P1}}(w_1) \\ y_{\Phi_{P1}}(w_1) \end{Bmatrix} = \begin{Bmatrix} a_{P1x}w_1^3 + b_{P1x}w_1^2 + c_{P1x}w_1 + d_{P1x} \\ a_{P1y}w_1^3 + b_{P1y}w_1^2 + c_{P1y}w_1 + d_{P1y} \end{Bmatrix} \tag{6}$$

$$\Phi_{P2} = \Phi_{P2}(w_2) = \begin{Bmatrix} x_{\Phi_{P2}}(w_2) \\ y_{\Phi_{P2}}(w_2) \end{Bmatrix} = \begin{Bmatrix} a_{P2x}w_2^3 + b_{P2x}w_2^2 + c_{P2x}w_2 + d_{P2x} \\ a_{P2y}w_2^3 + b_{P2y}w_2^2 + c_{P2y}w_2 + d_{P2y} \end{Bmatrix} \tag{7}$$

in which w_1 and w_2 are two independent variables regarding the order of points for each fitted contour, and the x- and y-components of each fitted contour are represented as cubic polynomials of w_1 or w_2 with their coefficients being solved by using the least-squares (LS) polynomial-fitting approach [35].

The BFC approach, illustrated in Figure 14l–o, is based on the application of the LS circle-fitting approach [36–38] in combination with a specific iterative procedure [34]. An initial group (set) of theoretical points on both fitted contours (denoted by $\{F\}_k = \{\Phi_{P1}, \Phi_{P2}\}_k$ for $k = 1$) can be calculated via Equations (6) and (7). The coordinate of each theoretical point can be represented as $\mathbf{F}_m = \{x_{Fm}\ y_{Fm}\}^T$ for $m = 1, 2, \ldots, n_{Fk}$, and n_{Fk} is the number of all points of $\{\mathbf{F}\}_k$. The LS circle-fitting approach is repeatedly executed with all points of $\{\mathbf{F}\}_k$ being involved and updated in every iteration. For the kth iteration, three coefficients $a_{F(k)}$, $b_{F(k)}$, and $c_{F(k)}$ of the equation of a fitted circle $C_{F(k)}$ can be solved by [38]

$$
\begin{Bmatrix} a_{F(k)} \\ b_{F(k)} \\ c_{F(k)} \end{Bmatrix} = \begin{bmatrix} \sum\limits_{m=1}^{n_{Fk}} x_{Fm}^2 & \sum\limits_{m=1}^{n_{Fk}} x_{Fm}y_{Fm} & \sum\limits_{m=1}^{n_{Fk}} x_{Fm} \\ \sum\limits_{m=1}^{n_{Fk}} x_{Fm}y_{Fm} & \sum\limits_{m=1}^{n_{Fk}} y_{Fm}^2 & \sum\limits_{m=1}^{n_{Fk}} y_{Fm} \\ \sum\limits_{m=1}^{n_{Fk}} x_{Fm} & \sum\limits_{m=1}^{n_{Fk}} y_{Fm} & \sum\limits_{m=1}^{n_{Fk}} 1 \end{bmatrix}^{-1} \begin{Bmatrix} -\sum\limits_{m=1}^{n_{Fk}} x_{Fm}\left(x_{Fm}^2 + y_{Fm}^2\right) \\ -\sum\limits_{m=1}^{n_{Fk}} y_{Fm}\left(x_{Fm}^2 + y_{Fm}^2\right) \\ -\sum\limits_{m=1}^{n_{Fk}} \left(x_{Fm}^2 + y_{Fm}^2\right) \end{Bmatrix} \tag{8}
$$

The center and diameter of the fitted circle $C_{F(k)}$, denoted by $O_{F(k)}$ and $D_{F(k)}$, respectively, can thus be determined by [38]

$$
\mathbf{O}_{F(k)} = \begin{Bmatrix} x_{O_{F(k)}} \\ y_{O_{F(k)}} \end{Bmatrix} = \begin{Bmatrix} -a_{F(k)}/2 \\ -b_{F(k)}/2 \end{Bmatrix} \tag{9}
$$

$$
D_{F(k)} = \sqrt{a_{F(k)}^2 + b_{F(k)}^2 - 4c_{F(k)}} \tag{10}
$$

The root-sum-square (RSS) error of the LS circle-fitting for the kth iteration is defined as

$$
\varepsilon_{F(k)} = \sqrt{\sum_{m=1}^{n_{Fk}} \left(x_{Fm}^2 + y_{Fm}^2 + a_{F(k)}x_{Fm} + b_{F(k)}y_{Fm} + c_{F(k)}\right)^2} \tag{11}
$$

Figure 14l shows the fitted circle $C_{F(1)}$ for the 1st iteration. Obviously, the major parts of theoretical points on contours Φ_{P1} and Φ_{P2} (i.e., all points of $\{\mathbf{F}\}_1$) are enclosed by circle $C_{F(1)}$. That is, circle $C_{F(1)}$ is too large to be a representative common tangent circle for measuring the CSWT. Therefore, partial points of $\{\mathbf{F}\}_1$ that are outside circle $C_{F(1)}$ are excluded for performing the next circle-fitting iteration. For the kth iteration, the group of theoretical points $\{\mathbf{F}\}_k$ is thus updated by

$$
\{\mathbf{F}\}_{k+1} = \{\mathbf{F}\}_k \setminus \left\{ (x_{Fm}, y_{Fm}) \,\middle|\, x_{Fm}^2 + y_{Fm}^2 - \left(D_{F(k)}^2/4\right) > 0 \right\} \tag{12}
$$

Thereby, the updated group $\{\mathbf{F}\}_2$ is within circle $C_{F(1)}$, as shown in Figure 14m. Subsequently, an updated fitted circle $C_{F(2)}$ can be obtained by using Equations (8)–(10) with all points of $\{\mathbf{F}\}_2$ being involved, as shown in Figure 14n. It can be observed that circle $C_{F(2)}$ is smaller than circle $C_{F(1)}$. Expectedly, the following circle-fitting iterations would lead to the fitted circles being gradually closer to a representative common tangent circle for measuring the CSWT. The iterative procedure is terminated if one of the following conditions is satisfied: (i) the RSS error $\varepsilon_{F(k)}$ is less than 0.5 pixels (or 0.1 μm) or (ii) the number of all points of $\{\mathbf{F}\}_k$ is less than 4. When the iterative procedure is terminated at the kth iteration, as shown in Figure 14o, the fitted circle $C_{F(k)}$ is regarded as the resultant best-fit circle (denoted by $C_{t(BF)}$), i.e., the representative common tangent circle for measuring the CSWT. As a result, the visually measured CSWT w can be obtained by

$$
w = D_{t(BF)} = D_{F(k)} \tag{13}
$$

in which $D_{t(BF)}$ is the diameter of the best-fit circle $C_{t(BF)}$. In addition, the center of the best-fit circle $C_{t(BF)}$, denoted by $O_{t(BF)}$, is equal to $O_{F(k)}$. When the data of obtained $D_{t(BF)}$ and $O_{t(BF)}$ are stored, this sub-process is stopped. The presented WTM sub-process with the BFC approach can be applied to measure the CSWTs of both ST and UC type microdrills.

3.5. Outer Diameter Measuring (ODM) Sub-Process

This sub-process is based on sequential image processing operations, similar to those in the WTM sub-process. An illustrative example of this sub-process is presented in Figure 15. As the sequential operations shown in Figure 15a–e are essentially identical to those shown in Figure 14a–e, their concepts will not be described again. The sequential operations shown in Figure 15f–h are also based on the BFC approach, and those shown in Figure 15i–l regard a specific approach called the fitted dividing line (FDL) approach; their fundamentals have been preliminarily studied by the authors [34] and are explained in this section.

According to the sequential operations shown in Figure 15a–e, an original axial-viewed image, obtained at the end of Step 9, is dealt with to obtain the exterior contour of the cross-section, Φ, in its modified binary image. Subsequently, as shown by the operations shown in Figure 15f–h, the BFC approach is again applied. That is, an initial group (set) of pixel points on the exterior contour Φ (denoted by $\{Q\}_l$ for $l = 1$) is extracted, for which the coordinate of each pixel point can be represented as $Q_m = \{x_{Rm}\ y_{Rm}\}^T$ for $m = 1, 2, \ldots,$ n_{Ql}, and n_{Ql} is the number of all points of $\{Q\}_l$. The LS circle-fitting is repeatedly executed with all points of $\{Q\}_l$ being involved and updated in every iteration. For the lth iteration, three coefficients $a_{Q(l)}$, $b_{Q(l)}$, and $c_{Q(l)}$ of the equation of a fitted circle $C_{Q(l)}$ can be solved by [38]

$$\begin{Bmatrix} a_{Q(l)} \\ b_{Q(l)} \\ c_{Q(l)} \end{Bmatrix} = \begin{bmatrix} \sum_{m=1}^{n_{Ql}} x_{Qm}^2 & \sum_{m=1}^{n_{Ql}} x_{Qm}y_{Qm} & \sum_{m=1}^{n_{Ql}} x_{Qm} \\ \sum_{m=1}^{n_{Ql}} x_{Qm}y_{Qm} & \sum_{m=1}^{n_{Ql}} y_{Qm}^2 & \sum_{m=1}^{n_{Ql}} y_{Qm} \\ \sum_{m=1}^{n_{Ql}} x_{Qm} & \sum_{m=1}^{n_{Ql}} y_{Qm} & \sum_{m=1}^{n_{Ql}} 1 \end{bmatrix}^{-1} \begin{Bmatrix} -\sum_{m=1}^{n_{Ql}} x_{Qm}\left(x_{Qm}^2 + y_{Qm}^2\right) \\ -\sum_{m=1}^{n_{Ql}} y_{Qm}\left(x_{Qm}^2 + y_{Qm}^2\right) \\ -\sum_{m=1}^{n_{Ql}} \left(x_{Qm}^2 + y_{Qm}^2\right) \end{Bmatrix} \quad (14)$$

The center and diameter of the fitted circle $C_{Q(l)}$, denoted by $O_{Q(l)}$ and $D_{Q(l)}$, respectively, can thus be determined by [38]

$$O_{Q(l)} = \begin{Bmatrix} x_{O_{Q(l)}} \\ y_{O_{Q(l)}} \end{Bmatrix} = \begin{Bmatrix} -a_{Q(l)}/2 \\ -b_{Q(l)}/2 \end{Bmatrix} \quad (15)$$

$$D_{Q(l)} = \sqrt{a_{Q(l)}^2 + b_{Q(l)}^2 - 4c_{Q(l)}} \quad (16)$$

Additionally, the RSS error of the LS circle-fitting for the lth iteration is defined as

$$\varepsilon_{Q(l)} = \sqrt{\sum_{m=1}^{n_{Ql}} (x_{Qm}^2 + y_{Qm}^2 + a_{Q(l)}x_{Qm} + b_{Q(l)}y_{Qm} + c_{Q(l)})^2} \quad (17)$$

The fitted circle $C_{Q(1)}$ for the 1st iteration is illustrated in Figure 15f. Obviously, the major parts of pixel points on contour Φ (i.e., all points of $\{Q\}_1$) are enclosed by circle $C_{Q(1)}$. In other words, circle $C_{Q(1)}$ is too small to overlap the pair of margin contours of the cross-section (as shown by contour Φ_M of Figure 2). Therefore, partial points of $\{Q\}_1$ that are inside circle $C_{Q(1)}$ are excluded for performing the next circle-fitting iteration. For the lth iteration, the group of pixel points $\{Q\}_l$ is thus updated by

$$\{Q\}_{l+1} = \{Q\}_l \setminus \left\{ (x_{Qm}, y_{Qm}) \middle| x_{Qm}^2 + y_{Qm}^2 - (D_{Q(l)}^2/4) < 0 \right\} \quad (18)$$

Thereby, the updated group $\{Q\}_2$ is out of circle $C_{Q(1)}$, and an updated fitted circle $C_{Q(2)}$ can be accordingly obtained by using Equations (14)–(16) with all points of $\{Q\}_2$ being involved, as illustrated in Figure 15g. Expectedly, the following circle-fitting iterations would lead to the fitted circles being gradually closer to the pair of margin contours. The iterative procedure is terminated if one of the following conditions is satisfied: (i) the RSS error $\varepsilon_{Q(l)}$ is less than 0.5 pixels (or 0.2 μm) or (ii) the number of all points of $\{Q\}_l$ is less than 15. When the iterative procedure is terminated at the *l*th iteration, as shown in Figure 15h, the fitted circle $C_{Q(l)}$ is regarded as the resultant best-fit circle (denoted by $C_{b(BF)}$) that overlaps the pair of margin contours. The center and diameter of the best-fit circle $C_{b(BF)}$, denoted by $O_{b(BF)}$ and $D_{b(BF)}$, can also be obtained.

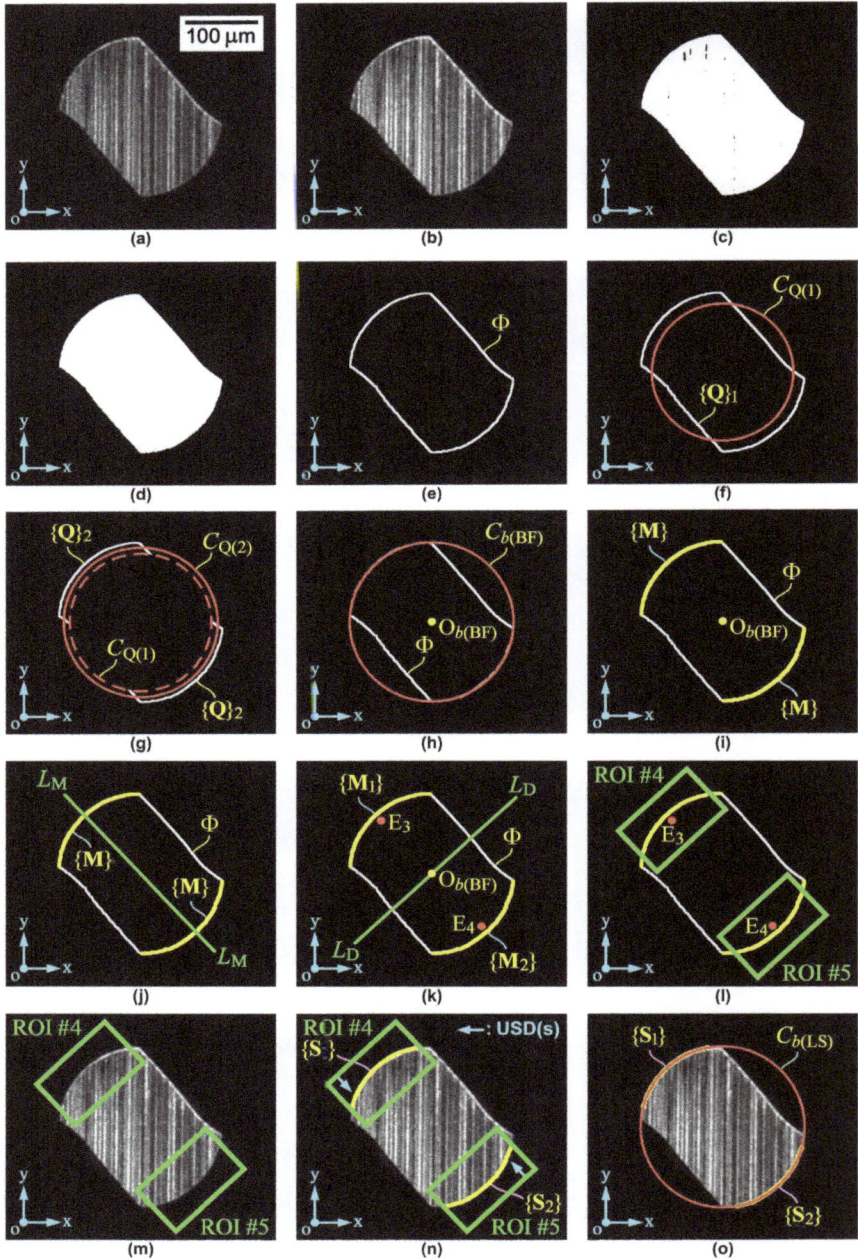

Figure 15. Illustration of the sequential operations of the ODM sub-process.

The FDL approach, illustrated in Figure 15i–l, is based on the following procedure. Firstly, pixel points on the pair of margin contours are extracted from the exterior contour Φ. To this end, the Euclidean distances between all points of $\{\mathbf{Q}\}_1$ and circular center $O_{b(\mathrm{BF})}$, denoted by l_m for $m = 1, 2, \ldots, n_{\mathrm{Q1}}$, are calculated by

$$l_m = \left\| \mathbf{Q}_m - \mathbf{O}_{b(\mathrm{BF})} \right\| = \sqrt{\left(x_{\mathrm{Q}m} - x_{\mathrm{O}_b(\mathrm{BF})} \right)^2 + \left(y_{\mathrm{Q}m} - y_{\mathrm{O}_b(\mathrm{BF})} \right)^2} \tag{19}$$

A criterion to judge whether a pixel point is on the pair of margin contours is

$$\left| l_m - D_{b(\mathrm{BF})}/2 \right| \le \varepsilon_{\mathrm{M}} \tag{20}$$

in which ε_{M} is a specified small constant which can be arbitrarily given as 3 pixels (or 0.5 μm). Thus, a pixel point is regarded as a margin contour point if its calculated distance l_m is quite close to the radius ($D_{b(\mathrm{BF})}/2$) of the best-fit circle $C_{b(\mathrm{BF})}$. As a result, a group (set) of pixel points on the pair of margin contours (denoted by $\{\mathbf{M}\}$) can be extracted from the exterior contour Φ, as shown in Figure 15i; the coordinate of each pixel point can be represented as $\mathbf{M}_m = \{x_{\mathrm{M}m} \; y_{\mathrm{M}m}\}^{\mathrm{T}}$ for $m = 1, 2, \ldots, n_{\mathrm{M}}$, and n_{M} is the number of all points of $\{\mathbf{M}\}$. Secondly, by using the LS line-fitting approach [35] (i.e., the well-known linear regression) with all points of $\{\mathbf{M}\}$ being involved, a fitted line (denoted by L_{M}) can be obtained via solving its slope a_{M} and intercept b_{M} by

$$\begin{Bmatrix} a_{\mathrm{M}} \\ b_{\mathrm{M}} \end{Bmatrix} = \begin{bmatrix} \sum\limits_{m=1}^{n_{\mathrm{M}}} x_{\mathrm{M}m}^2 & \sum\limits_{m=1}^{n_{\mathrm{M}}} x_{\mathrm{M}m} \\ \sum\limits_{m=1}^{n_{\mathrm{M}}} x_{\mathrm{M}m} & \sum\limits_{m=1}^{n_{\mathrm{M}}} 1 \end{bmatrix}^{-1} \begin{Bmatrix} \sum\limits_{m=1}^{n_{\mathrm{M}}} x_{\mathrm{M}m} y_{\mathrm{M}m} \\ \sum\limits_{m=1}^{n_{\mathrm{M}}} y_{\mathrm{M}m} \end{Bmatrix} \tag{21}$$

or

$$a_{\mathrm{M}} = \frac{n_{\mathrm{M}} \sum_{m=1}^{n_{\mathrm{M}}} x_{\mathrm{M}m} y_{\mathrm{M}m} - \sum_{m=1}^{n_{\mathrm{M}}} x_{\mathrm{M}m} \sum_{m=1}^{n_{\mathrm{M}}} y_{\mathrm{M}m}}{n_{\mathrm{M}} \sum_{m=1}^{n_{\mathrm{M}}} x_{\mathrm{M}m}^2 - \left(\sum_{m=1}^{n_{\mathrm{M}}} x_{\mathrm{M}m} \right)^2} \tag{22}$$

$$b_{\mathrm{M}} = \frac{\sum_{m=1}^{n_{\mathrm{M}}} x_{\mathrm{M}m}^2 \sum_{m=1}^{n_{\mathrm{M}}} y_{\mathrm{M}m} - \sum_{m=1}^{n_{\mathrm{M}}} x_{\mathrm{M}m} \sum_{m=1}^{n_{\mathrm{M}}} x_{\mathrm{M}m} y_{\mathrm{M}m}}{n_{\mathrm{M}} \sum_{m=1}^{n_{\mathrm{M}}} x_{\mathrm{M}m}^2 - \left(\sum_{m=1}^{n_{\mathrm{M}}} x_{\mathrm{M}m} \right)^2} \tag{23}$$

As shown in Figure 15j, the fitted line L_{M} intersects with the pair of margin contours. Thirdly, a line perpendicular to the fitted line L_{M} and passing through circular center $O_{b(\mathrm{BF})}$, denoted by L_{D}, can be obtained via calculating its slope a_{D} and intercept b_{D} by

$$a_{\mathrm{D}} = -1/a_{\mathrm{M}} \tag{24}$$

$$b_{\mathrm{D}} = y_{\mathrm{O}_b(\mathrm{BF})} - a_{\mathrm{D}} x_{\mathrm{O}_b(\mathrm{BF})} = y_{\mathrm{O}_b(\mathrm{BF})} + x_{\mathrm{O}_b(\mathrm{BF})}/a_{\mathrm{M}} \tag{25}$$

The obtained line L_{D} is called a dividing line, which can be used to divide the group of points $\{\mathbf{M}\}$ into two sub-groups $\{\mathbf{M}_1\}$ and $\{\mathbf{M}_2\}$, as indicated in Figure 15k. The criteria for judging whether the mth pixel point belongs to sub-group $\{\mathbf{M}_1\}$ or $\{\mathbf{M}_2\}$ are

$$\mathbf{M}_m = \begin{Bmatrix} x_{\mathrm{M}m} \\ y_{\mathrm{M}m} \end{Bmatrix} \in \{\mathbf{M}_1\} \text{ if } y_{\mathrm{M}m} - (a_{\mathrm{D}} x_{\mathrm{M}m} + b_{\mathrm{D}}) > 0 \tag{26}$$

$$\mathbf{M}_m = \begin{Bmatrix} x_{\mathrm{M}m} \\ y_{\mathrm{M}m} \end{Bmatrix} \in \{\mathbf{M}_2\} \text{ if } y_{\mathrm{M}m} - (a_{\mathrm{D}} x_{\mathrm{M}m} + b_{\mathrm{D}}) < 0 \tag{27}$$

Finally, the two centroids of $\{\mathbf{M}_1\}$ and $\{\mathbf{M}_2\}$ (i.e., the two average positions of all pixel points in individual sub-groups, denoted by E_3 and E_4, respectively) are calculated, as indicated in Figure 15k. Thereby, two rectangular ROIs centered at points E_3 and E_4 with their orientations being dependent on vector $\pm \mathbf{E}_3 \mathbf{E}_4$, numbered ROI #4 and ROI #5, respectively, are assigned to appropriately cover major parts of the pair of margin contours, as shown by the operation shown in Figure 15l. In addition, the length of each ROI (vertical to vector $\pm \mathbf{E}_3 \mathbf{E}_4$) is approximately 90 percent of the chord length of the margin contour, which can be estimated by the Euclidean distance between the first and last pixel points of $\{\mathbf{M}_1\}$ or $\{\mathbf{M}_2\}$, and the width of each ROI (parallel to vector $\pm \mathbf{E}_3 \mathbf{E}_4$) can be 50 to 90 percent of the estimated chord length.

Furthermore, the determined ROI #4 and ROI #5 are assigned to the enhanced grayscale image (see Figure 15m) for performing unidirectional edge detections within the two ROIs, as shown by the operation shown in Figure 15n. The inward direction parallel to vector $\pm \mathbf{E}_3 \mathbf{E}_4$ is defined as the USD of each ROI, and the concept of 1D edge detection [25,28–34] (described in Appendix A) is applied for the unidirectional edge detection. That is, multiple USD-parallel grayscale profiles (with an equal interval of 1 pixel vertical to the USD) are scanned along the USD of each ROI. Two groups (sets) of edge points on the pair of margin contours (denoted by $\{\mathbf{S}_1\}$ and $\{\mathbf{S}_2\}$, respectively) with a sub-pixel resolution (along their USDs) can be obtained. Subsequently, the LS circle-fitting approach [36–38], as shown in Equations (8)–(11) or Equations (14)–(17), is again executed once with all points of $\{\mathbf{S}_1\}$ and $\{\mathbf{S}_2\}$ being involved, and a resultant LS circle (denoted by $C_{b(LS)}$), i.e., the representative minimum bounding circle for measuring the CSOD, can be solved, as shown in Figure 15o. As a result, the visually measured CSOD D can be obtained by

$$D = D_{b(LS)} \tag{28}$$

in which $D_{b(LS)}$ is the diameter of the LS circle $C_{b(LS)}$. In addition, the center of the LS circle $C_{b(LS)}$, denoted by $O_{b(LS)}$, is also obtained. When the data of obtained $D_{b(LS)}$ and $O_{b(LS)}$ are stored, this sub-process is stopped. The presented ODM sub-process with the BFC and FDL approaches can be applied to measure the CSODs of both ST and UC type microdrills.

In addition, after the execution of both the WTM and ODM sub-processes, the concentricity of the representative common tangent and minimum bounding circles, $C_{t(BF)}$ and $C_{b(LS)}$, of a ground cross-section (denoted by ΔC) can be evaluated via calculating the eccentricity of circular centers $O_{t(BF)}$ and $O_{b(LS)}$, that is,

$$\Delta C = \left\| \mathbf{O}_{t(BF)} - \mathbf{O}_{b(LS)} \right\| \tag{29}$$

The amount of concentricity ΔC can be a rule of thumb to evaluate the radial symmetry of the ground cross-section.

4. Experimental Results and Discussion

In order to evaluate the actual effectiveness of the developed system, two series of experiments for measuring the CSWTs and CSODs of certain ST and UC type microdrill samples produced by several companies were conducted. The nominal outer diameter (denoted by D_n) and flute length (denoted by l_f) of each measured ST type sample were 0.30 and 5.5 mm, respectively. The measured UC type samples were of two specifications: (i) $D_n = 0.25$ mm and $l_f = 3.9$ mm and (ii) $D_n = 0.20$ mm and $l_f = 3.0$ mm. The nominal overall length and shank diameter of each sample were 38.1 and 3.175 mm (i.e., 1.5 and 0.125 inches), respectively.

4.1. Measuring Uncertainty and Accuracy Tests

This series of experiments aimed at testing the measuring uncertainty and accuracy that could be achieved by the developed system. A ST type microdrill sample (called sample ST1) and two UC type ones (called samples UC1 (with $D_n = 0.25$ mm) and UC2 (with $D_n = 0.20$ mm)) were measured. For each sample, its drill body was ground off to two specified axial positions of $l_c = 0.15l_f$ and $0.45l_f$, while its two ground cross-sections were repeatedly focused, captured, and measured 15 times. For performing the first measurement of each ground cross-section, the established AMP was completely executed (from Step 1 to Step 14) once. For performing the remaining 14 measurements of each ground cross-section, the established AMP was repeatedly executed with Step 4 to Step 7 (i.e., the DP and GO sub-processes) being skipped. As each measured sample was manually removed from and put into the microdrill fixture again after each measurement (according to Step 2 and Step 14), 15 axial-viewed images of each ground cross-section appearing in the FOV with different orientations (angular positions) could be focused and captured for measuring the CSGPs. In addition, for comparison purposes, a Sage Vision SG-1022 measuring microscope (in combination with RVS-250 metrology software), that can achieve a resolution of 0.01 μm and an accuracy of ±3 μm, was manually operated to repeatedly measure the CSGPs of each ground cross-section 15 times, as shown in Figure 16.

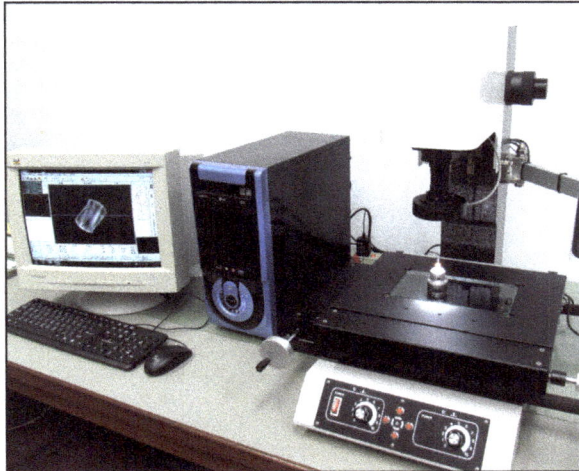

Figure 16. Measuring the CSGPs of a microdrill sample by a measuring microscope.

The statistical results of this series of experiments are listed in Tables 1–3, in which the CSWT, CSOD, and concentricity measured by the developed automation system are denoted by $w_{(A)}$, $D_{(A)}$, and $\Delta C_{(A)}$, respectively, and those measured by the measuring microscope are denoted by $w_{(M)}$, $D_{(M)}$, and $\Delta C_{(M)}$, respectively. For each CSGP, the maximum and minimum values, mean value (denoted by μ), standard deviation (denoted by σ), uncertainty (denoted by $\tilde{u}$), and statistical upper and lower bounds (denoted by $\beta_{(u)}$ and $\beta_{(l)}$, respectively) evaluated from its 15 measured values are presented. For each CSWT ($w_{(A)}$ or $w_{(M)}$) and CSOD ($D_{(A)}$ or $D_{(M)}$), the uncertainty $\tilde{u}$ is evaluated via the well-known 3σ-band approach [39], that is,

$$\tilde{u} = 3\sigma \tag{30}$$

for which the theoretical probability within the band of $\pm 3\sigma$ is 99.73% for a Gaussian distribution assumption. Thereby, its statistical upper and lower bounds are

$$\beta_{(u)} = \mu + \tilde{u} = \mu + 3\sigma \tag{31}$$

$$\beta_{(l)} = \mu - \tilde{u} = \mu - 3\sigma \tag{32}$$

The uncertainty range of $\pm \tilde{u}$ (or $\pm 3\sigma$) can be regarded as the measuring repeatability for each CSWT and CSOD. However, the 3σ-band approach is not suitable for evaluating the uncertainty of each concentricity ($\Delta C_{(A)}$ or $\Delta C_{(M)}$), since the resultant statistical lower bound may be a negative value that violates the essence of $\Delta C \geq 0$ based on Equation (29). Hence, a modified approach is used to evaluate the uncertainty of each concentricity. To this end, its 15 measured values are further divided into two data groups by comparing their individual magnitudes with the mean value μ, and the mean values and standard deviations of the group with smaller magnitudes (denoted by μ_S and σ_S) and of that with larger magnitudes (denoted by μ_L and σ_L) can be obtained. The condition of $0 \leq \mu_S < \mu_L$ essentially exists. Therefore, the modified statistical upper and lower bounds for each concentricity are estimated by

$$\beta_{(u)} = \mu_L + 1.6449\sigma_L \tag{33}$$

$$\beta_{(l)} = \mu_S - 1.6449\sigma_S \tag{34}$$

for which the theoretical probability within the band of $\pm 1.6449\sigma$ is 90% for a Gaussian distribution assumption. Accordingly, the modified uncertainty $\tilde{u}$ for each concentricity, as a rule of thumb, is defined as

$$\tilde{u} = \beta_{(u)} - \beta_{(l)} = (\mu_L - \mu_S) + 1.6449(\sigma_L + \sigma_S) \tag{35}$$

which is a conservatively estimated band based on the Gaussian distributions of the two divided data groups.

As can be observed in Tables 1–3, the representative statistical values (i.e., the results of $\mu \pm \tilde{u}$) of $w_{(A)}$ were 118.86 ± 0.538 and 169.11 ± 0.547 μm for sample ST1, 120.12 ± 0.619 and 149.86 ± 0.964 μm for sample UC1, and 125.74 ± 0.950 and 141.68 ± 0.701 μm for sample UC2, respectively; those of $w_{(M)}$ were 118.21 ± 1.771 and 169.29 ± 1.646 μm for sample ST1, 120.24 ± 2.196 and 149.01 ± 1.959 μm for sample UC1, and 126.02 ± 1.725 and 141.16 ± 1.829 μm for sample UC2, respectively. Thereby, the developed system could achieve a worst uncertainty range of ±1 μm for measuring the CSWTs, while that achieved by the measuring microscope was ±2.2 μm. It can also be found that, for each cross-section, the absolute difference between the mean values of $w_{(A)}$ and $w_{(M)}$ was significantly less than 1 μm. By calculating the maximum absolute differences between the two measured extremes of $w_{(A)}$ and the two of $w_{(M)}$ for each cross-section, the obtained results were 2.11 and 1.35 μm for sample ST1, 2.03 and 2.05 μm for sample UC1, and 1.26 and 1.62 μm for sample UC2. That is, for a conservative evaluation, the relative deviation between the measured CSWTs obtained by the developed system and by the measuring microscope could be less than 2.2 μm. As a result, based on the narrow uncertainty range of ±1 μm and the small relative deviation of 2.2 μm, the developed system could achieve good repeatability and accuracy for measuring the CSWTs.

Table 1. Statistical results for the measurements of sample ST1 ($D_n = 0.30$ mm and $l_f = 5.5$ mm).

| | Axial Position $l_c = 0.15l_f$ | | | | | | Axial Position $l_c = 0.45l_f$ | | | | | |
| | CSGP Measured by the Developed System | | | CSGP Measured by Measuring Microscope | | | CSGP Measured by the Developed System | | | CSGP Measured by Measuring Microscope | | |
Statistical Term	$w_{(A)}$ (μm)	$D_{(A)}$ (μm)	$\Delta C_{(A)}$ (μm)	$w_{(M)}$ (μm)	$D_{(M)}$ (μm)	$\Delta C_{(M)}$ (μm)	$w_{(A)}$ (μm)	$D_{(A)}$ (μm)	$\Delta C_{(A)}$ (μm)	$w_{(M)}$ (μm)	$D_{(M)}$ (μm)	$\Delta C_{(M)}$ (μm)
Measured maximum	119.17	295.31	1.94	119.07	295.93	18.12	169.56	292.32	2.23	170.10	292.87	19.10
Measured minimum	118.51	294.62	0.70	117.06	294.13	1.35	168.75	291.59	1.32	168.33	290.85	3.98
Mean value μ	118.86	294.89	1.13	118.21	295.09	9.11	169.11	291.94	1.92	169.29	291.70	9.57
Standard deviation σ	0.179	0.195	0.395	0.590	0.690	6.196	0.182	0.175	0.244	0.549	0.565	5.043
Uncertainty $\tilde{u}$	0.538	0.584	1.314	1.771	2.069	18.173	0.547	0.525	0.871	1.646	1.694	16.991
Upper bound $\beta_{(u)}$	119.40	295.47	1.94	119.98	297.16	18.79	169.65	292.46	2.28	170.94	293.40	20.15
Lower bound $\beta_{(l)}$	118.32	294.30	0.63	116.44	293.02	0.61	168.56	291.41	1.41	167.65	290.01	3.16

Table 2. Statistical results for the measurements of sample UC1 (D_n = 0.25 mm and l_f = 3.9 mm).

| Statistical Term | Axial Position $l_c = 0.15l_f$ | | | | | | Axial Position $l_c = 0.45l_f$ | | | | | |
| | CSGP Measured by the Developed System | | | CSGP Measured by Measuring Microscope | | | CSGP Measured by the Developed System | | | CSGP Measured by Measuring Microscope | | |
	$w_{(A)}$ (μm)	$D_{(A)}$ (μm)	$\Delta C_{(A)}$ (μm)	$w_{(M)}$ (μm)	$D_{(M)}$ (μm)	$\Delta C_{(M)}$ (μm)	$w_{(A)}$ (μm)	$D_{(A)}$ (μm)	$\Delta C_{(A)}$ (μm)	$w_{(M)}$ (μm)	$D_{(M)}$ (μm)	$\Delta C_{(M)}$ (μm)
Measured maximum	120.61	222.97	1.72	121.91	223.36	17.86	150.29	225.14	1.85	150.44	225.88	17.16
Measured minimum	119.88	222.44	0.94	119.34	221.26	1.38	149.37	224.08	0.62	148.24	223.87	1.46
Mean value μ	120.12	222.67	1.37	120.24	222.58	8.32	149.86	224.52	1.33	149.01	224.62	7.84
Standard deviation σ	0.206	0.152	0.216	0.732	0.653	5.655	0.321	0.328	0.416	0.653	0.547	4.551
Uncertainty $\tilde{u}$	0.619	0.456	0.770	2.196	1.958	19.304	0.964	0.985	1.332	1.959	1.641	14.995
Upper bound $\beta_{(u)}$	120.74	223.13	1.73	122.43	224.53	20.49	150.83	225.51	1.90	150.97	226.26	16.17
Lower bound $\beta_{(l)}$	119.50	222.22	0.96	118.04	220.62	1.18	148.90	223.54	0.56	147.05	222.98	1.18

Table 3. Statistical results for the measurements of sample UC2 (D_n = 0.20 mm and l_f = 3.0 mm).

| Statistical Term | Axial Position $l_c = 0.15l_f$ | | | | | | Axial Position $l_c = 0.45l_f$ | | | | | |
| | CSGP Measured by the Developed System | | | CSGP Measured by Measuring Microscope | | | CSGP Measured by the Developed System | | | CSGP Measured by Measuring Microscope | | |
	$w_{(A)}$ (μm)	$D_{(A)}$ (μm)	$\Delta C_{(A)}$ (μm)	$w_{(M)}$ (μm)	$D_{(M)}$ (μm)	$\Delta C_{(M)}$ (μm)	$w_{(A)}$ (μm)	$D_{(A)}$ (μm)	$\Delta C_{(A)}$ (μm)	$w_{(M)}$ (μm)	$D_{(M)}$ (μm)	$\Delta C_{(M)}$ (μm)
Measured maximum	126.35	201.85	3.32	126.64	202.71	18.74	141.93	187.22	2.66	141.94	187.83	18.64
Measured minimum	125.38	201.23	2.31	125.14	201.11	1.37	141.30	186.60	1.24	140.31	185.86	3.43
Mean value μ	125.74	201.46	2.69	126.02	201.88	6.91	141.68	186.93	1.63	141.16	186.66	10.25
Standard deviation σ	0.317	0.226	0.350	0.575	0.594	4.962	0.234	0.215	0.374	0.610	0.581	5.575
Uncertainty $\tilde{u}$	0.950	0.679	1.201	1.725	1.782	16.651	0.701	0.646	1.481	1.829	1.744	17.946
Upper bound $\beta_{(u)}$	126.69	202.14	3.42	127.74	203.66	17.10	142.38	187.57	2.72	142.99	188.40	20.17
Lower bound $\beta_{(l)}$	124.79	200.78	2.22	124.29	200.10	0.45	140.98	186.28	1.24	139.34	184.91	2.23

Furthermore, the representative statistical values of $D_{(A)}$ were 294.89 ± 0.584 and 291.94 ± 0.525 μm for sample ST1, 222.67 ± 0.456 and 224.52 ± 0.985 μm for sample UC1, and 201.46 ± 0.679 and 186.93 ± 0.646 μm for sample UC2, respectively; those of $D_{(M)}$ were 295.09 ± 2.069 and 291.70 ± 1.694 μm for sample ST1, 222.58 ± 1.958 and 224.62 ± 1.641 μm for sample UC1, and 201.88 ± 1.782 and 186.66 ± 1.744 μm for sample UC2, respectively. Thereby, the developed system could achieve a worst uncertainty range of ±1 μm for measuring the CSODs, while that achieved by the measuring microscope was ±2.1 μm. Additionally, for each cross-section, the absolute difference between the mean values of $D_{(A)}$ and $D_{(M)}$ was considerably less than 0.5 μm. By calculating the maximum absolute differences between the two measured extremes of $D_{(A)}$ and the two of $D_{(M)}$ for each cross-section, the obtained results were 1.31 and 1.47 μm for sample ST1, 1.71 and 1.80 μm for sample UC1, and 1.48 and 1.36 μm for sample UC2. In other words, the relative deviation between the measured CSODs obtained by the developed system and by the measuring microscope could be conservatively less than 2 μm. As a result, based on the narrow uncertainty range of ±1 μm and the small relative deviation of 2 μm, the developed system could achieve good repeatability and accuracy for measuring the CSODs.

In addition, all of the six mean values of $\Delta C_{(A)}$ ranged between 1.13 and 2.69 μm, but those of $\Delta C_{(M)}$ ranged between 6.91 and 10.25 μm; their corresponding absolute differences notably ranged between 4.22 and 8.62 μm. In addition, all of the six estimated uncertainty bands of $\Delta C_{(A)}$ ranged between 0.770 and 1.481 μm, however, those of $\Delta C_{(M)}$ ranged between 14.995 and 19.304 μm. The unreasonably large absolute differences (greater than 4 μm) and uncertainty bands of $\Delta C_{(M)}$ (greater than 14.5 μm) implied that the measuring microscope was not an adequate device to measure the amount of concentricity. Since the measuring microscope was manually operated to find the representative common tangent and minimum bounding circles of a ground cross-section, the positional stabilities of their circular centers sensitively depended on the skill and dexterity of the operator, although the uncertainty ranges of their diameters could be nonsensitively within ±2.2 μm. As compared with the used measuring microscope, an estimated uncertainty band less than 1.5 μm indicated that the developed system could achieve a narrower uncertainty band and a more confident measured value for the amount of concentricity.

In summary, from a conservative viewpoint, the developed system could achieve an uncertainty range of ±1.5 μm and an accuracy of 2.5 μm for the measurements of the CSWTs and CSODs of both ST and UC type microdrills. The developed system could also achieve a conservatively estimated uncertainty band of 2 μm, as well as a confident measured value, for the measurement of the concentricity, which could not be achieved by the used measuring microscope.

4.2. Applicability Tests for Sampling Inspection

This series of experiments aimed at evaluating the applicability of the developed system for automatically executing the sampling inspection procedure. An ST type microdrill sample (called sample ST2) and two UC type ones (called samples UC3 (with D_n = 0.25 mm) and UC4 (with D_n = 0.20 mm)) were measured. For each sample, five axial positions of l_c = 0.15l_f, 0.3l_f, 0.45l_f, 0.6l_f, and 0.75l_f were specified, and their corresponding cross-sections were sequentially ground and measured by executing the established AMP once; the measuring microscope (see Figure 16) was also employed to measure each ground cross-section. To this end, when a sample had been ground off to a specified axial position and measured, the AMP was immediately paused (at the end of Step 12), and the sample was temporarily removed from the microdrill fixture and then manually measured once by the use of the measuring microscope. Subsequently, the sample was thereupon put back into and held by the microdrill fixture, and the paused AMP was resumed for the remaining measuring cycles. Such operations were repeated until all the five axial positions of the drill body had been ground and measured. In addition, the required time for sequentially executing the DP and GO sub-processes (from Step 4 to Step 8) once was approximately 2.5 to 2.75 min, and that for sequentially executing the AF, WTM, and ODM sub-processes (from Step 9 to Step 12) once was approximately 1.5 to 1.75 min. Therefore, the developed system took approximately 4 to 4.5 min to destructively and visually measure the CSGPs at each specified axial position of a microdrill sample.

The statistical results of this series of experiments are listed in Tables 4–6. Based on the uncertainty and accuracy testing results presented in Section 4.1, the values of $\Delta C_{(M)}$ measured by the measuring microscope are not considered. For all the three samples, the 15 absolute differences between the measured values of $w_{(A)}$ and $w_{(M)}$ ranged between 0.37 and 1.83 μm, while those between the measured values of $D_{(A)}$ and $D_{(M)}$ ranged between 0.35 and 1.74 μm. In other words, for each cross-section, the developed system could measure its CSWT and CSOD with a good accuracy of less than 2 μm. The measured values of $\Delta C_{(A)}$ also reasonably ranged between 0.80 and 3.37 μm. Therefore, the established AMP could be executed successfully by the developed system.

Table 4. Statistical results for the measurements of sample ST2 (D_n = 0.30 mm and l_f = 5.5 mm).

Normalized Axial Position	CSGP Measured by the Developed System			CSGP Measured by Measuring Microscope		Absolute Difference	
l_c/l_f	$w_{(A)}$ (μm) [a]	$D_{(A)}$ (μm)	$\Delta C_{(A)}$ (μm)	$w_{(M)}$ (μm)	$D_{(M)}$ (μm)	$\mid w_{(A)} - w_{(M)} \mid$ (μm)	$\mid D_{(A)} - D_{(M)} \mid$ (μm)
0.15	117.97	294.78	1.82	117.23	295.85	0.74	1.07
0.30	144.53	293.36	3.37	145.46	294.95	0.93	1.59
0.45	169.38	291.33	1.43	170.60	290.55	1.22	0.78
0.60	191.54	291.40	2.22	191.17	293.14	0.37	1.74
0.75	214.24	289.95	0.80	213.79	288.60	0.45	1.35

[a] Fitted linear regression model: $w_{(A)} \approx 159.700(l_c/l_f) + 95.666$ μm (with a norm of residuals of 2.913 μm).

Table 5. Statistical results for the measurements of sample UC3 (D_n = 0.25 mm and l_f = 3.9 mm).

Normalized Axial Position	CSGP Measured by the Developed System			CSGP Measured by Measuring Microscope		Absolute Difference	
l_c/l_f	$w_{(A)}$ (μm) [a]	$D_{(A)}$ (μm)	$\Delta C_{(A)}$ (μm)	$w_{(M)}$ (μm)	$D_{(M)}$ (μm)	$\mid w_{(A)} - w_{(M)} \mid$ (μm)	$\mid D_{(A)} - D_{(M)} \mid$ (μm)
0.15	120.37	223.85	2.46	119.55	222.73	0.82	1.12
0.30	132.45	221.72	2.22	130.96	221.37	1.49	0.35
0.45	145.28	223.50	1.91	146.18	224.39	0.90	0.89
0.60	156.95	223.88	2.04	158.78	223.14	1.83	0.74
0.75	168.49	224.67	1.31	169.16	226.08	0.67	1.41

[a] Fitted linear regression model: $w_{(A)} \approx 80.490(l_c/l_f) + 108.490$ μm (with a norm of residuals of 0.749 μm).

Table 6. Statistical results for the measurements of sample UC4 (D_n = 0.20 mm and l_f = 3.0 mm).

Normalized Axial Position	CSGP Measured by the Developed System			CSGP Measured by Measuring Microscope		Absolute Difference	
l_c/l_f	$w_{(A)}$ (μm) [a]	$D_{(A)}$ (μm)	$\Delta C_{(A)}$ (μm)	$w_{(M)}$ (μm)	$D_{(M)}$ (μm)	$\mid w_{(A)} - w_{(M)} \mid$ (μm)	$\mid D_{(A)} - D_{(M)} \mid$ (μm)
0.15	114.77	207.74	2.76	115.40	208.94	0.63	1.20
0.30	129.94	188.23	1.79	130.54	186.91	0.60	1.32
0.45	135.78	187.03	2.03	134.37	186.39	1.41	0.64
0.60	141.16	186.81	1.85	142.25	187.99	1.09	1.18
0.75	145.11	185.64	1.31	144.08	184.23	1.03	1.41

[a] Fitted second-order regression model: $w_{(A)} \approx -72.724(l_c/l_f)^2 + 113.384(l_c/l_f) + 100.330$ μm (with a norm of residuals of 2.761 μm).

In addition, the measured values of $w_{(A)}$ of each sample obviously appeared a progressively increasing trend with the increased axial position, as shown in Figure 17, which can be fitted and formulated via some regression models. By using the LS polynomial-fitting approach [35], a first-order (linear) regression model for a set of measured values of $w_{(A)}$ can be represented as [24,25,28]

$$w_{(A)} \approx a_1\left(l_c/l_f\right) + a_0 \tag{36}$$

and a second-order regression model for it can be represented as

$$w_{(A)} \approx a_2\left(l_c/l_f\right)^2 + a_1\left(l_c/l_f\right) + a_0 \tag{37}$$

As can be observed in Figure 17, the measured values of $w_{(A)}$ of samples ST2 and UC3 both had a linearly increasing trend and are fitted by using Equation (36), while those of sample UC4 had a nonlinearly increasing trend and are fitted by using Equation (37). The resultant regression models are listed in the last rows (footnotes) of Tables 4–6, and the norm of residuals of each regression model was significantly less than 3 μm. The well-fitted regression models implied that the helical flutes of these samples were properly machined according to the web taper design of microdrills [5,23–25,28]. The measured values and fitted regression models of the CSWTs could be further used to compare with their corresponding theoretical designs, which should benefit the practical quality control tasks.

Accordingly, the applicability of the developed system for automating the sampling inspection procedure could be verified by the presented experimental results. Therefore, the actual effectiveness of the developed system for carrying out the destructive and visual measurements of the CSWTs and CSODs of both ST and UC type microdrills could be validated.

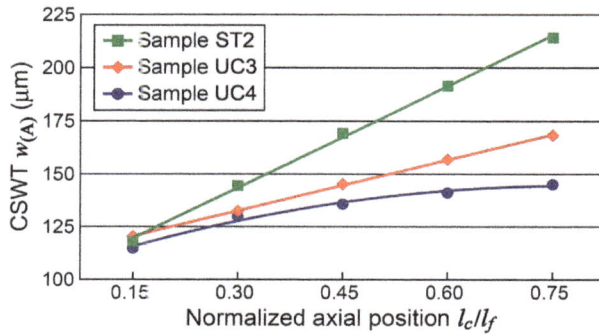

Figure 17. Measured values and fitted curves of the CSWTs of certain microdrill samples.

5. Conclusions

An automation system improved from the one proposed by Chang et al. [28], for carrying out the destructive and visual measurements of the CSGPs of both ST and UC type microdrills, has been presented in this paper. The primary conclusions are as follows:

(1) The major improvement of the hardware is characterized by the machine vision module consisting of several conventional machine vision components in combination with an innovative and lower cost optical subset formed by a set of PCA lenses and a reflection mirror. As a result, the essential functions of visually positioning the drilltip and visually measuring the CSGPs can both be achieved via the use of merely one machine vision module. As compared with the hardware design proposed by Chang et al. [28], a more compact and lower cost one can be realized.

(2) The major improvement of the AMP is characterized by the establishment of specific image processing operations for the AF sub-process based on 2D-DFT, for the WTM sub-process based on the BFC approach, and for the ODM sub-process based on integrated applications of the BFC and FDL approaches, respectively. As compared with the AMP proposed by Chang et al. [28] that can merely measure the CSWTs of ST type microdrills and does not have the functions of auto-focusing and measuring the CSODs, the execution of the AF sub-process enables the ground cross-sections of a microdrill to be automatically and clearly focused, and the execution of the WTM and ODM sub-processes enables the CSWTs and CSODs of the ground cross-sections of both ST and UC type microdrills to be automatically and visually measured.

(3) The hardware of the improved automation system had been constructed. Two series of experiments for measuring the CSGPs of certain microdrill samples had been conducted. The measuring uncertainty and accuracy of the developed system, as well as its applicability for automatically executing the sampling inspection procedure, had been tested and evaluated. It was found that the established AMP could be executed successfully by the developed system to yield stable and reliable measurement results.

(4) The experimental results showed that, from a conservative viewpoint, the developed system could achieve an uncertainty range of ±1.5 μm and an accuracy of 2.5 μm for the measurements of the CSWTs and CSODs of both ST and UC type microdrills. The developed system could also achieve a confident value for the measurement of the concentricity, which could not be achieved by the used measuring microscope.

Accordingly, the actual effectiveness of the developed system had been validated, since it could actually provide good repeatability and accuracy for carrying out the destructive and visual measurements of the CSGPs of both ST and UC type microdrills. Therefore, the developed system could effectively and comprehensively automate the conventional sampling inspection procedure.

Author Contributions: Conceptualization, W.-T.C.; methodology, W.-T.C.; software, W.-T.C. and Y.-Y.L.; validation, W.-T.C. and Y.-Y.L.; formal analysis, W.-T.C.; investigation, W.-T.C. and Y.-Y.L.; resources, W.-T.C.; data curation, W.-T.C. and Y.-Y.L.; writing—original draft preparation, W.-T.C.; visualization, W.-T.C. and Y.-Y.L.; funding acquisition, W.-T.C. All authors have read and agreed to the published version of the manuscript.

Funding: This research is funded by National Science and Technology Council of Taiwan under Grant Nos. 101-2218-E-019-003, 110-2221-E-019-059, and 111-2221-E-019-073, and is also funded by National Taiwan Ocean University under Grant Nos. NTOU-RD-AA-2012-104011 and NTOU-RD-AA-2013-104041.

Data Availability Statement: Not applicable.

Conflicts of Interest: The authors declare no conflict of interest.

Appendix A

The fundamental of the 1D edge detection used in the DP, WTM, and ODM sub-processes, which is based on the concept of 1D grayscale profile scanning [29–33] and has also been adopted by several studies [25,28,34], is described here. By setting a local coordinate system o_r-$x_r y_r$ on an assigned rectangular ROI with its x_r-axis parallel to one of the two pairs of opposite sides of the ROI (i.e., parallel to the search direction) and its origin o_r coinciding with a corner of the ROI, an arbitrary 1D grayscale profile within the ROI, denoted by g_p, can be represented as

$$g_p = g_p\left(x_r, y_{r(S)}\right) \tag{A1}$$

in which x_r is a positional variable (that is, $x_r = 0, 1, \ldots, n_{rx}$ (in unit of pixels) with n_{rx} being the total number of pixels of the x_r-directional side of the ROI), and $y_{r(S)}$ is a specified y_r-directional position (a constant in unit of pixels). The criterion applied to evaluate whether an edge point $(x_{r(E)}, y_{r(S)})$ exists is given as follows:

$$\max_{x_{r(E)} \in (0, n_{rx})} \Delta g_p\left(x_{r(E)}, y_{r(S)}\right) \geq C_E \tag{A2}$$

in which C_E is a given contrast grayscale value [29], and the criterion value Δg_p is defined as

$$\Delta g_p\left(x_{r(E)}, y_{r(S)}\right) = \frac{\left|\sum_{x_r = x_{r(E)} - (W_E + S_E/2)}^{x_{r(E)} - S_E/2} g_p\left(x_r, y_{r(S)}\right) - \sum_{x_r = x_{r(E)} + S_E/2}^{x_{r(E)} + (W_E + S_E/2)} g_p\left(x_r, y_{r(S)}\right)\right|}{W_E}$$
$$\text{for} \begin{cases} W_E + S_E/2 \leq x_{r(E)} \leq n_{rx} - (W_E + S_E/2) \\ 2W_E + S_E < n_{rx} \\ 0 < S_E \leq W_E \end{cases} \tag{A3}$$

where W_E and S_E are given filter width and steepness values (both in pixels) [29], respectively. As referred to Equations (A2) and (A3), an edge point would be located at $(x_{r(E)}, y_{r(S)})$ when the absolute difference between the two average grayscale values calculated from $x_r = x_{r(E)} - (W_E + S_E/2)$ to $x_r = x_{r(E)} - S_E/2$ and from $x_r = x_{r(E)} + S_E/2$ to $x_r = x_{r(E)} + (W_E + S_E/2)$, respectively, is greater than or equal to the given contrast grayscale value C_E. The location of $(x_{r(E)}, y_{r(S)})$ can be determined via Equation (A2) to find a local maximum of Δg_p, and a fitted 1D grayscale profile, denoted by $\hat{g}_p$, can be further obtained as

$$\hat{g}_p = \hat{g}_p\left(x_r, y_{r(S)}\right) \text{ for } x_{r(E)} - S_E/2 \leq x_r \leq x_{r(E)} + S_E/2 \tag{A4}$$

To this end, curve-fitting approaches [35,40,41] with cubic polynomials or spline curves can be applied. Subsequently, a refined edge point $\left(x_{r(E)}^*, y_{r(S)}\right)$ can be determined by numerically solving the following equation:

$$\hat{g}_p\left(x_{r(E)}^*, y_{r(S)}\right) - C_E = 0 \tag{A5}$$

The refined edge point $\left(x^*_{r(E)}, y_{r(S)}\right)$ can achieve a sub-pixel resolution along the search direction (i.e., the x_r-direction). As a result, multiple refined edge points with an equal y_r-directional interval of 1 pixel (within the ROI) can all be detected.

Finally, each refined edge point is represented in a global coordinate system o-xy (that is consistently set on the captured digital image) via the well-known homogeneous transformation matrix [42], that is,

$$\left\{\begin{matrix} x_{(E)} \\ y_{(E)} \\ 1 \end{matrix}\right\} = \begin{bmatrix} \cos\Delta\theta & -\sin\Delta\theta & \Delta x \\ \sin\Delta\theta & \cos\Delta\theta & \Delta y \\ 0 & 0 & 1 \end{bmatrix} \left\{\begin{matrix} x^*_{r(E)} \\ y_{r(S)} \\ 1 \end{matrix}\right\} \tag{A6}$$

in which Δx and Δy are the x- and y-components of a vector from origin O to origin O_r, respectively, and $\Delta\theta$ is the angle of the positive x_r-axis counterclockwise measured from the positive x-axis. The transformed edge coordinate ($x_{(E)}$, $y_{(E)}$) is the refined edge point represented in the global coordinate system o-xy. Multiple transformed edge coordinates can therefore be obtained and stored for further usage.

References

1. Haney, T. Precision interconnect drilling. In *Printed Circuits Handbook*, 6th ed.; Coombs, C.F., Jr., Ed.; McGraw-Hill: New York, NY, USA, 2008; pp. 25.1–25.19.
2. Vandervelde, H. Drilling processes. In *Printed Circuits Handbook*, 6th ed.; Coombs, C.F., Jr., Ed.; McGraw-Hill: New York, NY, USA, 2008; pp. 24.1–24.22.
3. Shi, H.; Liu, X.; Lou, Y. Materials and micro drilling of high frequency and high speed printed circuit board: A review. *Int. J. Adv. Manuf. Technol.* **2019**, *100*, 827–841. [CrossRef]
4. Stephonson, D.A.; Agapiou, J.S. *Metal Cutting Theory and Practice*; Marcel Dekker: New York, NY, USA, 1997; pp. 205–236.
5. DeGarmo, E.P.; Black, J.T.; Kohser, R.A. *Materials and Processes in Manufacturing*, 9th ed.; Wiley: Hoboken, NJ, USA, 2003; pp. 583–597.
6. Groover, M.P. *Principles of Modern Manufacturing*, 4th ed.; Wiley: Hoboken, NJ, USA, 2011; pp. 561–564.
7. Imran, M.; Mativenga, P.T.; Gholinia, A.; Withers, P.J. Evaluation of surface integrity in micro drilling process for nickel-based superalloy. *Int. J. Adv. Manuf. Technol.* **2011**, *55*, 465–476. [CrossRef]
8. Imran, M.; Mativenga, P.T.; Withers, P.J. Assessment of machining performance using the wear map approach in micro-drilling. *Int. J. Adv. Manuf. Technol.* **2012**, *59*, 119–126. [CrossRef]
9. Lei, X.; Shen, B.; Sun, F. Optimization of diamond coated microdrills in aluminum alloy 7075 machining: A case study. *Diam. Relat. Mater.* **2015**, *54*, 79–90. [CrossRef]
10. Zheng, L.J.; Wang, C.Y.; Qu, Y.P.; Song, Y.X.; Fu, L.Y. Interaction of cemented carbide micro-drills and printed circuit boards during micro-drilling. *Int. J. Adv. Manuf. Technol.* **2015**, *77*, 1305–1314. [CrossRef]
11. Hyacinth Suganthi, X.; Natarajan, U.; Ramasubbu, N. A review of accuracy enhancement in microdrilling operations. *Int. J. Adv. Manuf. Technol.* **2015**, *81*, 199–217. [CrossRef]
12. Mittal, R.K.; Yadav, S.; Singh, R.K. Mechanistic force and burr modeling in high-speed microdrilling of Ti6Al4V. *Procedia CIRP* **2017**, *58*, 329–334. [CrossRef]
13. Guo, H.; Wang, X.; Liang, Z.; Zhou, T.; Jiao, L.; Liu, Z.; Teng, L.; Shen, W. Drilling performance of non-coaxial helical flank micro-drill with cross-shaped chisel edge. *Int. J. Adv. Manuf. Technol.* **2018**, *99*, 1301–1311. [CrossRef]
14. Chang, D.Y.; Lin, C.H. High-aspect ratio mechanical microdrilling process for a microhole array of nitride ceramics. *Int. J. Adv. Manuf. Technol.* **2019**, *100*, 2867–2883. [CrossRef]
15. Huang, G.; Wan, Z.; Yang, S.; Li, Q.; Zhong, G.; Wang, B.; Liu, Z. Mechanism investigation of micro-drill fracture in PCB large aspect ratio micro-hole drilling. *J. Mater. Process. Technol.* **2023**, *316*, 117962. [CrossRef]
16. Liu, Y.; Zhang, D.; Geng, D.; Shao, Z.; Zhou, Z.; Sun, Z.; Jiang, Y.; Jiang, X. Ironing effect on surface integrity and fatigue behavior during ultrasonic peening drilling of Ti-6Al-4V. *Chin. J. Aeronaut.* 2023; *in press*. [CrossRef]
17. Tien, F.C.; Yeh, C.H.; Hsieh, K.H. Automated visual inspection for microdrills in printed circuit board production. *Int. J. Prod. Res.* **2004**, *42*, 2477–2495. [CrossRef]
18. Tien, F.C.; Yeh, C.H. Using eigenvalues of covariance matrices for automated visual inspection of microdrills. *Int. J. Adv. Manuf. Technol.* **2005**, *26*, 741–749. [CrossRef]
19. Huang, C.K.; Liao, C.W.; Huang, A.P.; Tarng, Y.S. An automatic optical inspection of drill point defects for micro-drilling. *Int. J. Adv. Manuf. Technol.* **2008**, *37*, 1133–1145. [CrossRef]
20. Su, J.C.; Huang, C.K.; Tarng, Y.S. An automated flank wear measurement of microdrills using machine vision. *J. Mater. Process. Technol.* **2006**, *180*, 328–335. [CrossRef]

21. Duan, G.; Chen, Y.W.; Sukegawa, T. Automatic optical flank wear measurement of microdrills using level set for cutting plane segmentation. *Mach. Vis. Appl.* **2010**, *21*, 667–676. [CrossRef]
22. Huang, C.K.; Wang, L.G.; Tang, H.C.; Tarng, Y.S. Automatic laser inspection of outer diameter, run-out and taper of micro-drills. *J. Mater. Process. Technol.* **2006**, *171*, 306–313. [CrossRef]
23. Chuang, S.F.; Chen, Y.C.; Chang, W.T.; Lin, C.C.; Tarng, Y.S. Nondestructive web thickness measurement of micro-drills with an integrated laser inspection system. *Nondestruct. Test. Eval.* **2010**, *25*, 249–266. [CrossRef]
24. Chang, W.T.; Lu, S.Y.; Chuang, S.F.; Shiou, F.J.; Tang, G.R. An optical-based method and system for the web thickness measurement of microdrills considering runout compensation. *Int. J. Precis. Eng. Manuf.* **2013**, *14*, 725–734. [CrossRef]
25. Chang, W.T.; Wu, J.H. An innovative optical-based method and automation system for rapid and non-destructive measurement of the web thickness of microdrills. *Measurement* **2016**, *94*, 388–405. [CrossRef]
26. Jaini, S.N.B.; Lee, D.W.; Kim, K.S.; Lee, S.J. Measurement of cemented carbide-PCD microdrill geometry error based on computer vision algorithm. *Measurement* **2022**, *187*, 110186. [CrossRef]
27. Beruvides, G.; Quiza, R.; Rivas, M.; Castaño, F.; Haber, R.E. Online detection of run out in microdrilling of tungsten and titanium alloys. *Int. J. Adv. Manuf. Technol.* **2014**, *74*, 1567–1575. [CrossRef]
28. Chang, W.T.; Chuang, S.F.; Tsai, Y.S.; Tang, G.R.; Shiou, F.J. A vision-aided automation system for destructive web thickness measurement of microdrills. *Int. J. Adv. Manuf. Technol.* **2014**, *71*, 983–1003. [CrossRef]
29. National Instrument Corp. *NI Vision Concepts Manual*; National Instrument Corp.: Austin, TX, USA, 2007; pp. 3.1–3.18, 7.1–7.12, 8.1–8.9, 11.1–11.22.
30. Jain, R.; Kasturi, R.; Schunck, B.G. *Machine Vision*; McGraw-Hill: New York, NY, USA, 1995; pp. 76–86, 140–185, 309–405.
31. Gonzalez, R.C.; Woods, R.E. *Digital Image Processing*, 2nd ed.; Prentice-Hall: Upper Saddle River, NJ, USA, 2002; pp. 147–156, 519–585.
32. Gonzalez, R.C.; Woods, R.E.; Eddins, S.L. *Digital Image Processing Using MATLAB*; Prentice-Hall: Upper Saddle River, NJ, USA, 2003.
33. McAndrew, A. *Introduction to Digital Image Processing with MATLAB*; Brooks/Cole: Pacific Grove, CA, USA, 2004.
34. Chang, W.T.; Lu, Y.Y. Machine vision-based methods for measuring the cross-sectional geometric accuracy of microdrills and their comparisons. In *Machine Vision and Human-Machine Interface: Technologies, Applications and Challenges*; Yates, S.T., Ed.; Nova Science Publishers: New York, NY, USA, 2016; pp. 41–78.
35. Faires, J.D.; Burden, R. *Numerical Methods*, 3rd ed.; Thomson Learning: Pacific Grove, CA, USA, 2003; pp. 33–38, 64–110, 340–356.
36. Fan, K.C.; Lee, M.Z.; Mou, J.I. On-line non-contact system for grinding wheel wear measurement. *Int. J. Adv. Manuf. Technol.* **2002**, *19*, 14–22. [CrossRef]
37. Su, J.C.; Tarng, Y.S. Measuring wear of the grinding wheel using machine vision. *Int. J. Adv. Manuf. Technol.* **2006**, *31*, 50–60. [CrossRef]
38. Chang, W.T.; Chen, T.H.; Tarng, Y.S. Measuring characteristic parameters of form grinding wheels used for microdrill fluting by computer vision. *Trans. Can. Soc. Mech. Eng.* **2011**, *35*, 383–401. [CrossRef]
39. Beckwith, T.G.; Marangoni, R.D.; Lienhard, J.H. *Mechanical Measurements*, 5th ed.; Pearson Education Taiwan Ltd.: Taipei, Taiwan, 2004; pp. 45–125.
40. Yamaguchi, F. *Curves and Surfaces in Computer Aided Geometric Design*; Springer: Berlin/Heidelberg, Germany, 1988.
41. Zeid, I. *CAD/CAM Theory and Practice*; McGraw-Hill: New York, NY, USA, 1991.
42. Litvin, F.L. *Theory of Gearing*; NASA Reference Publication 1212: Washington, DC, USA, 1989; pp. 1–19.

machines

MDPI

Article

Improving Industrial Robot Positioning Accuracy to the Microscale Using Machine Learning Method

Vytautas Bucinskas [ID], Andrius Dzedzickis *[ID], Marius Sumanas [ID], Ernestas Sutinys, Sigitas Petkevicius, Jurate Butkiene, Darius Virzonis and Inga Morkvenaite-Vilkonciene [ID]

Department of Mechatronics, Robotics and Digital Manufacturing, Vilnius Gediminas Technical University, 10223 Vilnius, Lithuania
* Correspondence: andrius.dzedzickis@vilniustech.lt

Abstract: Positioning accuracy in robotics is a key issue for the manufacturing process. One of the possible ways to achieve high accuracy is the implementation of machine learning (ML), which allows robots to learn from their own practical experience and find the best way to perform the prescribed operation. Usually, accuracy improvement methods cover the generation of a positioning error map for the whole robot workspace, providing corresponding correction models. However, most practical cases require extremely high positioning accuracy only at a few essential points on the trajectory. This paper provides a methodology for the online deep Q-learning-based approach intended to increase positioning accuracy at key points by analyzing experimentally predetermined robot properties and their impact on overall accuracy. Using the KUKA-YouBot robot as a test system, we perform accuracy measurement experiments in the following three axes: (i) after a long operational break, (ii) using different loads, and (iii) at different speeds. To use this data for ML, the relationships between the robot's operating time from switching on, load, and positioning accuracy are defined. In addition, the gripper vibrations are evaluated when the robot arm moves at various speeds in vertical and horizontal planes. It is found that the robot's degrees of freedom (DOFs) clearances are significantly influenced by operational heat, which affects its static and dynamic accuracy. Implementation of the proposed ML-based compensation method resulted in a positioning error decrease at the trajectory key points by more than 30%.

Keywords: online machine learning; deep-q learning; positioning accuracy; industrial robot; vibrations

check for updates

Citation: Bucinskas, V.; Dzedzickis, A.; Sumanas, M.; Sutinys, E.; Petkevicius, S.; Butkiene, J.; Virzonis, D.; Morkvenaite-Vilkonciene, I. Improving Industrial Robot Positioning Accuracy to the Microscale Using Machine Learning Method. *Machines* **2022**, *10*, 940. https://doi.org/10.3390/machines10100940

Academic Editors: Raul D.S.G. Campilho, Francisco J. G. Silva and Dan Zhang

Received: 11 August 2022
Accepted: 13 October 2022
Published: 17 October 2022

1. Introduction

Robots are used in various processes, including manufacturing, entertainment, services, and scientific research. To maintain a technical edge and thereby remain competitive, more and more businesses are applying advanced technology and programming solutions to their operational processes [1]. Such a wide application encourages the development of universal robotic systems and requires research of their capabilities and performance characteristics.

In general, industrial robots provide high-level static and dynamic positioning accuracy. Nevertheless, they must be maintained to ensure that they continue to meet the conditions for which they have been programmed and in which they operate [2]. Therefore, for each specific task, it is important to determine the following: (i) positioning accuracy (positioning error between stated and the real position of arm end effector); (ii) repeatability (positioning error between real positions of arm end effector performing repeating movements); (iii) other parameters, which are considered to be unique characteristics of the particular machine [3]. Positioning accuracy depends on a number of actions including, but not restricted to the following: (i) the parameters of the drives guiding the robot's movements; (ii) tolerances in manufacturing parts of the machinery; (iii) tolerances due to the articulation of the robot's chains [4]; (iv) control algorithms [5,6]; (v) dynamic properties of the mobile platform [7,8], and (vi) robot arm properties [9,10]. Each of the factors may become important for the accuracy of the robot, depending on its type, lifting capacity, and operational conditions [4]. Due to the complex nature of robots, the increasing positioning accuracy of robotic installations develops challenging tasks. The positioning errors in offline mode can be determined using special algorithms [11] and mathematical models [12–15]. Common positioning accuracy problems in path and trajectory planning can be resolved by improving offline programming software when the code for a robot is generated automatically or by implementing online feedback [16–20]. However, typical compensation methods are limited by the chaotic nature of the positioning errors. Therefore, more adaptive methods, such as machine learning (ML), can increase positioning accuracy by compensating for positioning errors in particular cases [21,22]. Advances in ML with respect to simple error compensation are in the accountancy of feature-chaotic robot error distribution about positioning points. ML ensures respect for robot fluctuation and fits into the prescribed tolerance field. There are a few cases [23–26] of ML approaches compensating for the positioning errors. Chen and Zhang developed an ML kinematics-based positioning error compensation method for high-precision mechanical machining operations [26]. This positioning error compensation method provided good results, but it contains sophisticated procedures due to its complexity and use of excessive data. More to say, it combines the following: analytical modeling; extended Kalman filtering; spatial interpolation algorithm; an adaptive mesh division algorithm, and an inverse distance weighted interpolation algorithm. To use it at a few trajectory points, this method becomes too costly.

Moreover, ML could be implemented to improve the trajectory-planning process [27–29] or reduce vibrations [30,31].

The main aim of our research is to create a methodology to increase industrial robot positioning accuracy and minimize robot end-of-arm vibrations by applying ML in online mode. This methodology will be used to compensate for positioning errors by shifting the destination point or altering the moving velocity of the end-of-arm reference point before defining the forward kinematic task. The method is focused on destination point approach accuracy in defined arbitrary robot positions. Moreover, our proposed approach suits well for industrial robots with typically closed controllers since position correction is performed externally with respect to the robot controller; thus, modified motion commands to target coordinates are processed in a standard way.

2. Concept of Research

This paper focused on creating a universal methodology for most types of industrial robots, which can increase positioning accuracy at the robot trajectory endpoints. One of the pillars of ML is mathematical optimization, which involves the numerical computation of parameters for a system designed to make decisions based on unseen data [32–34]. While the ML procedure is based on existing data collection, such a method all the time remains retrospective and reflects errors in previous applications with corresponding load cases. The use of the robot for precise machining or assembly operations requires data about the vibration level and settling time. This will be critically important in addition to static positioning error values and directions. The ML procedure enables a decrease in positioning error values, and end-of-arm vibration influences the resulting accuracy. Nevertheless, the robot workspace mapping procedure, which provides error values and direction at any point of the workspace, all the time remains inaccurate for each particular case.

It is possible to calculate optimal parameters for a given learning problem using currently available data [35]. The collection of data required for ML is a complex process, defined by the robot design, its characteristics, and the aim of ML implementation.

The procedure of implementing our proposed ML method is divided into the following three phases: (i) initial preparation, (ii) positioning task formulation, and (iii) optimization (Figure 1).

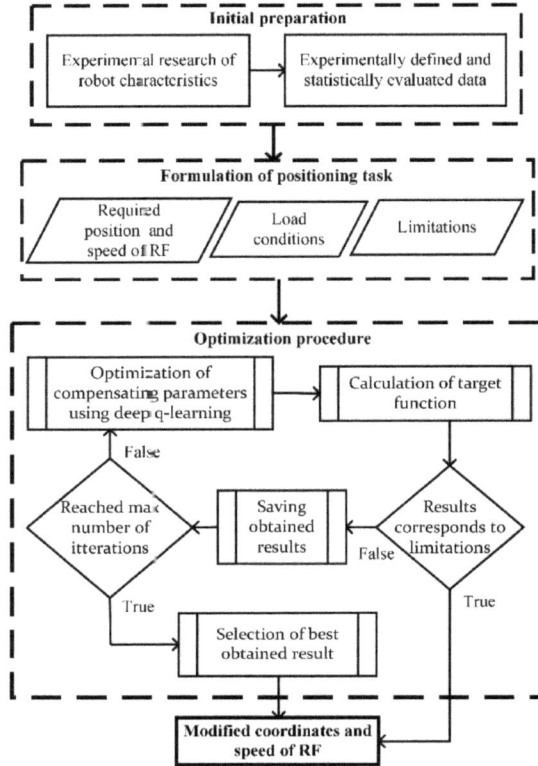

Figure 1. Simplified block diagram of positioning accuracy increase procedure by implementing ML method. RF—reference point.

The success of the ML procedure depends on the quantity and quality of data generated during initial preparation. This phase includes experiments for the definition of the main robot's characteristics, such as positioning accuracy, reference point vibration level, statistical evaluation, and analysis of obtained data. The initial preparation procedure should be performed for all robot trajectory points of interest. This procedure must be repeated in cases of mechanical wear, change of tool configuration, or essential variation of environmental conditions.

The obtained data shows the technical conditions of the robot since these data consist of a set of dependencies between positioning errors, vibration level, reference point position, operating time, robot speed, and load. In the general case, these dependencies can be expressed as follows:

$$\begin{cases} \Delta_{x,y,z} = F(x, y, z, V, M, t_{op}) \\ \varepsilon_{x,y,z} = F(x, y, z, V, M) \end{cases} \tag{1}$$

where: $\Delta_{x,y,z}$ is the overall positioning error, $\varepsilon_{x,y,z}$ is the overall level of reference point vibrations, $x\ y\ z$ are coordinates of robot reference point, V is the speed of reference point, M is load, t_{op} is the time from the moment when a robot was switched on.

The second phase of our proposed ML method is the Formulation of the positioning task. In this phase, it is necessary to define the required position and speed of the reference point ($x\ y\ z$, V) and load (M). Moreover, it is necessary to specify the following limitations:

acceptable overall positioning error Δ_{min}, and overall vibration level of reference point ε_{min}. In general, case limitations can be identified as follows:

$$\begin{cases} \Delta_{x,y,z} \leq \Delta_{min} \\ \varepsilon_{x,y,z} \leq \varepsilon_{min} \end{cases} \tag{2}$$

Moreover, to ensure stable operation of the ML algorithm and avoid looping in the algorithm (when the desired accuracy cannot be achieved), it is necessary to include a parameter specifying the maximum number of iterations k_{max}.

The third phase is the Optimization procedure, which contains the cycle during which the optimization algorithm is used to obtain the most suitable compensation parameters in order to achieve the required positioning accuracy and an acceptable level of reference point vibrations.

The ML algorithm runs with each new positioning task; ML selects compensation parameters and predicts the expected positioning error $\Delta_{x,y,z}$, and reference point vibration level. This prediction is based on experimentally defined data obtained during the Initial preparation phase. After defining the expected positioning error $\Delta_{x,y,z}$ and reference point vibration level $\varepsilon_{x,y,z}$, their values are compared with the limitations. If positioning error and reference point vibration level correspond to the limitations of Equation (2), then modified coordinates of the required reference point position and speed are transferred to the robot controller. In case if results do not correspond to the limitations, the data is saved and checked if the maximum number of iterations k_{max} are reached. The optimization of compensation parameters is performed many times until the limit of iterations is reached. In case if the maximum number of iterations is not achieved, the program selects the best result from all previous iterations.

Deep q-Learning Algorithm

The deep Q-learning algorithm was implemented to realize the mentioned ML procedure. The advantage of this algorithm in comparison to others is emphasized due to its efficiency in similar problems and its relatively simple implementation [36–38]. The deep Q-learning algorithm combines the Q-learning algorithm and a deep artificial neural network. The idea of Q-learning is based on the perception of the environment and state to take respective actions in order to achieve the maximum reward. A neural network enables the algorithm to operate in a much larger environment and optimize calculations procedure by enabling approximation features. It allows the algorithm to observe the pattern in the environment and discover the optimal sequences of actions instead of calculating and evaluating each state and the value of each action at each point in the environment. The same principle was used to optimize the definition of the algorithm loss function. We used the stochastic gradient descent method (SGD) [35] to define the function by the theoretical gradient calculated from randomly selected data points instead of calculating the actual gradient from the entire dataset. Such an approach diminishes the dependency of obtained data from strong dataset influence on functional distribution. Other parameters of the implemented algorithm were selected experimentally by evaluating their impact on positioning accuracy after many trials in a separate study. The configuration of the algorithm used in this study is provided in Table 1.

Table 1. Parameters of the implemented algorithm.

Parameter	Activation Function	Optimizer	Hidden Neurons	Hidden Layers	Replay Memory	Temperature
Value	Softsign	SGD	37	1	100,000	70

The parameters within Table 1 were defined through the experimental running of the algorithm. Process activation through the Softsign procedure corresponds to the robot learning mode, while there are possibilities for other functions. The activation margin (sensitivity to the activation condition) needs redefinition; the ML process converges correctly and, within 1000 iterations, achieves learning process influence saturation. The implemented optimizer—SGD—corresponds to our purposes and achieves the prescribed result. A number of neurons is chosen according to the desired process parameter resolution, and the number of hidden layers is naturally chosen as one according to the dimensionality of the ML process. The amount of replay memory is defined by the experimental test, and it occurs outside our paper scope as well as the temperature parameter.

Furthermore, this paper provides a detailed methodology for experimental research and data collection for ML.

3. Experimental Research

3.1. Experimental Setup

The experiments were performed using the KUKA-YouBot robot (KUKA, Augsburg, Germany) as an industrial robot testbench; it was fixed to a special stable base. The robot's geometric parameters and main characteristics are provided in Table 2 [39]. Positions of the robot's gripper were detected using two USB cameras with a resolution of 1920×1080, a checkerboard matrix of 4×8 mm size, and a user-defined function, implementing the detect Checkerboard Points procedure in MatLab (Figure 2). According to [40], the use of such a measurement method can ensure an average measurement deviation better than 0.0004 mm.

Table 2. General characteristics KUKA-YoubBot arm [39].

General Information		Axes Motion Range, Speed	
Serial kinematics	5 axes	Axis 0 (A0)	$+/-169°, 90°/s$
Height	655 mm	Axis 1 (A1)	$+90°/-65°, 90°/s$
Work envelope	0.513 m^3	Axis 2 (A2)	$+146°/-151°, 90°/s$
Weight	6.3 kg	Axis 3 (A3)	$+/-102°, 90°/s$
Payload	0.5 kg	Axis 4 (A4)	$+/-167°, 90°/s$

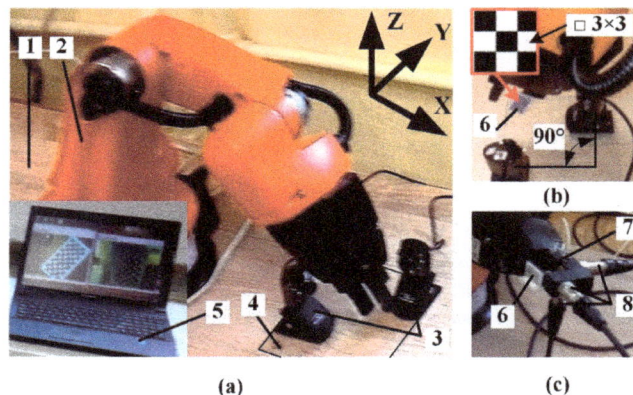

Figure 2. Experimental setup for measurements of positioning accuracy: (**a**) general view; (**b**) position of the cameras; (**c**) accelerometers mounted on robot gripper. 1—base; 2—robot; 3—USB cameras; 4—based on which are mounted cameras; 5—personal computer; 6—optical checkerboard matrix; 7—holder for cameras; 8—accelerometers.

The cameras were fixed by a special holder positioned at a $90°$ angle relative to each other to resolve spatial coordinates of the end-of-arm reference point (Figure 2b). The holder with attached cameras was screwed to the same base as the robot. Two special checkerboard patterns (identification marks for cameras) were attached to the gripper (Figure 2b,c).

Absolute vibration accelerations of the end-of-arm reference point were measured using industrial accelerometers Ini 603C01 (PCB Piezotronics, Depew, NY, USA) with a measurement range of 0.5~10,000 Hz and an acceleration limit of 51 g. Accelerometers were attached to the gripper in three perpendicular directions, as shown in Figure 2c. Signals from accelerometers were collected using the data acquisition system USB-4432 (National Instruments, Austin, TX, USA).

Ambient temperature was measured using a digital thermometer MWF-DT-616CT (CEM Corporation, Matthews, NC, USA), with a resolution of 0.1 C. Loads were weighted using Silver crest HG01025 (Silvercrest, Corona, CA, USA) scales, with a resolution of 0.001 kg. For the determination of the required warm-up time, the unloaded robot moves at 50% of the max joint speed (which is 90°/s for all axes) to the trajectory endpoint (Figure 3), rotating joints by the following angles: φ_0—90°, φ_1—15.2°, φ_2—45.7°, φ_3—44.2°, φ_4—20°, and backward. Then the robot returns to the start position, waits for 3 s, and the image of the checkerboard matrix is captured. The experiment trials took place after long operation breaks of 12, 2, and 1-h. In the case of a 2- and 1-h operation break, the warming procedure was applied before the experiment. Firstly, the robot was warmed up using a 30 min warming program and 2- and 1-h operation breaks afterward, correspondingly.

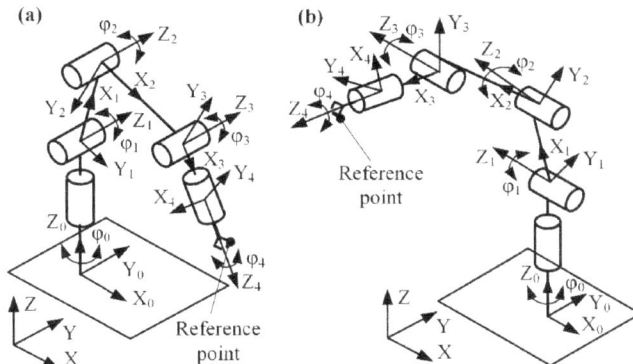

Figure 3. Position of the end-of-arm reference point: (**a**) start (measuring) position; (**b**) endpoint of the trajectory.

The positioning accuracy measurements were performed in the same conditions, but in this case, experiments were carried out with 10%, 50%, and 100% of the maximum joint speed, without load, and loaded by 0.250 kg and 0.500 kg weights. Each measurement was repeated 20 times.

The proposed methodology for the online ML procedure was tested for 2 different robot configurations, imitating the pick-and-place task. In each individual case, using our online ML method, the procedure can be performed at any chosen position in the robot workspace.

3.2. Evaluation of the End-of-Arm Reference Point Vibrations

The dynamic parameters of the robot—the end-of-arm reference point vibration level and settling time when the robot stops at the desired positions—were evaluated using accelerometers (Figure 2c). There are the following three parameters for measuring vibrations: displacement, velocity, and acceleration. The evaluation of vibration acceleration provides the best sensitivity in higher frequencies and suits well for evaluating impacts caused by bearing damage, abnormal gears, and noise. Moreover, this method is recommended when it is necessary to evaluate forces and stresses acting on or rotating parts.

Moreover, measurements of acceleration do not require reference points. Therefore, it is more convenient for industrial robots.

The measurements were carried out at the following three different speeds of the joints of the robot: at 10%, 50%, and 100% of the maximum (90°/s) speed of the joint, using the same robot movement trajectory as shown in Figure 3. The measurement was repeated three times.

The separate DOF effect on the robot end-of-arm vibrations was evaluated by rotating each robot joint separately. For this purpose, joints I-V at coordinates φ_0–φ_4 were rotated clockwise by a 60° angle at maximum (90°/s) speed from the start position Figure 4. The end-of-arm reference point vibrations were measured after the movement of each joint by 60° at a maximum (90°/s) speed. After 10 s, the joints were rotated back to the start position. Measurements were performed when the robot moved in forward and backward directions, and the procedure was repeated three times.

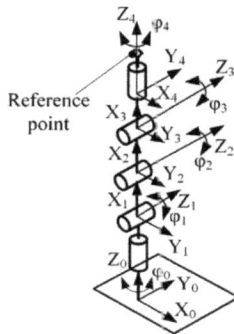

Figure 4. The start position of the robot for the measurements of separate joint impact to robot vibrations.

To define the impact of joint I (coordinate φ_0, Figure 5) on robot end-of-arm vibrations, joint I was rotated at 60° from the start position, and the vibrations were measured. The vibrations at the following two start positions were evaluated: (i) fully extended position (Figure 5a); (ii) compacted position (Figure 5b).

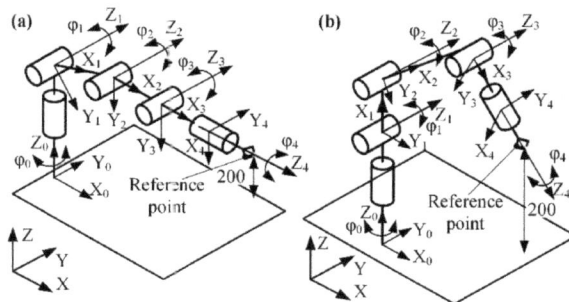

Figure 5. Start positions of the robot used for the measurements of gripper vibrations: (**a**) maximum peripheral reach on extended position, (**b**) compacted position. In both cases, the reference point was distanced from the z-axis origin by 200 mm.

In a fully extended position robot's own weight fully loads joint I. Other joints remain unloaded and stay in a singularity position. The compacted robot position is represented by the situation when joint I is loaded by half of the own weight of other links.

All vibration measurements were performed according to the flowchart provided in Figure 6.

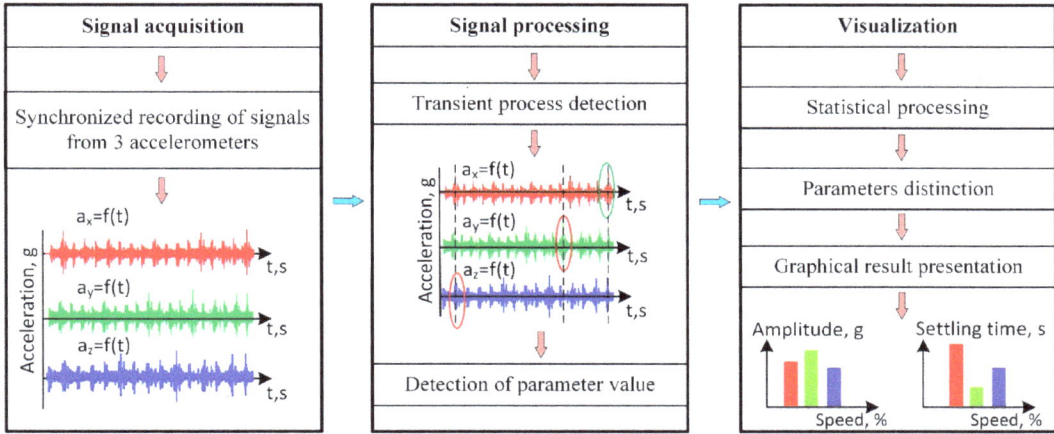

Figure 6. Flowchart of the vibration measurement process.

Signals from three accelerometers were recorded continuously during robot movement between the endpoints of the trajectory. Later acquired data were processed offline (Figure 7), defining transient processes due to robot stops at the trajectory endpoints.

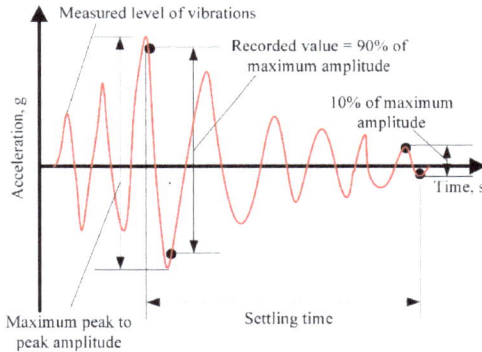

Figure 7. Definition of vibration parameters from experimental data.

Main parameters (end-of-arm reference point maximum peak-to-peak vibration amplitude, settling time) were extracted from raw data, processed statistically, grouped according to testing conditions, and represented in graphical form.

3.3. Calculations

Overall positioning accuracy from measurements in separate axes was evaluated according to the methodology provided by ISO 9283:1998 [41]. It defines positioning accuracy AP_p as the deviation between a reference position and the mean of attained positions while approaching reference positions from the same direction as follows:

$$AP_p = \sqrt{(\bar{x} - x_c)^2 + (\bar{y} - y_c)^2 + (\bar{z} - z_c)^2},$$ (3)

where x_c y_c z_c are coordinates of the prescribed end-of-arm reference point position, and $\bar{x}$, $\bar{y}$, $\bar{z}$ are mean values of the resulting position.

General robot positioning error consists of accumulated values from all errors, and it is an object of ML-obtained compensation. General error from three-axis measurements for the experiments in 1, 2, and 12 h evaluated using the simplified equation, since target position defined as $x_c = 0$, $y_c = 0$ and $z_c = 0$ as follows:

$$AP_p = \sqrt{(\overline{x})^2 + (\overline{y})^2 + (\overline{z})^2}. \tag{4}$$

where $\overline{x}\ \overline{y}\ \overline{z}$ are calculated from experimental data as follows:

$$\overline{x} = \frac{1}{n}\sum_{j-1}^{n} x_j \ , \ \overline{y} = \frac{1}{n}\sum_{j-1}^{n} y_j \ , \ \overline{z} = \frac{1}{n}\sum_{j-1}^{n} z_j, \tag{5}$$

where: $x_j\ y_j\ z_j$ are coordinates of the resulting position, and n is the measurement cycle number.

Robot positioning repeatability RP_l, also was evaluated using the methodology proposed in ISO 9283:1998. According to it, RP_l value is the radius of the sphere that defines the closeness of an attained position after n attempts to achieve the same position from the same direction as follows:

$$RP_l = \overline{l} + 3S_l, \tag{6}$$

here:

$$\overline{l} = \frac{1}{n}\sum_{j=1}^{n} l_j; \quad l_j = \sqrt{(\overline{x} - x_j)^2 + (\overline{y} - y_j)^2 + (\overline{z} - z_j)^2}; \quad S_l = \sqrt{\frac{\sum_{j-1}^{n}(l_j - \overline{l})^2}{n - 1}}. \tag{7}$$

Robot positioning repeatability was evaluated using the same experimental data as used for the positioning accuracy measurement; values obtained after the 14th minute of the experiment were used.

3.4. Statistical Evaluation of Research Data

The confidence of results for robot gripper vibration level and settling time measurements were evaluated by statistical parameters. Measurement data was assessed by correlation-regression analysis. Arithmetic averages, their standard deviations, and confidence intervals at a 0.95 probability level were calculated according to [42].

In correlation-regression analysis, the difference between the measured result m_v and real measured parameter value r_v is defined as absolute measurement error as follows [42]:

$$\Delta m_v = m_v - r_v. \tag{8}$$

The measured parameter m_v have a prescribed probability, representing the real value of positioning error if the exact measurements are repeated n times. From several measurements, we calculated the arithmetic mean $\overline{m_v}$ as follows:

$$\overline{m_v} = \frac{1}{n}\sum_{i=1}^{n} m_{vi} = \frac{m_{v1} + m_{v2} + \dots m_{vn}}{n} \tag{9}$$

Absolute measurement error consists of systematic, random, and random errors of deduction. Using correlation-regression analysis, only random errors were estimated in statistical data evaluations. In our case, we evaluated only random errors, as if the methodology and measurement devices are far more accurate than expected error, systematic and accidental deduction errors are not significant and therefore cannot be considered [42]. To evaluate random error, it is necessary to calculate the experimental standard deviation σ of each individual measurement as follows:

$$\sigma = \sqrt{\frac{\sum_{i=1}^{n}(m_v - \overline{m_v})^2}{n - 1}}. \tag{10}$$

The experimental standard mean deviation S_{md} is calculated by the following:

$$S_{md} = \frac{\sigma}{\sqrt{n}}. \tag{11}$$

The random error of the measured parameter is calculated by the following:

$$\Delta m_{v,n,P} = S_c \cdot S_{md}. \tag{12}$$

where: S_c—the value of the Student Criterion selected according to the number of experiment variables $(n-1)$ and the probability level ($\alpha = 0.95$).

The final result of the measured (n times) value m_v is expressed as the sum of the arithmetic mean $\overline{m_v}$ and the random error $\Delta_{mv,n,P}$ as follows:

$$\overline{m_v} \pm \Delta m_{v,n,P}. \tag{13}$$

4. Results

4.1. Robot's Accuracy after a Long Operation Break

The robot's accuracy after a long operation break degenerates because of the temperature regime change and lubricant film disappearance from joint clearances. Thus, the measurement of the positioning error of the robot after a long operation break is important for determining the accuracy of restoring to operating time. To determine the robot's positioning accuracy in time, the positioning error measurement was performed in the x, y, and z-axes when the robot starts to operate after a 12-h break (Figure 8a). In all axes, positioning error increases in the first 14 min of operation time. Positioning errors in the x and y-axes are almost identical as follows: after 14 min, errors are 0.09 mm in both the x and y-axis, while in the z-axis, the error is 0.12 mm.

Regression approximation parameters are provided in Table 3.

Results of experiments performed after a 1-h and 12-h operation break showed that the required warm-up period is 14 min (Figure 8b). However, with a 1-h break, positioning errors decreased compared to a 12-h break, causing the following errors: 0.05 mm in the x-axis, 0.06 mm in the y-axis, and 0.09 mm in the z-axis. This phenomenon can be explained by the fact that 1 h is not enough for all joints to cool down to the initial temperature and to sediment the lubricant layer. The uneven distribution of positioning errors in the x and y axes requires an increase in the operational break. To check this assumption, the same experiments were performed after a 2 h break (Figure 8c).

It was observed that the time until positioning errors become stable is 12.5 min for the y and z-axes and 11.2 min for the x-axis. The stable positioning error in the x-axis is 0.07 mm, 0.08 mm in the y-axis, and 0.12 mm in the z-axis. This situation is quite similar to the case when the robot starts operating after a 12-h break. Small differences can be explained by assuming that after a 2-h break, the robot almost cools down, but some elements in separate joints may still have higher temperatures.

From the dependencies between positioning error fluctuations over time, we can define the minimum required warm-up period for the robot or predict the positioning accuracy in respect of operating time. Moreover, the obtained data allow us to define the robot's overall positioning accuracy and repeatability.

Results presented in Figure 8d correspond to results presented in Figure 8a–c where it is seen that after a lengthy break (12 h), overall positioning precision depends upon time. In the first 13 min of operation after the 1-h break, positioning error slightly increases up to 0.12 mm. After a 2-h break, the error reaches 0.16 mm, and finally 0.17 mm after a 12-h break. This allows us to state that the minimum warm-up time required for the robot to reach stable positioning accuracy values is more than 14 min.

The repeatability error of the robot was specified using experimental data provided in Figure 8d. The overall evaluation of repeatability error after data processing is 0.0146 ± 0.00526 mm (Equations (6) and (7)).

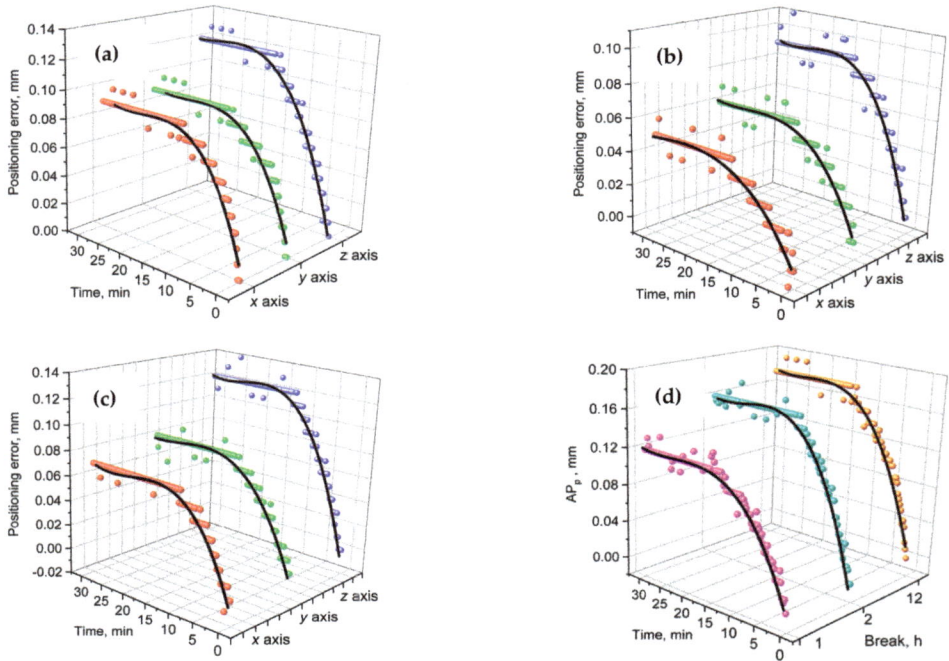

Figure 8. Positioning error dependency on time after: (**a**) 12 h operation break; (**b**) 1-h operation break; (**c**) 2-h operation break; (**d**) general positioning errors *APp*, calculated by Equation (4) after 1, 2, and 12 h breaks. Trend lines—regression approximation.

Table 3. Coefficients for the regression estimated robot positioning error.

Operation Break	Axis	a_0	a_1	a_2	a_3
1-h	x	2.8464×10^{-4}	5.5500×10^{-3}	-1.9554×10^{-4}	2.1655×10^{-6}
	y	1.6700×10^{-3}	7.7700×10^{-3}	-3.3957×10^{-4}	4.9016×10^{-6}
	z	-4.6000×10^{-3}	1.3590×10^{-2}	-6.3174×10^{-4}	9.5758×10^{-6}
	AP_p	8.5400×10^{-3}	2.2560×10^{-2}	-1.0000×10^{-3}	1.4723×10^{-5}
2-h	x	-7.3200×10^{-3}	1.2260×10^{-2}	-6.1872×10^{-4}	9.9370×10^{-6}
	y	-5.8300×10^{-3}	1.1690×10^{-2}	-5.0854×10^{-4}	7.0670×10^{-6}
	z	-9.7000×10^{-3}	1.8300×10^{-2}	-8.2850×10^{-4}	1.2126×10^{-5}
	AP_p	-1.1120×10^{-2}	2.4470×10^{-2}	-1.1200×10^{-3}	1.6540×10^{-5}
12-h	x	8.0000×10^{-3}	1.2340×10^{-2}	-6.1009×10^{-4}	9.8658×10^{-6}
	y	7.5200×10^{-3}	1.1750×10^{-2}	-5.5304×10^{-4}	8.5685×10^{-6}
	z	-2.7500×10^{-3}	1.5180×10^{-2}	-6.0424×10^{-4}	7.8195×10^{-6}
	AP_p	4.5125×10^{-4}	1.5930×10^{-2}	-6.9624×10^{-4}	9.9989×10^{-6}

The collected data suggest that, if the robot stays idle for an hour or more, in order to achieve a higher level of accuracy and repeatability, it is recommended that within the first 14 min after the break, do not perform any operational action, but rather use some warm-up programs. In other cases, we can provide coordinates of the prescribed position, which should be adjusted according to the dependencies obtained during this research. The value and direction of defined positioning error become important parameters for compensation used in ML-based robot trajectory correction. The ML process will use the value and direction of the defined positioning error as parameters and use them as the target line for further process of ML. The orientation of the error vector also has meaning for error compensation. In the case of a simulation procedure rather than an experimental ML approach, there is a chance to significantly decrease learning time. Obtained data will have a value not only as an absolute achieved point but also as a beacon for further learning, especially when another loading configuration or load occurs.

We observed the positioning error dependencies over time and after a long operation break. These dependencies can be included in ML algorithms to compensate for positioning errors at different periods of the warming process operating times.

4.2. Definition of Robot Positioning Accuracy

The accuracy of robot positioning was examined when the robot was moved at various speeds with different loads. Positioning errors were measured in the following three perpendicular directions: x, y, and z (Figure 9). The average distribution of positioning errors was calculated using Equation (12) when the robot moves at the speed values of 10%, 50%, and 100% of maximum speed for every joint being unloaded or with 0.25 kg and 0.50 kg loads.

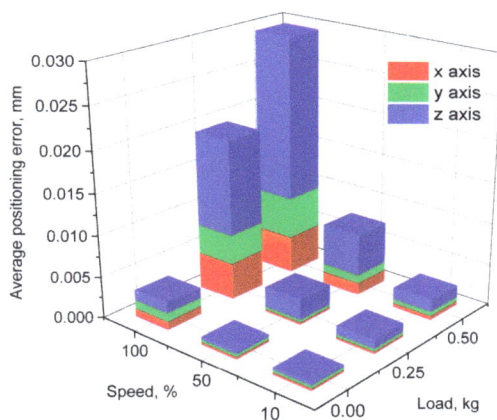

Figure 9. Positioning errors of the robot when it operates at various speeds and loads.

It was defined that in all researched cases, the highest positioning error value appears along the z-axis operating at maximum speed (90°/s). When the robot moves in unloaded mode, the average positioning error in the z-axis is 0.0014 mm. When the robot moves with a 0.50 kg load, this error increases up to 0.02 mm. A similar tendency is noticed when analyzing the accuracy of the x and y-axis, but in those cases, when the values of positioning errors are lower compared with values of the z-axis as follows: 0.001 mm when the robot moves unloaded and 0.005 mm when the robot moves with a 0.50 kg load.

The results of the measurements allow us to state that if the robot is warmed up, positioning accuracy of ±0.03 mm can be achieved with a maximum (0.50 kg) weight at the highest speed (90°/s). At lower loads or at lower speeds, it is possible to achieve a positioning accuracy of up to ±0.01 mm (Figure 9).

Results obtained from accuracy and repeatability measurements show that at desired conditions, the robot achieves better parameters compared to the ones declared by the manufacturer [39]. The achieved values of positioning error with prescribed load and speed let us compensate key-point coordinates through ML procedure and achieve actual positioning accuracy higher than guaranteed by the robot producer.

Moreover, ML can compensate for the dynamic error of the robot, but for this purpose, it is necessary to analyze the actual vibrations during the robot's movement. Measurements of robot vibrations allow us to fulfill information about positioning accuracy by evaluating such parameters as maximum vibration level and settling time when the robot stops at the desired position. The results of these researches are presented in the next subchapter. The error values are derived from experimental research as error ranges, and the direction of error known as well; therefore, the compensation is performed in advance for defined load and robot configuration.

4.3. Measurement of the Robot Vibrations

Data obtained from robot vibration measurements allowed us to define the average vibration level of the end-of-arm reference point and settling time. These characteristics describe the robot's dynamics and have a great influence on its accuracy and operating speed. In addition, these characteristics allow indirect evaluation of position overshoots and show the minimum required time to reach a stable position after movement. This is extremely important in robotics for the assembly of precise objects and the cooperation of several robots.

Statistically evaluated results show the average level of the reference point absolute vibration acceleration values (Figure 10).

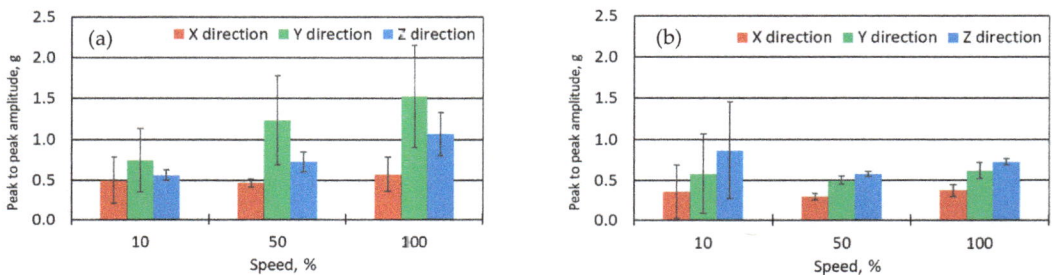

Figure 10. Peak-to-peak vibrations amplitudes: (**a**) at the start (measuring) position (Figure 3a), (**b**) at the endpoint of the trajectory (Figure 3b). Statistical evaluation was performed according to Equations (8)–(13).

From Figure 10a, it is seen that the average amplitude of the vibrations increases when the speed of the robot increases practically on all measured axes. The highest values of vibration levels were recorded at the highest speed. Comparing vibration levels in separate axes, we noticed that the highest values (0.75 g at speed 10% and 1.6 g at speed 100%) were defined on the *y*-axis.

Results presented in Figure 10b show the following inverse tendency: here, the highest values of vibration level occur when the robot moves at 10% speed. By analyzing the vibration level in the *z*-axis direction, it was 0.87 g at 10% speed, 0.6 g at 50% speed, and 0.75 g at 100% speed. A similar tendency was also defined by analyzing the *x* and *y*-axes.

Controversial results regarding Figure 10a,b can be explained by the fact that when the robot moves up from the start position, the gravitational force acts as an additional load; when the robot moves down from the endpoint of the trajectory, gravity acts in the same direction as robot displacement (Figure 2). This analysis was useful to evaluate the settling time of vibrations (Figure 11).

Settling time for all axes increases with movement speed; however, the resulting settling time of the *x*-axis at 50% speed does not follow this tendency (Figure 11a). At the ends of the trajectory (Figure 3), the lowest defined settling time is 0.1 s, whilst the highest value is equal to 0.5 s. Comparing the results presented in Figure 11a,b, the impact of the direction of gravity on the direction of the robot's movement is demonstrated. Settling time in the case when the robot moves from its starting point is about 0.07 s. If the robot moves from its trajectory endpoint, it does not exceed 0.2 s. Results presented in Figures 10 and 11 allow us to state that the vibration level when the robot stops depends not only on the movement speed but also on the movement direction. The results of all experimental research revealed the settling time did not exceed 0.5 s.

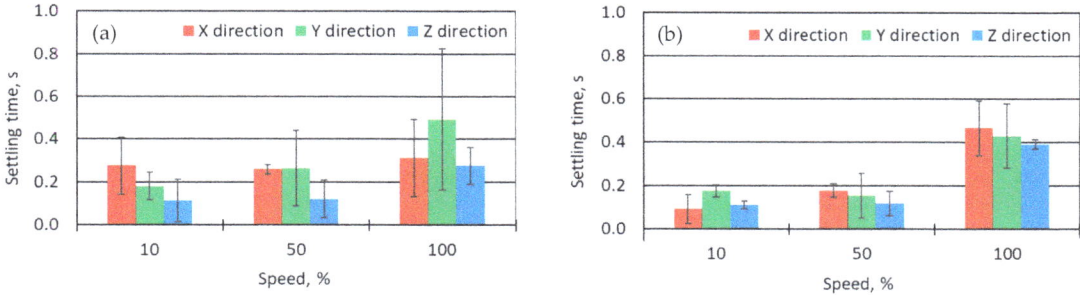

Figure 11. Settling time: (**a**) at the start (measuring) position (Figure 3a) and (**b**)—at the endpoint of the trajectory (Figure 3b). Statistical evaluation processed according to Equations (8)–(13).

4.4. Separate Joint Vibrations Evaluation

To evaluate the behavior of each joint, we performed a separate experiment, which allows defining the influence of each joint on the general vibrations of the entire robot (Figure 12).

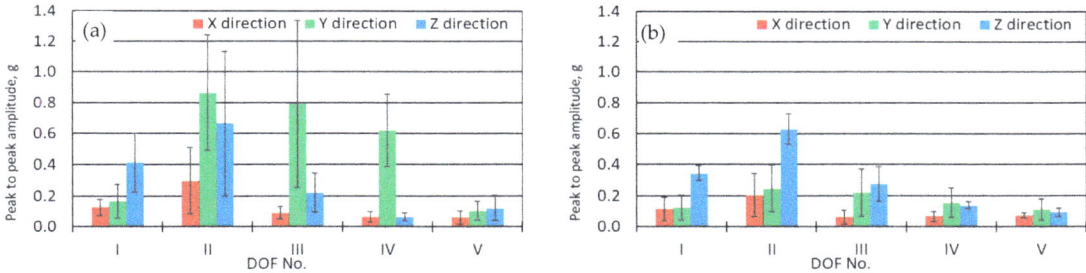

Figure 12. Peak-to-peak vibrations amplitudes evaluated after rotation of separate joints in (**a**) clockwise direction; (**b**) anti-clockwise direction. Statistical evaluation processed according to Equations (8)–(13). Robot start position marked in Figure 3a.

Peak-to-peak vibration amplitudes were evaluated after the rotation of separate joints (Figure 4). The largest measured vibration amplitude was recorded at the x and y axes. The largest amplitude of all measurements was recorded when rotating joint II in a clockwise direction (0.6 g), and the smallest amplitude was recorded when rotating joint V (in a counterclockwise direction, 0.05 g). Probably some variations in clockwise and counterclockwise directions were produced by gravity, which was affecting the robot's movement. Results of measured settling time when separate joints were moved are presented in Figure 13.

From Figure 13a, it is seen that settling time had the greatest value in the x and y directions when joint II was rotated. When the robot joints move clockwise (Figure 13b), the longest settling time was recorded when rotating the second and third joints. The shortest settling time was detected while moving joint I and joint V in both cases. Results obtained from this experiment show that dynamic robot characteristics are mostly affected by the characteristics of the second and third joints, and the settling time at various positions can be more than 0.5 s.

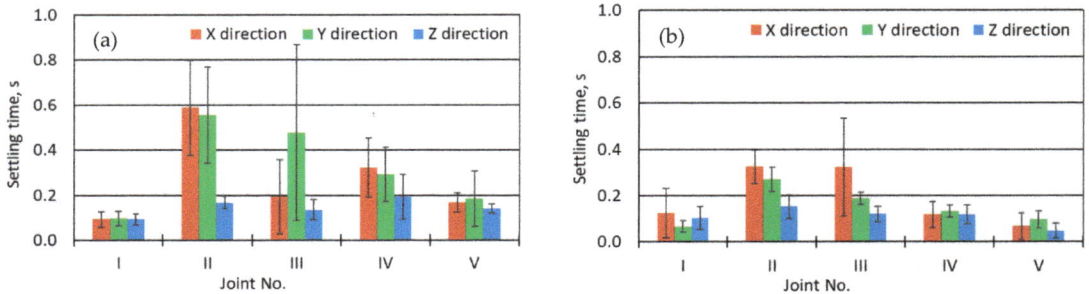

Figure 13. Settling time evaluated after rotation of separate joints in: (**a**) clockwise direction; (**b**) counterclockwise direction. Statistical evaluation was performed by Equations (8)–(13). Robots start position is shown in Figure 3a.

4.5. Extended Position Vibration Measurement

The impact of the first joint characteristics on the robot vibrations when this joint is loaded by maximum force due to an extended robot arm was evaluated (Figure 5). Under such conditions, the maximum vibration level is expected to be influenced by the gravity force direction.

From Figure 14, the highest level of vibration, 0.46 g, was detected in the x-axis when the robot is at an extended position; in the compact position case, this value was 0.27 g. The lower levels of vibrations were detected in the y and z-axes. The moderately high level of vibrations in the z-axis allows us to state that during the movement of the first joint, the whole structure of the robot is kinematically excited. In the extended position, the longest settling time of 0.56 s was defined in the z-axis, while in the compact position, the longest settling time of 0.44 s was detected in the y-axis. This can be explained by the fact that in various configurations, due to the distribution of mass centers, the variable loads and inertia forces in all joints can be detected. Thus, it could be concluded that in most uncomfortable configurations, settling time will not exceed 0.6 s.

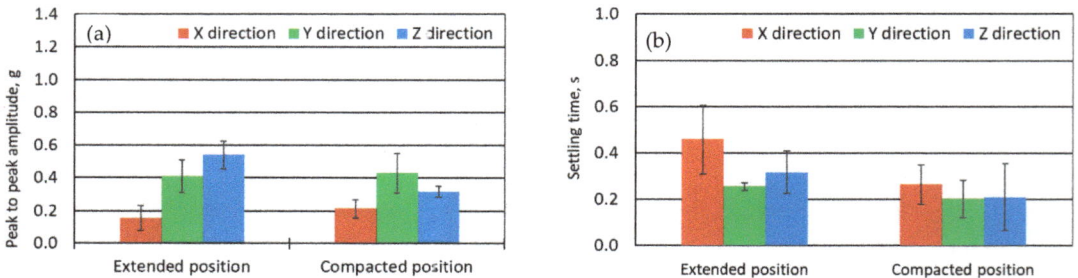

Figure 14. Influence of the robot positions (extended position and compact position) in to (**a**) peak-to-peak vibrations amplitudes; (**b**) settling times. Statistical evaluation was performed by Equations (8)–(13).

4.6. Experiment with Implemented Correction

The correction was applied to the real (Kuka YouBot) robot using processed experimental data. The flowchart of the performed procedure is presented in Figure 15. During the first run, algorithms take experimentally defined data representing robot characteristics and previous compensation values as input parameters and provide an output as a natural number that is later transformed into a new compensation value.

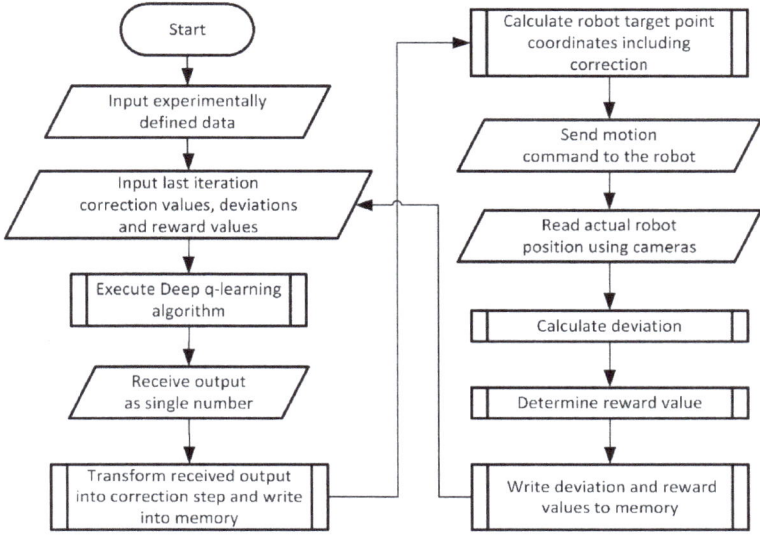

Figure 15. Flowchart of machine learning based positioning errors compensation procedure.

The compensation parameters were applied by shifting the stated start and endpoint positions of the trajectory provided in Figure 3. The efficiency of the proposed methodology was evaluated by measuring positioning accuracy at the start and endpoint of the robot trajectory (Figure 16). During the experiment, the robot was loaded with a 0.5 kg payload and moved at maximum velocity.

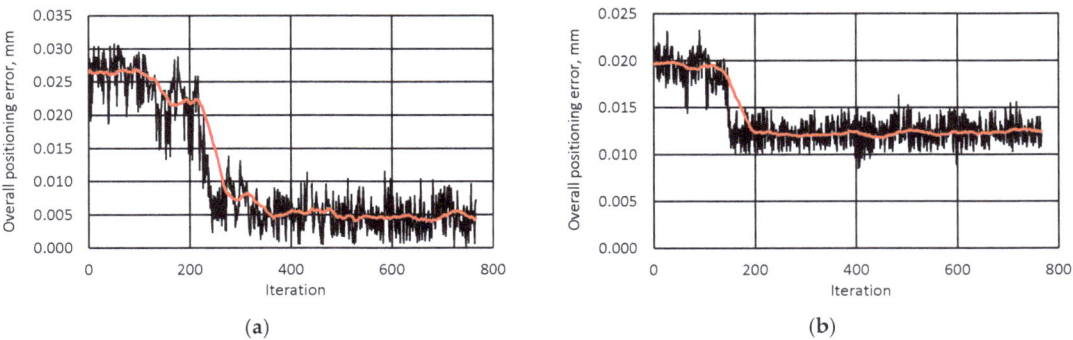

Figure 16. Comparison of robot positioning accuracy before applying correction procedure and after. (**a**) Experiment at the start point (Figure 3a); (**b**) experiment at the endpoint (Figure 3b).

The machine learning process was terminated after 780 iterations by the program because the general positioning error stops to decrease. The ML process has different progress for overall positioning error in the start (Figure 3a) and endpoints (Figure 3b) of the trajectory. The learning process at the start point of the trajectory progresses at the 180th iteration, while the process at the endpoint of the trajectory starts approaching the learning target at the 180th iteration and ends at the 240th iteration. Differences in the learning process are influenced by the gravity force depending on the robot links' configuration.

From the obtained results, we defined the following: the overall positioning error at the start point decreases by 40%: from 0.021 mm to 0.0125 mm; the overall positioning error at the endpoint decreases by 34%: from 0.028 mm to 0.0185 mm.

The proposed method deals with the complex improvement of robot accuracy in comparison with classic methods. Comparison of the proposed method to the existing classic ML methodologies has some scale problems. The online ML process, as proposed in the paper, can minimize robot positioning errors at the interesting point by compensating them in the operation online. Available offline learning procedures are focused on preparatory modes of robot's implementation. Offline methods, as well as statistical ones (workspace mapping), are not sensitive to the fluctuations of the robot operating process. In such a case, robot installation in the production enterprise begs for readjusting for the time being, while the proposed procedure is valid for active robot operation. The complexity of the evaluated and compensated errors is hidden behind the process—we achieve a goal as a diminishing of robot error regardless of the actual coordinate position and configuration of the robot in the target point position. The distribution of the task components along the axes as it is performed for some methods here is not applicable. The disadvantage of the proposed method is some operational time for the ML procedure to set up compensation values, so actual disturbances of the robotic system can cause some periods of increased error values. The good point of the proposed method—initial data collection is possible to optimize. There are available synthetic (simulated) data implementations into the learning process and the use of advanced technical equipment for automated data collection.

5. Conclusions

The dynamic behavior of a robot is to a large extent affected by the robot's warming-up time, the load, the speed of movement, and the trajectory in the working space. The experimental research on warming-up time showed that the highest positioning error was 0.12 mm if the robot operated without warm-up. The optimal time for warming up is 14–15 min. The completed research allowed a conclusion that when the manipulator stays idle for 1 h or more, it is recommended to allow the manipulator to operate initially without performing any work. Where the use of a warm-up program is inconvenient, higher positioning accuracy can be achieved by adjusting the prescribed positions according to experimentally defined dependencies between positioning errors and operating time. Implementing ML here could compensate for positioning errors during the robot warming period if the learning procedure can be taken for many operation breaks. Robot repeatability measured after the warming-up process was 0.0146 ± 0.00526 mm.

Research performed when the robot moves at various speeds with various loads showed that the highest 0.019 mm positioning error appeared in the z-axis when the robot moves at its highest speed with a 0.50 kg load. The obtained results significantly exceed the tolerance of ± 1.0 mm declared by the manufacturer. This allows us to state that ML can improve the accuracy of industrial robots with known operating and environmental conditions. ML is suitable for positioning static and dynamic error compensation for steady-state trajectory and fast accuracy improvement during initial or test operation. The obtained accuracy error value is better than that declared by the manufacturer, and it can even be improved for precision operations; the manufacturer could also implement this option using ML procedures.

Results obtained from vibration measurements show that gripper vibration amplitude mostly depends on the movement trajectory, speed, and direction; the highest measured peak-to-peak vibration amplitude, 1.6 g, was observed at maximum (90°/s) speed. Additional research performed with separate joints showed that the highest impact on the overall vibration level is on joint II and joint III if the robot operates in common configurations. When the robot operates in the closed position, overall vibration levels are mostly affected by joint I.

The results of the settling time measurements, in most cases, correspond to the results of vibration measurements. It was noticed that settling time in most of the tests depends on movement speed, while vibration amplitude is additionally affected by movement direction. The longest settling time was defined as 0.55 s at maximum speed. Therefore, this allows us to state that when operating at maximum speed, the settling time for the robot can last up to 0.6 s.

The set of results obtained from our research allows us to predict the behavior of the analyzed robot and to define its characteristics under various conditions. These results can be used to improve robot parameters for precise operations. Moreover, this result and methodology can be used as a basic database for creating and implementing in practice various ML algorithms. To compensate for positioning errors, the data about the current robot operation must be collected and processed in a way suitable to further use in ML. First, positioning accuracy and vibrations under various conditions in positions that are important for the performing task and in which the accuracy is diminished (for example, near the singularity) should be determined experimentally. Second, the measurements of vibrations on the end-of-arm reference point, including measurements of separate joint impact to general vibration level, should be performed. Then the collected data can be processed using ML algorithms.

Author Contributions: Conceptualization, V.B., A.D. and D.V.; methodology, M.S. and I.M.-V.; software, S.P. and E.S.; validation, M.S., I.M.-V. and A.D.; formal analysis, I.M.-V. and E.S.; investigation, M.S.; resources, M.S.; data curation, D.V.; writing—original draft preparation, A.D., I.M.-V. and J.B.; writing—review and editing, J.B., A.D. and I.M.-V.; visualization, A.D. and S.P.; supervision, V.B.; project administration, M.S.; funding acquisition, V.B. All authors have read and agreed to the published version of the manuscript.

Funding: This work is part of the AI4DI project, receiving funding from the Electronic Components and Systems for European Leadership Joint Undertaking in collaboration with the European Union's H2020 Framework Programme (H2020/2014-2020) and National Authorities, under grant agreement n° 826060.

Data Availability Statement: The data presented in this study are available on request from the corresponding author. The data are not publicly available due to an internal project consortium agreement.

Conflicts of Interest: The authors declare no conflict of interest.

Nomenclature

g	Gravitational acceleration equivalent to ~9.81 m/s^2
φ	Actual joint rotation angle, deg
AP_p	Overall positioning accuracy
RP_i	Robot positioning repeatability
m_v	Measured result
r_v	The real measured parameter value
$\triangle mv$	Absolute measurement error
σ	Experimental standard deviation
S_{md}	Experimental standard mean deviation
$\Delta m_{v,n,P}$	The random error of the measured parameter
$\Delta_{x,y,z}$	Overall positioning error
V	Speed of reference point
M	Load
t_{op}	Time from the moment when the robot was switched on
k_{max}	Maximum number of iterations
$\varepsilon_{x,y,z}$	The overall level of reference point vibrations

References

1. Lu, Y. Industry 4.0: A survey on technologies, applications and open research issues. *J. Ind. Inf. Integration* **2017**, *6*, 1–10. [CrossRef]
2. Jin, Y.; Chen, I.-M.; Yang, G. Kinematic design of a family of 6-DOF partially decoupled parallel manipulators. *Mech. Mach. Theory* **2009**, *44*, 912–922. [CrossRef]
3. Bhangale, P.; Agrawal, V.; Saha, S. Attribute based specification, comparison and selection of a robot. *Mech. Mach. Theory* **2004**, *39*, 1345–1366. [CrossRef]
4. Park, K.-J. Flexible robot manipulator path design to reduce the endpoint residual vibration under torque constraints. *J. Sound Vib.* **2004**, *275*, 1051–1068. [CrossRef]
5. Wu, K.; Krewet, C.; Kuhlenkötter, B. Dynamic performance of industrial robot in corner path with CNC controller. *Robot. Comput. Manuf.* **2018**, *54*, 156–161. [CrossRef]
6. Sintov, A.; Shapiro, A. Dynamic regrasping by in-hand orienting of grasped objects using non-dexterous robotic grippers. *Robot. Comput. Manuf.* **2018**, *50*, 114–131. [CrossRef]
7. Ponce-Hinestroza, A.N.; Castro-Castro, J.A.; Guerrero-Reyes, H.I.; Parra-Vega, V.; Olguin-Diaz, E. Cooperative redundant omnidirectional mobile manipulators: Model-free decentralized integral sliding modes and passive velocity fields. In Proceedings of the IEEE International Conference on Robotics and Automation, Stockholm, Sweden, 16–21 June 2016; Volume 2016, pp. 2375–2380. [CrossRef]
8. Adamov, B.I. Influence of mecanum wheels construction on accuracy of the omnidirectional platform navigation (on example of KUKA youBot robot). In Proceedings of the 25th Saint Petersburg International Conference on Integrated Navigation Systems, ICINS 2018-Proceedings, Petersburg, Russia, 28–30 May 2018; pp. 1–4. [CrossRef]
9. Huckaby, J.; Christensen, H.I. *Dynamic Characterization of KUKA Light-Weight Robot Manipulators Technical Report GT-RIM-CR-2012-001*; Georgia Institute of Technology: Atlanta, GA, USA, 2012.
10. Mohamed, K.T.; Ata, A.; El-Souhily, B.M. Dynamic analysis algorithm for a micro-robot for surgical applications. *Int. J. Mech. Mater. Des.* **2011**, *7*, 17–28. [CrossRef]
11. Urrea, C.; Pascal, J. Design, simulation, comparison and evaluation of parameter identification methods for an industrial robot. *Comput. Electr. Eng.* **2018**, *67*, 791–806. [CrossRef]
12. Abele, E.; Weigold, M.; Rothenbücher, S. Modeling and Identification of an Industrial Robot for Machining Applications. *CIRP Ann.* **2007**, *56*, 387–390. [CrossRef]
13. Sharifi, M.; Chen, X.; Pretty, C.; Clucas, D.; Cabon-Lunel, E. Modelling and simulation of a non-holonomic omnidirectional mobile robot for offline programming and system performance analysis. *Simul. Model. Pr. Theory* **2018**, *87*, 155–169. [CrossRef]
14. Li, R.; Zhao, Y. Dynamic error compensation for industrial robot based on thermal effect model. *Measurement* **2016**, *88*, 113–120. [CrossRef]
15. Shang, D.; Li, Y.; Liu, Y.; Cui, S. Research on the Motion Error Analysis and Compensation Strategy of the Delta Robot. *Mathematics* **2019**, *7*, 411. [CrossRef]
16. Borisov, O.I.; Gromov, V.S.; Pyrkin, A.A.; Vedyakov, A.A.; Petranevsky, I.; Bobtsov, A.; Salikhov, V.I. Manipulation Tasks in Robotics Education. *IFAC-Pap.* **2016**, *49*, 22–27. [CrossRef]
17. Wang, L.; Chen, L.; Shao, Z.; Guan, L.; Du, L. Analysis of flexible supported industrial robot on terminal accuracy. *Int. J. Adv. Robot. Syst.* **2018**, *15*, 1729881418793022. [CrossRef]
18. Mueggler, E.; Faessler, M.; Fontana, F.; Scaramuzza, D. Aerial-guided navigation of a ground robot among movable obstacles. In Proceedings of the 12th IEEE International Symposium on Safety, Security and Rescue Robotics, SSRR 2014-Symposium Proceedings, Hokkaido, Japan, 27–30 October 2014. [CrossRef]
19. Sharma, S.; Kraetzschmar, G.K.; Scheurer, C.; Bischoff, R. Unified closed form inverse kinematics for the KUKA youBot. In Proceedings of the 7th German Conference on Robotics, ROBOTIK 2012, Munich, Germany, 21–22 May 2012; pp. 139–144.
20. Li, B.; Tian, W.; Zhang, C.; Hua, F.; Cui, G.; Li, Y. Positioning error compensation of an industrial robot using neural networks and experimental study. *Chin. J. Aeronaut.* **2021**, *35*, 346–360. [CrossRef]
21. Cao, C.-T.; Do, V.-P.; Lee, B.-R. A Novel Indirect Calibration Approach for Robot Positioning Error Compensation Based on Neural Network and Hand-Eye Vision. *Appl. Sci.* **2019**, *9*, 1940. [CrossRef]
22. Yuan, P.; Chen, D.; Wang, T.; Cao, S.; Cai, Y.; Xue, L. A compensation method based on extreme learning machine to enhance absolute position accuracy for aviation drilling robot. *Adv. Mech. Eng.* **2018**, *10*, 2018. [CrossRef]
23. Zhu, W.; Li, G.; Dong, H.; Ke, Y. Positioning error compensation on two-dimensional manifold for robotic machining. *Robot. Comput. Manuf.* **2019**, *59*, 394–405. [CrossRef]
24. Marchal, P.C.; Sörnmo, O.; Olofsson, B.; Robertsson, A.; Ortega, J.G.; Johansson, R. Iterative Learning Control for Machining with Industrial Robots. *IFAC Proc. Vol.* **2014**, *47*, 9327–9333. [CrossRef]
25. Pane, Y.P.; Nageshrao, S.P.; Kober, J.; Babuška, R. Reinforcement learning based compensation methods for robot manipulators. *Eng. Appl. Artif. Intell.* **2018**, *78*, 236–247. [CrossRef]
26. Qi, J.; Chen, B.; Zhang, D. A calibration method for enhancing robot accuracy through integration of kinematic model and spatial interpolation algorithm. *J. Mech. Robot.* **2021**, *13*, 061013. [CrossRef]
27. Jing, W.; Goh, C.F.; Rajaraman, M.; Gao, F.; Park, S.; Liu, Y.; Shimada, K. A Computational Framework for Automatic Online Path Generation of Robotic Inspection Tasks via Coverage Planning and Reinforcement Learning. *IEEE Access* **2018**, *6*, 54854–54864. [CrossRef]

28. Schaal, S.; Atkeson, C.G. Learning Control in Robotics. *IEEE Robot. Autom. Mag.* **2010**, *17*, 20–29. [CrossRef]
29. Sumanas, M.; Bucinskas, V.; Vilkonciene, I.M.; Dzedzickis, A.; Lenkutis, T. Implementation of Machine Learning Algorithms for Autonomous Robot Trajectory Resolving. In Proceedings of the 2019 Open Conference of Electrical, Electronic and Information Sciences (eStream), Vilnius, Lithuania, 25 April 2019; pp. 1–3. [CrossRef]
30. A Fahmy, A.; Kalyoncu, M.; Castellani, M. Automatic design of control systems for robot manipulators using the bees algorithm. *Proc. Inst. Mech. Eng. Part I J. Syst. Control Eng.* **2011**, *226*, 497–508. [CrossRef]
31. Liu, K.P.; Li, Y.C. Vibration suppression for a class of flexible manipulator control with input shaping technique. In Proceedings of the 2006 International Conference on Machine Learning and Cybernetics, Dalian, China, 13–16 August 2006; Volume 2006, pp. 835–839. [CrossRef]
32. Jordan, M.I.; Mitchell, T.M. Machine learning: Trends, perspectives, and prospects. *Science* **2015**, *349*, 255–260. [CrossRef]
33. Ghahramani, Z. Probabilistic machine learning and artificial intelligence. *Nature* **2015**, *521*, 452–459. [CrossRef]
34. Dimiduk, D.M.; Holm, E.A.; Niezgoda, S.R. Perspectives on the Impact of Machine Learning, Deep Learning, and Artificial Intelligence on Materials, Processes, and Structures Engineering. *Integr. Mater. Manuf. Innov.* **2018**, *7*, 157–172. [CrossRef]
35. Bottou, L.; Curtis, F.E.; Nocedal, J. Optimization Methods for Large-Scale Machine Learning. *SIAM Rev.* **2018**, *60*, 223–311. [CrossRef]
36. Adam, B.; Smith, I.F. Reinforcement Learning for Structural Control. *J. Comput. Civ. Eng.* **2008**, *22*, 133–139. [CrossRef]
37. Qiu, J.; Wu, Q.; Ding, G.; Xu, Y.; Feng, S. A survey of machine learning for big data processing. *EURASIP J. Adv. Signal Process.* **2016**, *2016*, 67. [CrossRef]
38. Arimoto, S.; Kawamura, S.; Miyazaki, F. Bettering operation of Robots by learning. *J. Robot. Syst.* **1984**, *1*, 123–140. [CrossRef]
39. YouBot Detailed Specifications. Available online: http://www.youbot-store.com/wiki/index.php/YouBot_Detailed_Specifications (accessed on 14 October 2022).
40. Wang, Z.; Wu, Z.; Zhen, X.; Yang, R.; Xi, J.; Chen, X. A two-step calibration method of a large FOV binocular stereovision sensor for onsite measurement. *Measurement* **2015**, *62*, 15–24. [CrossRef]
41. *ISO 9283:1998*; Manipulating Industrial Robots—Performance Criteria and Related Test Methods; ISO/TC 299 Robotics. Technical Committee: Geneva, Switzerland, 2003.
42. Cowan, G. *Statistical Data Analysis*; Clarendon Press: Oxford, UK, 1998.

![machines logo] *machines*

MDPI

Article

A GAN-BPNN-Based Surface Roughness Measurement Method for Robotic Grinding

Guojun Zhang 🔟, Changyuan Liu, Kang Min, Hong Liu and Fenglei Ni *

Department of State Key Laboratory of Robotics and System, Harbin Institute of Technology, Harbin 150001, China
* Correspondence: flni@hit.edu.cn; Tel.: +86-186-0451-3569

Abstract: Existing machine vision-based roughness measurement methods cannot accurately measure the roughness of free-form surfaces (with large curvature variations). To overcome this problem, this paper proposes a roughness measurement method based on a generative adversarial network (GAN) and a BP neural network. Firstly, this method takes images and curvature of free-form surfaces as training samples. Then, GAN is trained for roughness measurement through each game between generator and discriminant network by using real samples and pseudosamples (from generator). Finally, the BP neural network maps the image discriminant value of GAN and radius of curvature into roughness value (Ra). Our proposed method automatically learns the features in the image by GAN, omitting the independent feature extraction step, and improves the measurement accuracy by BP neural network The experiments show that the accuracy of the proposed roughness measurement method can measure free-form surfaces with a minimum roughness of 0.2 μm, and measurement results have a margin of 10%.

Keywords: robotic belt grinding; surface roughness measurement; generative adversarial network

✅ **check for updates**

Citation: Zhang, G.; Liu, C.; Min, K.; Liu, H.; Ni, F. A GAN-BPNN-Based Surface Roughness Measurement Method for Robotic Grinding. *Machines* **2022**, *10*, 1026. https://doi.org/10.3390/machines10111026

Academic Editors: Raul D. S. G. Campilho and Francisco J. G. Silva

Received: 21 September 2022
Accepted: 31 October 2022
Published: 4 November 2022

Publisher's Note: MDPI stays neutral with regard to jurisdictional claims in published maps and institutional affiliations.

1. Introduction

Grinding is a widespread but essential process in manufacturing products. It is widely used in aerospace, automobile manufacturing, rail transit and other industries [1]. Traditionally, grinding is done by manual operation or multi-axis CNC machine tools. The former is time-consuming and labor-intensive, while the latter is limited by operating space. Recently, robotic belt grinding has become an alternative due to its low cost, efficiency and large operating space [2]. For small workpiece, the robot generally uses appropriate contact force to ensure the stability of the workpiece [3]. The robot holds the workpiece to complete the grinding operation and uses its dexterity to move to the best inspection position on the surface of the workpiece [4]. Multiple sampling points are planned according to the curvature of the workpiece surface, and the robot moves to each sampling point in turn to complete the surface image sampling.

As an important indicator of the quality of machined surfaces, surface roughness has a significant impact on product life and reliability [5]. The traditional probe-based measurement method [6–9] uses a probe that slides over the surface of the workpiece for measurement. This method has been proven to be highly reliable and highly accurate. However, it has several limitations, such as long measurement times, high environmental requirements and complex steps.

With the development of optical technology, a variety of noncontact measurement methods has been developed to address the limitations of the contact measurement method. For example, optical microscopy, confocal laser scanning microscopy and white light interferometry are used to measure surface roughness, but these methods are mainly used in laboratories, because their high cost and complicated operation are not suitable for real industrial production. Currently, the most popular noncontact measurement method is based on machine vision for roughness measurements of the machined workpiece surface [10–15]; this is widely used because of its advantages, which include unrestricted size of the measured object, low cost, high speed, and easy factory automation deployment.

Usually, vision-based measurement methods can be divided into four stages: image acquisition, image processing, feature extraction and prediction strategy [16]. Most vision-based methods for surface roughness assessments rely on feature extraction. Suganandha et al. [17] used a single-point laser to irradiate the surface of an object and analyzed the gray-level covariance matrix (GLCM) features of the scatter pattern, such as contrast, correlation, energy, entropy homogeneity and maximum probability, in order to investigate the correlation between 3D surface roughness and scatter images. Shanta et al. [18] obtained roughness measures by using a singular value decomposition method to analyze contrast, as well as light and dark pixels of binarized scatter patterns. Samie et al. [19] proposed a method to transform the image of a surface into an unweighted and undirected network graph. The graph-theoretic invariants and Fiedler number were estimated for use as discriminative factors for the surface roughness of the workpiece, avoiding the filtering and segmentation of complex images. Liu et al. [20] proposed an improved method based on microscopic vision to detect the surface roughness of R surfaces in valves. The method analyzed the surface morphology images of R surfaces by the gray-level covariance matrix (GLCM) method and used a support vector machine (SVM) model to describe the relationship between GLCM features and the actual surface roughness. Huaian et al. [21] used a specific of color light source to illuminate the surface and measured the surface roughness of the object based on the color distribution statistical matrix for features such as texture in the image.

Vision-based surface roughness assessment methods have been proven to be reliable and accurate, but there are still some limitations, especially in the feature extraction process. Visual assessment methods that rely on feature extraction are difficult to reapply to different datasets. Feature extraction is a highly intensive process that requires a high degree of computation and expert decision making in selecting the appropriate surface features. As a result, the process can lead to long processing times, which are not conducive to the rapid diffusion of the technique.

Some researchers had accomplished roughness measurement by using intelligent algorithms for the self-learning of image features. For example, Du-Ming [22] used the two-dimensional Fourier transform to extract quantitative measures of surface roughness in the spatial frequency domain. The roughness features were used as input to build an artificial neural network to determine the surface roughness. Gürcan [23] converted the surface image into a binary image as input data. The log-sigmoid function was chosen as the transfer function and the neural network model was trained using the scaled conjugate gradient algorithm. Kaixuan et al. [24] proposed a roughness classification method. The method expanded and preprocessed the images, and then trained an AlexNet-based surface roughness classification model for milled samples.

Jamal et al. [25] proposed that a convolutional neural network (CNN) is used as a regressor in order to obtain steel surface roughness, and that a CNN based on spatial pooling pyramid is applied for roughness classification. Achmad et al. [26] proposed a deep learning model containing convolutional layer, ReLU, pooling layer and two FC layers to automatically learn image features using a gradient descent method to optimize hyperparameters in the model.

GAN is an emerging self-supervised learning technique that provides a method to learn deep representations without the use of extensively labeled training data [27,28].

In this paper, we design a surface roughness prediction model based on GAN + BP neural network. Firstly, we consider the difference in surface images with different curvatures and use the images and curvatures as training samples. Then, the GAN network performs model training in such a way that the generative network and the discriminative network play with each other to automatically learn the features in the images, omitting the feature extraction step. Finally, the BP neural network represents the correlation between the GAN discriminant value and the roughness (Ra).

2. Generative Adversarial Network

A GAN consists of two parts: the generative and discriminative networks, which use an adversarial game between generators and discriminators to achieve self-supervised learning. The main difference from the traditional network model is that the data training process contains both consistent and adversarial data. The generators and discriminators each optimize in different directions and form a competitive relationship with each other, but they also depend on each other to form a unified whole. In the adversarial training mode, the generator no longer learns directly from the training dataset, but learns iteratively in an indirect way through the optimization directions output by the discriminator, generating pseudosamples to mix the spurious with the genuine. A GAN computes faster and has greater expansion flexibility than a traditional network. In this paper, we propose using a GAN to discriminate the roughness of the surface of workpieces using the structure shown in Figure 1.

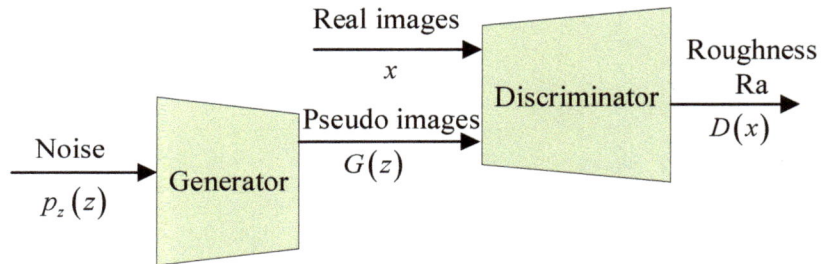

Figure 1. The structure of the generative adversarial network (GAN).

$p_z(z)$ represented a prior on input noise variables, $G(z)$ was generator of GAN, and $D(x)$ represented the probability that x comes from real images. GAN's goal is training $D(x)$ to maximize the probability of assigning the correct label to both training real samples and pseudosamples from $G(z)$, at the same time training $G(z)$ to minimize $\log(1 - D(G(z)))$. In other words, the loss function of GAN is as follows:

$$\min_G \max_D V(D, G) = E_{x \sim p_{data}(x)}[\log D(x)] + E_{z \sim p_z(z)}[\log(1 - D(G(z)))] \tag{1}$$

The disadvantages of GAN:

(1) GAN training is unstable, and the training degree of generator and discriminator should be carefully balanced.

(2) When the generator learns that some features of the real data successfully deceive the discriminator, it will not update. As a result, the generated samples lack diversity, and a collapse mode occurs in GAN.

(3) When the overlap between the real and generated distributions is negligible, the JS divergence between the real and generated distributions is minimized to a constant log2. When the discriminator is optimal, the loss of the minimized generator is also closer to log2, and G is no longer updated, resulting in the disappearance of the generator gradient.

In order to overcome the above shortcomings, the generator and discriminator network structures of the original GAN network are redesigned. In this paper, a deep convolutional neural network is used as the network structure of the generator and discriminator:

The structure of the generative network is shown in Figure 2, and the corresponding discriminator network structure is shown in Figure 3.

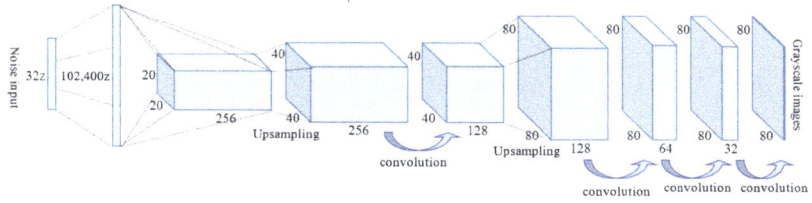

Figure 2. The structure of the generative network.

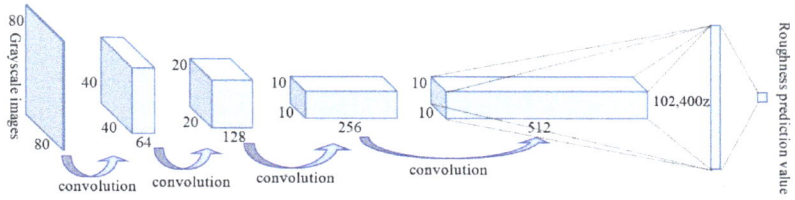

Figure 3. The structure of the discriminant network.

From Equation (1), it can be seen that the objective of the discriminator is to obtain the maximum value of $V(G, D)$. Then, it is necessary to maximize $D(x)$, which means maximizing the probability that real samples and generated samples will be assigned the correct label. At the same time, the generator G is trained to minimize $\log(1 - D(G(z)))$, which means minimizing the difference between the generated samples and the true samples.

Therefore, the value function of GAN can be decomposed into two optimization problems:

(1) Fix the generator G, train the discriminator D, so that it can maximize the correct determination of whether the sample is from the real sample or from the sample generated by G. The objective function of D is,

$$\max_D V(D, G) = E_{x \sim p_{data}}[\log D(x)] + E_{x \sim p_g}[\log(1 - D(x))] \tag{2}$$

Finding the derivative of Equation (2) with respect to $D(x)$, such that its derivative 0 gives,

$$\frac{P_{data}(x)}{D(x)} - \frac{P_g(x)}{1 - D(x)} = 0 \tag{3}$$

Rewriting Equation (3), the optimal discriminator is obtained as follows:

$$D^*(x) = \frac{P_{data}(x)}{P_{data}(x) + P_g(x)} \tag{4}$$

(2) Fix the discriminator D and train the generator G, such that $L = \max_D V(D, G)$ minimizes the difference between the generated samples and the real samples. The objective function of G is given by,

$$\min_G L = E_{x \sim p_{data}}[\log D^*(x)] + E_{x \sim p_g}[\log(1 - D^*(x))] \tag{5}$$

Substituting Equation (4) into Equations (2) and (5), we obtain,

$$\min_G L = E_{x \sim p_{data}}\left[\log \frac{P_{data}(x)}{P_{data}(x) + P_g(x)}\right] + E_{x \sim p_g}\left[\log\left(\frac{P_g(x)}{P_{data}(x) + P_g(x)}\right)\right] \tag{6}$$

Introducing two important similarity metrics KL divergence (Kullback–Leibler divergence) and JS divergence (Jensen–Shannon divergence), Equation (6) is rewritten as,

$$\min_G L = KL\left(P_{data} \parallel \frac{P_{data}+P_g}{2}\right) + KL\left(P_z \parallel \frac{P_{data}+P_g}{2}\right) - 2 \cdot \log 2$$
$$= 2 \cdot JSD\left(P_{data} \parallel P_g\right) - 2 \cdot \log 2 \tag{7}$$

From Equation (7), it can be seen that fixing the discriminator, the goal of training the generator is to have a JS dispersion of 0 between P_g and P_{data}, i.e., $P_g = P_{data}$. At this point, the discriminator $D^*(G) = \frac{1}{2}$. Thus, the equilibrium between generator and discriminator is realized by reciprocally training the GAN.

3. Measurement of Surface Roughness of Workpieces

In this method, a neural network is used to discriminate the surface roughness of a grinded workpiece. A small number of data images are used to train a generative adversarial network (GAN) to generate the required large dataset for training the discriminative network, which outputs the surface roughness measurements of the workpiece.

3.1. Image Acquisition and Data Preprocessing Methods Subsection

The radius of the curvature and roughness of the workpiece surface directly determine the refractive index of light, a factor which affects the imaging effect of the workpiece surface.

Therefore, 72 images of workpieces with different curvature radii and roughness were collected in this study. The 12 images with the roughest and smoothest surfaces were selected as the original training dataset for the GAN generator, with 6 smooth and 6 rough images, as shown in Figure 4. However, the amount of data in the original dataset of 12 images was too small to be used directly for training the GAN discriminator, which would lead to severe overfitting of the discriminator. Therefore, the original images were first rotated and panned to generate a sufficient number of derived images (32,000 images were generated in the experiment).

Ra = 0.19 R = 50   Ra = 0.23 R = 1000   Ra = 0.25 R = ∞

Figure 4. *Cont.*

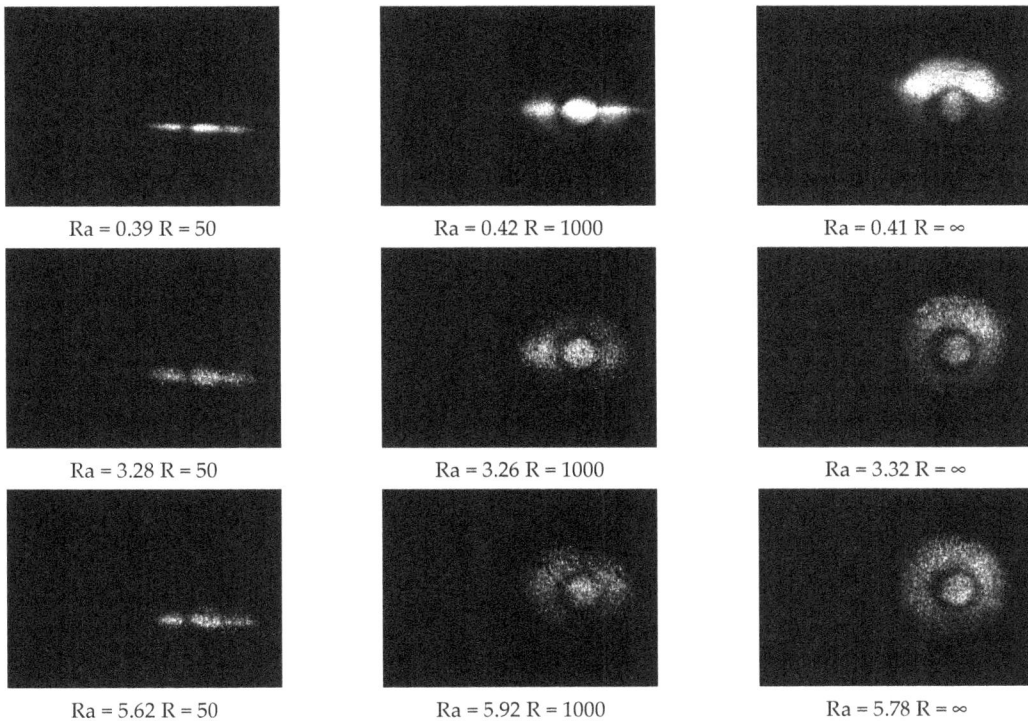

Figure 4. The raw data used for training in this study, where Ra denotes the degree of roughness (the larger the Ra, the rougher the surface) and R is the radius of curvature (the larger the R, the flatter the surface).

3.2. GAN-Based Surface Roughness Discrimination Method

The generative network attempts to create images that the discriminant network cannot distinguish from real images during the training process, while the discriminant network is equivalent to a binary classification network to distinguish between the real training images and the images created by the generator. In this way, the model parameters are continuously optimized by the adversarial training between the two networks. Over multiple iterations, the images created by the generative network become closer to the real images, and the discriminant network increasingly understands the meaning of the roughness represented by the images, and finally gives the roughness value.

The process of training the GAN is shown in Figure 5.

Step 1: The original datasets (surface roughness images) are classified in a binary fashion and labeled as "smooth" and "rough" according to a certain threshold.

Step 2: Real surface roughness images marked by the classification in step 1 are rotated and panned to expand datasets.

Step 3: The GAN is trained using the "rough" category data in the dataset in Step 2.

Step 4: The GAN is trained using the "smooth" category data in the dataset in Step 2.

Step 5: The image data is generated from the generation network trained in steps 3 and 4 to complete the data set expansion.

Step 6: The discriminative network is retrained for the expanded smooth and rough datasets generated by the generative network in step 5 to obtain the final roughness recognition network model.

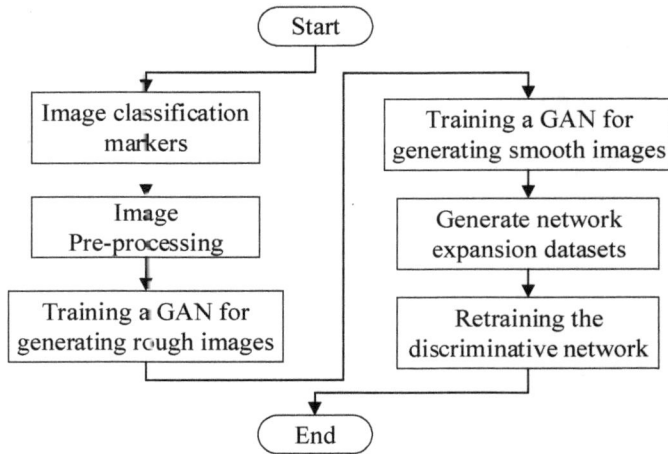

Figure 5. The training process of the GAN.

3.3. Training Results

To teach the GAN the meaning of "rough," real smooth images are added as fake data to train the discriminator along with the output of the generator, as shown in Figure 6a. Figure 6b shows examples of the surface roughness of workpieces generated by the GAN.

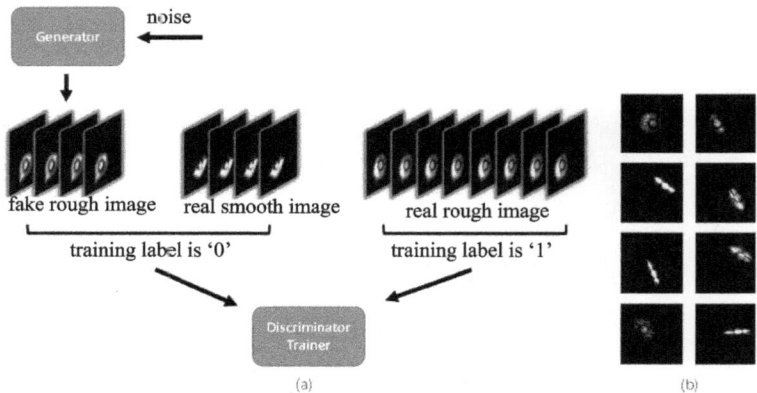

Figure 6. (a) Training method for the discriminator in the GAN used to generate rough images **(b)** Data generated by the GAN.

The GAN is trained using the expanded datasets, and then the GAN is able to generate a series of smooth and rough images as the training set. The problem is treated as a binary classification problem using the sigmoid function as the output activation function and the binary cross-entropy as the loss function for training, with the rough images labeled '1' and the smooth images labeled '0'. In this method, the same network structure of the classifier with the discriminator of the GAN is used, as shown in Table 1.

Table 1. The network structure of the generator and the discriminator.

Generator		Discriminator	
Input	(32)	Input	(80, 80, 1)
Dense	(102400)	Conv2D	(40, 40, 64)
BatchNormalization	(102400)	LeakyReLU	(40, 40, 64)
Activation-ReLU	(102400)	Dropout	(40, 40, 64)
Reshape	(20, 20, 256)	Conv2D	(20, 20, 128)
Dropout	(20, 20, 256)	LeakyReLU	(20, 20, 128)
UpSampling2D	(40, 40, 256)	Dropout	(20, 20, 128)
Conv2DTranspose	(40, 40, 128)	Conv2D	(10, 10, 256)
BatchNormalization	(40, 40, 128)	LeakyReLU	(10, 10, 256)
Activation-ReLU	(40, 40, 128)	Dropout	(10, 10, 256)
UpSampling2D	(80, 80, 128)	Conv2D	(10, 10, 512)
Conv2DTranspose	(80, 80, 64)	LeakyReLU	(10, 10, 512)
BatchNormalization	(80, 80, 64)	Dropout	(10, 10, 512)
Activation-ReLU	(80, 80, 64)	Flatten	(51200)
Conv2DTranspose	(80, 80, 32)	Dense	(1)
BatchNormalization	(80, 80, 32)	Activation-Sigmoid	(1)
Activation-ReLU	(80, 80, 32)		
Conv2DTranspose	(80, 80, 1)		
Activation-Sigmoid	(80, 80, 1)		

In the method described above, a dataset with only roughness Ra of 0.2, 0.4, 3.2, 6.3 and radius of curvature R of 50, 1000, and infinity are used as the training set. However, the dataset in the actual test contains more cases of roughness and radius of curvatures. A qualified classifier should be able to accurately distinguish the cases which are not encountered in the training set. Some of the actual test results are shown in Table 2, which visually reflects the classifier's scoring of images with different roughness levels. The smaller score means the smoother surface. The test results show that the method can accurately determine the untrained intermediate roughness (0.8 and 1.6 are the untrained cases), which indicates that the network can understand the concept of "roughness."

Table 2. Surface roughness prediction based on GAN.

Radius of Curvature (mm)	Roughness Ra (µm)	Predictive Value
	0.2	−22.61
	0.4	−8.75
150	0.8	9.86
	1.6	12.40
	3.2	28.27
	6.3	30.37
	0.2	−27.65
	0.4	−14.10
700	0.8	9.75
	1.6	19.65
	3.2	39.47
	6.3	41.16
	0.2	−48.25
	0.4	−31.05
∞	0.8	37.82
	1.6	45.70
	3.2	54.27
	6.3	62.00

4. Discriminating Method of Workpiece Surface Roughness

Roughness (Ra) is a small-distance (usually less than 1 mm) peak–valley that forms from a microgeometric shape error of the surface of the part. As shown above, the real surface image can obtain a discriminant value, nonlinearly related to roughness Ra through the trained GAN. In order to establish a mapping relationship between the discrimination value of GAN and surface roughness Ra, this paper proposes a surface roughness Ra discrimination method based on the BP neural network, which is shown in Figure 7.

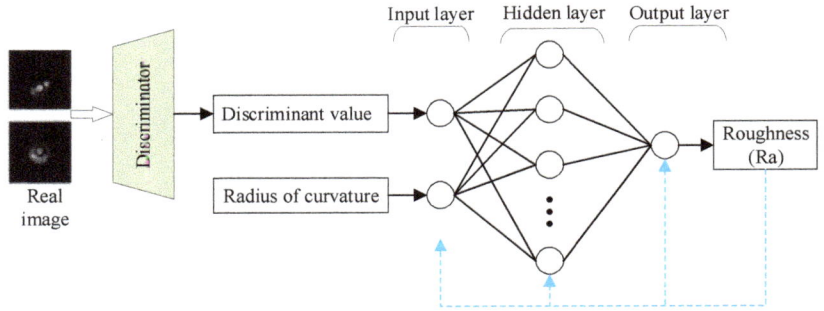

Figure 7. Back-propagation neural network structure to evaluate grinding roughness, according to discriminant value.

The input layer of the BP neural network for image discriminant value and radius of curvature is denoted as $x_i (i = 1, 2)$, and the output surface roughness is denoted as z. The number of neurons in the hidden layer is s, denoted as $y_i (1, 2, \cdots, s)$. The bias value of the hidden layer is γ, and the bias of the output layer is θ. w denotes the connection weights from the input layer to the hidden layer, v denotes the connection weights from the hidden layer to the output layer, while f_y and f_z are the activation functions of the hidden and output layers, respectively. The forward propagation process of the BP neural network is expressed as:

$$y_i = f_y \left(\sum_{i=1}^{2} wx_i - \gamma \right)$$
$$z = f_z \left(\sum_{j=1}^{s} vy_j - \theta \right) \tag{8}$$

Equation (8) completes the mapping from $x_i (i = 1, 2)$ to z. α denotes the target value of the output, and the error of the BP neural network is:

$$\min \delta = z - \alpha \tag{9}$$

The error back-propagation process of the BP neural network is implemented by minimizing the objective function by Equation (9). The weights and biases of each layer are adjusted for each propagation as follows:

$$
\begin{aligned}
\hat{v} &= v - \eta \frac{\partial \delta(v,\theta)}{\partial v} & \hat{w} &= w - \eta \frac{\partial \delta(w,\gamma)}{\partial w} \\
\hat{\theta} &= \theta - \eta \frac{\partial \delta(v,\theta)}{\partial \theta} & \gamma' &= \gamma - \eta \frac{\partial \delta(w,\gamma)}{\partial \gamma}
\end{aligned} \tag{10}
$$

From Equation (8), it can be seen that the data transfer process between the layers is transformed and connected by the activation function. Meanwhile, in the error back-propagation process in the BP neural network in Equation (10), the derivatives of the activation function adjust the connection weights of each layer to reduce the error to the desired range. Considering the requirement that the activation function be continuously differentiable, the sigmoid function is chosen as the output layer activation function, and the tanh function is chosen as the output layer activation function.

$$f_y = \frac{1}{1 + e^{-x}}, \quad f_z = \frac{e^x - e^{-x}}{e^x + e^{-x}} \tag{11}$$

To ensure accuracy of roughness estimation, this study determines the number of neurons in the hidden layer according to the Hecht–Nelson method [29] because the hidden layer affects the stability of the network. The number of input layer neurons is n, while the number of hidden layer neurons is 2n + 1. That is, the number of hidden layer neurons in the network is 5. Networks with different numbers of neurons have also been tested, as shown in Table 3.

Table 3. BP neural network errors for different network structures.

Structure of the BP Neural Network	Average Error (%)
	Roughness
2-3-1	12.42
2-5-1	6.24
2-7-1	8.79
2-9-1	14.63

After determining that there should be one hidden layer for the BP neural network model, the number of neurons in the hidden layer must be determined. For each neural network model in Table 3, the average error between the actual and predicted values were recorded in 10 training sessions. It is clear from Table 3 that as the number of neurons in the hidden layer increases, the accuracy does not necessarily follow. When the number of hidden layer neurons increases from 3 to 5, the average error decreases from 12.42% to 6.24%. However, as the number of hidden layer neurons continues to increase to 7, the average error increases from 6.24% to 14.63%. The 2-5-1 neural network structure has the smallest error, and so this network structure is used in the real-time surface roughness model in this study.

5. Experiments

In this study, a robot is used to hold the workpiece while grinding a complex, curved item. The robotic grinding platform is built as shown in Figure 8, consisting of a belt grinder, a lightweight robot with seven degrees of freedom and a camera.

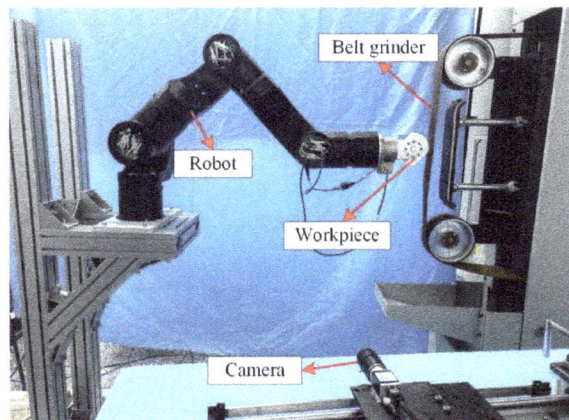

Figure 8. Robotic grinding system.

Surface roughness is a key feature of the surface texture of a workpiece because it has a significant impact on the service life and reliability of a mechanical product. This paper proposes a vision-based roughness measurement method for comprehensively considering the cost, usability, and efficiency. This method consists of three stages: image acquisition, self-supervised learning and result discrimination. A narrow-angle industrial camera is used to build a vision-based surface roughness measurement system, as shown in Figure 9.

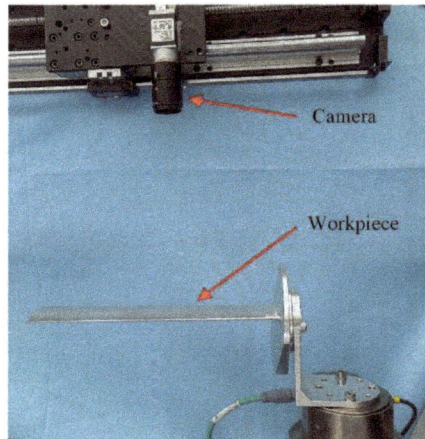

Figure 9. Surface roughness measurement system based on vision.

The narrow-angle Balser ace 2 industrial camera meets the needs of this method for surface roughness measurement of curved workpieces. Its detailed parameters are shown in Table 4.

Table 4. Parameters of Balser ace 2 narrow angle camera.

	Parameters
Resolution ratio	2.3 MP
Chip	IMX392
Frame rate	51.0 fps
Port	GigE
Noise	$2.6e^{-}$
Dynamic range	71.7 dB
Signal-to-noise ratio	40.2 dB
Photosensitive chip type	CMOS
Photosensitive chip size	6.6 mm × 4.1 mm

Balser ACE 2 samples images of standard grinding surface roughness contrasts as shown in Figure 10a, and images of standard milled surface roughness contrasts as shown in Figure 10b. There is no significant difference between images Ra 0.2 and Ra 0.1. The camera can distinguish surfaces with surface roughness Ra greater than 0.2. The image samples in Figure 10 are added to the training set of the proposed GAN to enrich the samples and increase the measurement accuracy of the model.

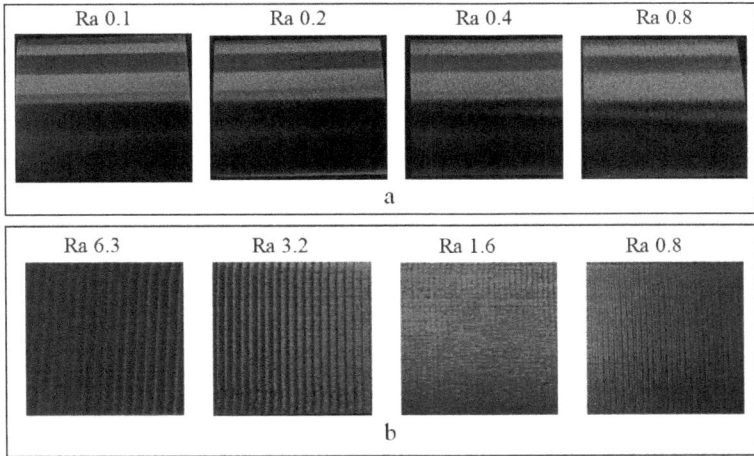

Figure 10. Images of the roughness standard samples were obtained by Balser ace 2. (**a**) Images of the roughness standard grinding surface roughness contrasts; (**b**) Images of the roughness standard milled surface roughness contrasts.

The blade-grinding process is divided into three processing stages according to the requirements of the grinding process: rough grinding, semi-finishing grinding and finishing grinding. The rough grinding stage removes a large amount of the material using the grinding parameters belt line speed 5 m/s, feed speed 3 mm/s, belt mesh 120 and contact force 15 N. The semi-finishing stage removes a small amount of material with the grinding parameters of belt line speed 5 m/s, feed speed 3 mm/s, belt mesh 320 and contact force 10 N. The finishing stage is mainly for finishing the surface of the workpiece with the grinding parameters belt line speed 5 m/s, feed speed 3 mm/s, belt mesh 320, and contact force 5 N. After each grinding stage, the surface roughness is measured and analyzed.

In this study, an industrial camera was used to take pictures of the workpiece surface under a specific light environment to measure the surface roughness conditions, as shown in Figure 11. In order to compare the measurement accuracy of different learning algorithms, ANN and CNN are added to train the surface roughness measurement model in this paper. The GAN-BPNN, ANN, CNN were used to judge the smoothness of the polished processing surface in Figure 11, as shown in Table 5. The proposed GAN-BPNN has a margin of error of 10%. However, the prediction error of ANN and CNN were more than 10%.

Table 5. Identify workpiece surface roughness with GAN-BPNN, ANN, CNN.

	Real Ra (μm)	GAN-BPNN (μm)	Error (%)	ANN (μm)	Error (%)	CNN (μm)	Error (%)
Before grinding	1.75	1.65	6.1	1.61	8.0	1.9	8.6
Rough grinding stage	1.29	1.36	5.1	1.49	15.5	1.45	12.4
Semi-finishing stage	0.93	0.87	6.9	0.84	9.7	1.03	10.8
Finishing stage	0.59	0.64	7.8	0.5	15.2	0.66	11.9

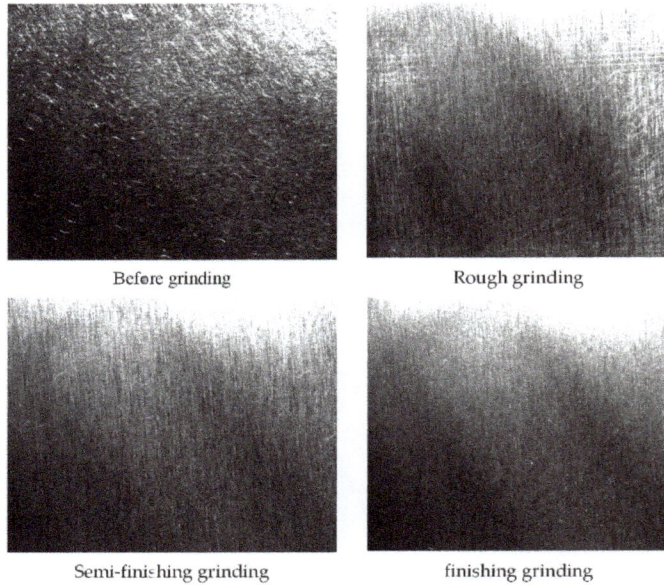

Before grinding

Rough grinding

Semi-finishing grinding

finishing grinding

Figure 11. Roughness measurement based on machine vision.

In order to measure blade surface roughness more accurately, this paper designs a sampling point distribution method based on constant chord height. The method realizes a dense distribution of sampling points in the region with a large curvature variation and a sparse distribution in the region with a small curvature variation, as shown in Figure 12.

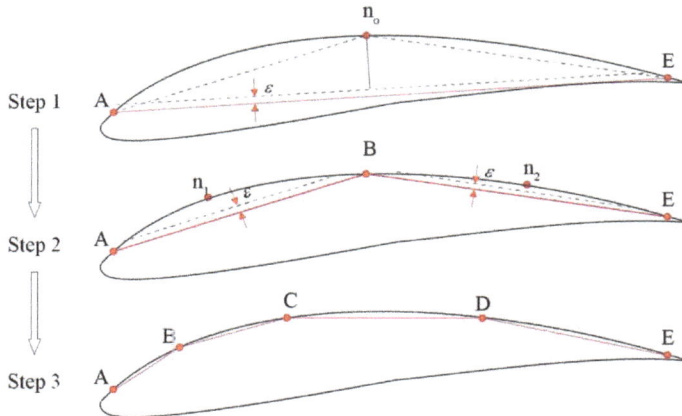

Figure 12. Sampling point distribution method.

(1) Set a chord height ε.

(2) The baseline AE was obtained by connecting the end points of the curve.

(3) Determine whether there exists a point on the curve $\overset{\frown}{AE}$ whose distance to the datum line is greater than ε. If there is a point on $\overset{\frown}{AE}$, whose distance from the datum line AE is the farthest point is n_0, then delete the original datum line AE.

(4) Set n_0 to B and connect the AB and BE as the new datum line.

(5) Determine whether there exists a point on the curve $\overset{\frown}{AB}$ whose distance to the datum line is greater than ε, and if there is a point on $\overset{\frown}{AB}$ whose distance from the datum line AB is the farthest point is n_1. Similarly, point n_2 on curve $\overset{\frown}{AB}$ is obtained. Rewrite A, n_1, B, n_2, E as A, B, C, D, E.

Repeat the above process until all points on the curve to the datum line position are less than the threshold ε. In this paper, five sampling points (A, ... , E) marked in the figure were taken as examples to compare the roughness changes of blades before and after grinding, as shown in Figure 13. The robot planed the motion trajectory according to the position and curvature of the sampling points (20 mm × 20 mm area centered on this point), collecting images of the five sampling points in turn at the best angle, as shown in Figure 14.

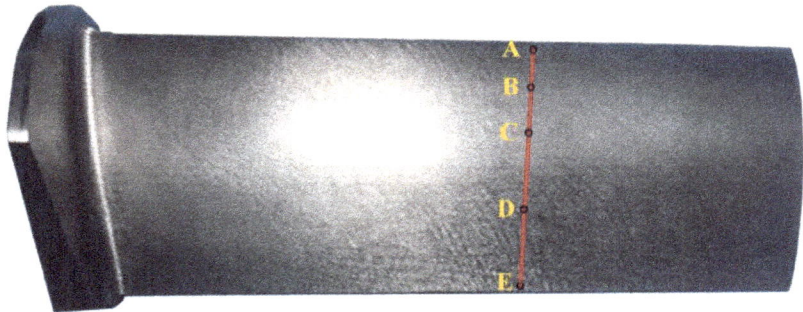

Figure 13. Distribution of sampling points for blade surface roughness.

The image data of the sampling points before and after polishing can obtain the corresponding predicted value by the roughness discrimination model of the GAN + BP neural network, as shown in Table 6. It can be seen that the relative error between the predicted roughness and the real roughness (measured by SJ-210) has a margin of error of 10%, which meets the needs of industrial detection.

Table 6. The relative error between the roughness value predicted by GAN + BP and the real roughness.

Compressor Blade		Roughness Ra (µm)				
Before	Real Ra	1.64	1.59	1.65	1.68	1.62
	GAN Ra	1.75	1.69	1.79	1.74	1.72
	Error (%)	6.7	6.2	8.5	3.5	6.2
After	Real Ra	0.67	0.64	0.65	0.68	0.61
	GAN Ra	0.72	0.67	0.71	0.74	0.66
	Error (%)	7.4	4.6	9.2	8.8	8.2

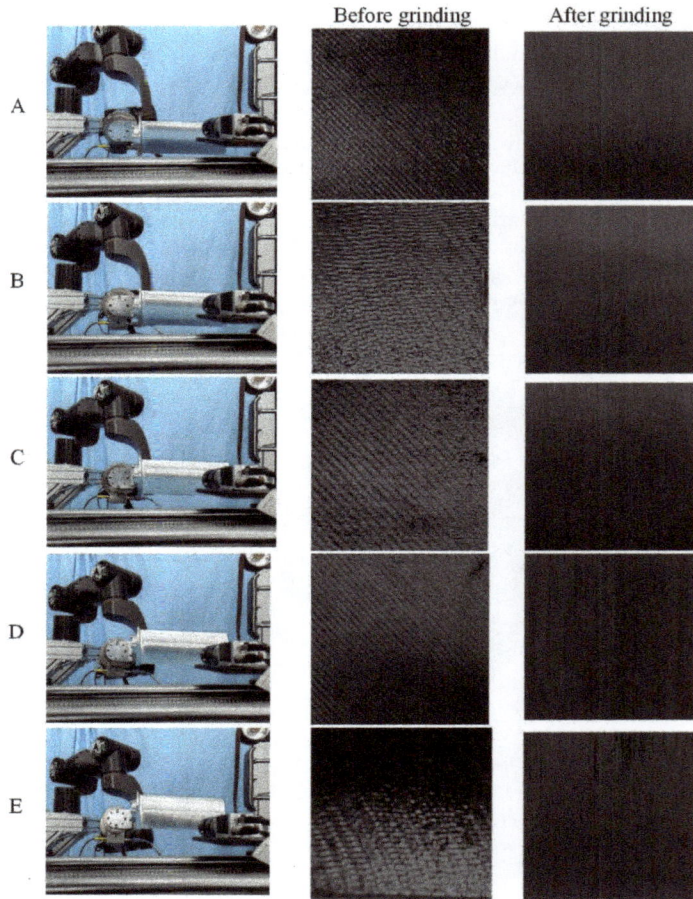

Figure 14. The roughness images at the five sampling points. (**A**) The roughness image at point A; (**B**) The roughness image at point B; (**C**) The roughness image at point C; (**D**) The roughness image at point D; (**E**) The roughness image at point E.

6. Conclusions

In this paper, we propose a roughness evaluation method combining generative an adversarial network (GAN) and a BP neural network, which avoids the influence of surface image differences with different curvatures on the accuracy of roughness measurements. The method automatically learns features in the image by generators playing with discriminators, eliminating the independent feature extraction step. This does not merely shorten the prediction time but reduces the complexity of model training as well. Experiments show that the proposed method can measure a free-form surface with a minimum roughness of 0.2 μm, and measurement results have a margin of error of 10%. Since the proposed method does not require the operator to have the field knowledge of feature recognition, the method is easier to apply in a factory. However, the proposed roughness evaluation method using GAN takes a long time for model training due to its convolution process and depth structure characteristics. To overcome this problem, the adaptive model can be considered in future research to automatically adjust the hyperparameters.

Author Contributions: G.Z. was responsible for methodology and manuscript writing. C.L. was responsible for software. K.M. was responsible for experiments. F.N. was responsible for reviewing papers and providing funding support. H.L. provided resources. All authors have read and agreed to the published version of the manuscript.

Funding: This work is supported by Self-Planned Task (NO.SKLRS202204B) of State Key Laboratory of Robotics and System (HIT).

Institutional Review Board Statement: Not applicable.

Data Availability Statement: Not applicable.

Conflicts of Interest: The authors declare no conflict of interest.

References

1. Wei, J.; Wang, L. Industrial Robotic Machining: A Review. *Int. J. Adv. Manuf. Technol.* **2019**, *103*, 1239–1255.
2. Zhu, D.; Feng, X.; Xu, X.; Yang, Z.; Li, W.; Yan, S.; Ding, H. Robotic grinding of complex components: A step towards efficient and intelligent machining–challenges, solutions, and applications. *Robot. Comput. Integr. Manuf.* **2020**, *65*, 101908. [CrossRef]
3. Liu, Z.; Jiang, L.; Yang, B. Task-Oriented Real-Time Optimization Method of Dynamic Force Distribution for Multi-Fingered Grasping. *Int. J. Hum. Robot.* **2022**, 2250013. [CrossRef]
4. Xu, X.; Zhu, D.; Zhang, H.; Yan, S.; Ding, H. TCP-based calibration in robot-assisted belt grinding of aero-engine blades using scanner measurements. *Int. J. Adv. Manuf. Technol.* **2017**, *90*, 635–647. [CrossRef]
5. Liu, J.; Lu, E.; Yi, H.; Wang, M.; Ao, P. A new surface roughness measurement method based on a color distribution statistical matrix. *Measurement* **2017**, *103*, 165–178. [CrossRef]
6. Launhardt, M.; Wörz, A.; Loderer, A.; Laumer, T.; Drummer, D.; Hausotte, T.; Schmidt, M. Detecting surface roughness on SLS parts with various measuring techniques. *Polym. Test.* **2016**, *53*, 217–226. [CrossRef]
7. Hiziroglu, S. Surface roughness analysis of wood composites: A stylus method. *For. Prod. J.* **1996**, *46*, 67.
8. Poon, C.Y.; Bhushan, B. Comparison of surface roughness measurements by stylus profiler, AFM and non-contact optical profiler. *Wear* **1995**, *190*, 76–88. [CrossRef]
9. Haitjema, H. Uncertainty analysis of roughness standard calibration using stylus instruments. *Precis. Eng.* **1998**, *22*, 110–119. [CrossRef]
10. Kiran, M.B.; Ramamoorthy, B.; Radhakrishnan, V. Evaluation of surface roughness by vision system. *Int. J. Mach. Tools Manuf.* **1998**, *38*, 685–690. [CrossRef]
11. Patel, D.R.; Kiran, M.B. Vision based prediction of surface roughness for end milling. *Mater. Today: Proc.* **2021**, *44*, 792–796. [CrossRef]
12. John, J.G.; Arunachalam, N. Illumination compensated images for surface roughness evaluation using machine vision in grinding process. *Procedia Manuf.* **2019**, *34*, 969–977. [CrossRef]
13. Joshi, K.; Patil, B. Prediction of surface roughness by machine vision using principal components based regression analysis. *Procedia Comput. Sci.* **2020**, *167*, 382–391. [CrossRef]
14. Gandla, P.K.; Inturi, V.; Kurra, S.; Radhika, S. Evaluation of surface roughness in incremental forming using image processing based methods. *Measurement* **2020**, *164*, 108055. [CrossRef]
15. Sanjeevi, R.; Nagaraja, R.; Krishnan, B.R. Vision-based surface roughness accuracy prediction in the CNC milling process (Al6061) using ANN. *Mater. Sci.* **2020**, *2214*, 7853. [CrossRef]
16. Rifai, A.P.; Aoyama, H.; Tho, N.H.; Dawal, S.Z.M.; Masruroh, N.A. Evaluation of turned and milled surfaces roughness using convolutional neural network. *Measurement* **2020**, *161*, 107860. [CrossRef]
17. Jayabarathi, S.B.; Ratnam, M.M. Comparison of Correlation between 3D Surface Roughness and Laser Speckle Pattern for Experimental Setup Using He-Ne as Laser Source and Laser Pointer as Laser Source. *Sensors* **2022**, *22*, 6003. [CrossRef]
18. Patil, S.H.; Kulkarni, R. Surface roughness measurement based on singular value decomposition of objective speckle pattern. *Opt. Lasers Eng.* **2022**, *150*, 106847. [CrossRef]
19. Tootooni, M.S.; Liu, C.; Roberson, D.; Donovan, R.; Rao, P.K.; Kong, Z.J.; Bukkapatnam, S.T. Online non-contact surface finish measurement in machining using graph theory-based image analysis. *J. Manuf. Syst.* **2016**, *41*, 266–276. [CrossRef]
20. Liu, W.; Tu, X.; Jia, Z.; Wang, W.; Ma, X.; Bi, X. An improved surface roughness measurement method for micro-heterogeneous texture in deep hole based on gray-level co-occurrence matrix and support vector machine. *Int. J. Adv. Manuf. Technol.* **2013**, *69*, 583–593. [CrossRef]
21. Huaian, Y.I.; Jian, L.I.U.; Enhui, L.U.; Peng, A.O. Measuring grinding surface roughness based on the sharpness evaluation of colour images. *Meas. Sci. Technol.* **2016**, *27*, 025404. [CrossRef]
22. Tsai, D.M.; Chen, J.J.; Chen, J.F. A vision system for surface roughness assessment using neural networks. *Int. J. Adv. Manuf. Technol.* **1998**, *14*, 412–422. [CrossRef]
23. Samtaş, G. Measurement and evaluation of surface roughness based on optic system using image processing and artificial neural network. *Int. J. Adv. Manuf. Technol.* **2014**, *73*, 353–364. [CrossRef]

24. Tang, K.; Chen, F.; Chang, F. Roughness Classification of End Milling Based on Machine Vision. In Proceedings of the 3rd World Conference on Mechanical Engineering and Intelligent Manufacturing (WCMEIM), Shanghai, China, 4–16 December 2020; p. 292.
25. Saeedi, J.; Dotta, M.; Galli, A.; Nasciuti, A.; Maradia, U.; Boccadoro, M.; Giusti, A. Measurement and inspection of electrical discharge machined steel surfaces using deep neural networks. *Mach. Vis. Appl.* **2021**, *32*, 1–15. [CrossRef]
26. Rifai, A.P.; Fukuda, R.; Aoyama, H. Surface roughness estimation and chatter vibration identification using vision-based deep learning. *J. Jpn. Soc. Precis. Eng.* **2019**, *85*, 658–666. [CrossRef]
27. Goodfellow, I.; Pouget-Abadie, J.; Mirza, M.; Xu, B.; Warde-Farley, D.; Ozair, S.; Courville, A.; Bengio, Y. *Generative Adversarial Nets, Neural Information Processing Systems*; MIT Press: Cambridge, MA, USA, 2014.
28. Creswell, A.; White, T.; Dumoulin, V.; Arulkumaran, K.; Sengupta, B.; Bharath, A.A. Generative adversarial networks: An overview. *IEEE Signal Process. Mag.* **2018**, *35*, 53–65. [CrossRef]
29. Hecht-Nielsen, R. *Kolmogorov's Mapping Neural Network Existence Theorem*; IEEE Press: Piscataway, NJ, USA, 1987.

![machines logo] machines

MDPI

Article

Managing Delays for Realtime Error Correction and Compensation of an Industrial Robot in an Open Network

Seemal Asif *[iD] and Phil Webb

School of Aerospace, Transport and Manufacturing, Cranfield University, Cranfield, Bedford MK43 0AL, UK;
p.f.webb@cranfield.ac.uk
* Correspondence: s.asif@cranfield.ac.uk

Abstract: The calibration of articulated arms presents a substantial challenge within the manufacturing domain, necessitating sophisticated calibration systems often reliant on the integration of costly metrology equipment for ensuring high precision. However, the logistical complexities and financial burden associated with deploying these devices across diverse systems hinder their widespread adoption. In response, Industry 4.0 emerges as a transformative paradigm by enabling the integration of manufacturing devices into networked environments, thereby providing access through cloud-based infrastructure. Nonetheless, this transition introduces a significant concern in the form of network-induced delays, which can significantly impact realtime calibration procedures. To address this pivotal challenge, the present study introduces an innovative framework that adeptly manages and mitigates network-induced delays. This framework leverages two key components: controller and optimiser, specifically the MPC (Model Predictive Controller) in conjunction with the Extended Kalman Filter (EKF), and a Predictor, characterised as the Dead Reckoning Model (DRM). Collectively, these methodologies are strategically integrated to address and ameliorate the temporal delays experienced during the calibration process. Significantly expanding upon antecedent investigations, the study transcends prior boundaries by implementing an advanced realtime error correction system across networked environments, with particular emphasis on the intricate management of delays originating from network traffic dynamics. The fundamental aim of this research extension is twofold: firstly, it aims to enhance realtime system performance on open networks, while concurrently achieving an impressive level of error correction precision at 0.02 mm. The employment of the proposed methodologies is anticipated to effectively surmount the intricacies and challenges associated with network-induced delays. Subsequently, this endeavour serves to catalyse accurate and efficient calibration procedures in the context of realtime manufacturing scenarios. This research significantly advances the landscape of error correction systems and lays a robust groundwork for the optimised utilisation of networked manufacturing devices within the dynamic realm of Industry 4.0 applications.

Keywords: robotics calibration; network delays management; resource sharing; temporal delays; realtime calibration; realtime application over the network

![check for updates]

Citation: Asif, S.; Webb, P. Managing Delays for Realtime Error Correction and Compensation of an Industrial Robot in an Open Network. *Machines* **2023**, *11*, 863. https://doi.org/10.3390/machines11090863

Academic Editor: Dan Zhang

Received: 18 June 2023
Revised: 11 August 2023
Accepted: 16 August 2023
Published: 28 August 2023

1. Introduction

Large-scale, complex, and low-volume manufacturing systems, particularly in the aerospace industry, rely heavily on robotics for automation. The precision required for tasks like position accuracy, module assembly, inspection, and fastening poses unique challenges due to robot kinematics and environmental factors. Existing packages provid-ed by robot manufacturers for error correction and compensation suffer from cost and lack of realtime capabilities, resulting in static correction and dedicated resources. The research addressed these challenges by developing a dynamic and realtime error correction system, achieving error correction in the range of 0.02 mm. [1] The study extends the existing re-search by implementing the proposed system over the network and addressing the chal-lenge of managing delays caused by network traffic.

Considering the implications of connecting the tracker or measurement device to a singular network switch, its utility is limited to a specific robotic cell as a result. In such a scenario, the tracker's functionality would be exclusive to that particular cell, rendering it inaccessible to other cells that might require its services. The tracker's mobility, which allows it to orient itself towards different cells as needed, would be compromised. The cost-effectiveness also comes into play when optimizing resource utilisation. Employing a single tracker for multiple cells, numbering around 10 or more, and coordinating this shared resource via a resource management system, emerges as a strategic approach. This allows the tracker's capabilities to be harnessed across multiple cells while minimizing redundant costs. The paper could further emphasise these considerations to underscore the rationale behind the chosen network configuration.

The issue of delay represents a prominent challenge within Network Control Systems (NCS) and other systems reliant on network infrastructure. This delay stems from two primary factors: network traffic and the limited bandwidth of the communication channel. The use of Kalman filter [2,3] and MPC for NCS can help to eliminate the noise and both types of delays but still there is a need to implement a strategy to handle network delays to obtain a realtime effect in the system [4,5]. The system in question is the distributed application.

There are lots of techniques available to handle network delays for distributed Interactive applications (DIAs) [6]. Some of the techniques can be benchmarked; for example, the DRM [7] is a popular technique in positioning systems. DRM is widely used in DIA for the predictive contract agreement mechanism in managing network latencies [8]. A study shows that the use of DRM can enhance the Network speed and optimise the performance [9].

The same DRM technique can be used for NCS to predict the information about the position of the robot arm using extrapolation and the smoothing function can help to smooth the extrapolated position and real position [9,10].

MPC has also been used to deal with variable data losses and time delays in a realtime environment [4], which proves itself the most suitable method to be induced in realtime error compensation over the network. To handle delays in the NCS, the observer algorithm was proposed [11]. It proposed the two feedback loops one for the control system the other for the observer. They further used two more controllers for the anticipated and non-anticipated data. The Kalman filter [2] was proven to be useful for the environment in which data losses are common, and data are received intermittently [12]. This approach is particularly useful for a chaotic (where multiple cells are running at the same time on the network) and realtime sensitive environment (where time is crucial).

To make the system extendable, configurable, manageable, and distributed, some techniques need to be used, for example, Service Oriented Architecture (SOA) on the application level [13]. SOA is becoming popular for developing distributed and manageable manufacturing systems [14,15].

The gap here is to combine the realtime error correction with the network so the resources can be shared amongst other operations, significantly reducing cost and increasing flexibility. This study will need to deal with network issues while using the robot and measurement device connected to the network. The study will be more gelled with Industry 4.0 [16], which will be automating the error correction procedure and also allowing the resources to be shared and available over the network. The data about the resources, their scheduling, and their programs will be stored in the databases [17] which can be used as a knowledge base in the future [18].

Another study addresses compliance modeling and error compensation for an industrial robot's application in ship hull welding [19]. The Cartesian stiffness matrix is obtained through the virtual-spring approach, a method that considers factors like actuation and structural stiffness, arm gravity, and external loads. While this approach enhances accuracy, it's important to note that it doesn't operate in real-time error correction on the network. The derived stiffness model offers the foundation for error compensation. This compensation method is demonstrated using an industrial robot executing a welding trajectory. The outcomes reveal that this compensation approach effectively enhances the robot's operational accuracy. It aligns the robot's actual trajectory, even when influenced by auxiliary loads, to closely match the intended trajectory.

In precision-demanding industrial robotic applications, a novel hybrid computational method is introduced for error compensation [20]. Combining Local POE calibration and Gaussian Process Regression, it reduces positioning errors by up to 37.2% compared to existing methods. While it proposes an innovative hybrid method for error compensation in precision industrial robotic tasks, it doesn't address network integration or real-time adjustments, which are crucial aspects in modern manufacturing environments.

The novelty of the present study lies in addressing the intricate calibration challenges faced by articulated arms within the manufacturing domain. Traditionally, calibrating these arms necessitates costly and physically attached metrology equipment, impeding a widespread adoption due to cost and practical limitations. Leveraging Industry 4.0 principles, this study introduces an innovative approach by incorporating manufacturing devices into a networked ecosystem, thereby enabling cloud accessibility. However, the subsequent issue of network delays emerges as a significant hurdle, particularly concerning realtime calibration procedures.

To surmount this predicament, this study introduces a pioneering solution. It integrates an optimiser, the EKF, in tandem with a Predictor, the DRM. These techniques collectively combat and resolve network-induced delays encountered during calibration processes. A significant aspect of novelty lies in the integration of the DRM into the realm of robotics, with a specific focus on estimating position and error. This innovative application of DRM directly addresses the challenge associated with managing delays linked to the reception of measurements from a tracking device. By incorporating DRM, the system ensures realtime compensation by predicting and adapting to potential errors, thus enhancing the accuracy and efficiency of the compensation process. Furthermore, building upon earlier research efforts [1], this study extends the boundaries by implementing a realtime error correction system across a network. Notably, it homes in on mitigating delays stemming from network congestion.

The choice to adopt a realtime approach stems from catering to sectors characterised by a demand for substantial variability, such as shipbuilding and aircraft manufacturing. In these industries, the imperative for precision, quality, and efficiency is paramount. The offline provision of information and training, while potentially time-consuming, could lead to inefficiencies in meeting the stringent requirements of these sectors. Considering the inherent variability in components and robot paths, opting for realtime solutions becomes a pragmatic choice. The realtime approach aligns well with the need to swiftly adapt to diverse scenarios and ensure the precision demanded by these industries, making it a viable solution within this context.

This paper employs a three-section structure to systematically present its content. Section 2 comprehensively elucidates the methodology behind Realtime Error Correction over the network, encompassing aspects such as error prediction and estimation, the management of network delays, system development, and resource allocation. In Section 3, a thorough exploration of results and discussions unfolds, offering insights into the conducted tests aimed at assessing the system's capabilities. Finally, Section 4 delves into a meticulous analysis of these findings, fostering a detailed discourse that culminates in the paper's conclusive remarks.

2. Realtime Error Correction over the Network—Problem Realisation

The field of static Error Correction encompasses various technologies and systems, as highlighted in the Introduction. However, an unexplored area of research lies in addressing realtime error correction over regular or open networks. While the preceding section discussed systems that have been developed for realtime error correction, they lack the suitability to operate effectively over standard networks due to their stringent feedback requirements, which cannot be met within the required realtime constraints. The problem at hand encompasses several dimensions, including Realtime Error Correction on the Network, Handling Network Traffic and delays and making the resources sharable.

Figure 1 illustrates the resolution of the issues pertaining to static calibration and resource wastage through the implementation of a realtime networked control system, dynamic calibration techniques, and flexible resource management. The diagram visually depicts the problem area encapsulated within a red rectangle, while the corresponding solution is represented by the green region. The green arrows superimposed on the black arrows symbolise the specific approaches and algorithms employed to address the problem. The final rectangle, distinguished by a white background and a green outline, signifies the complete resolution of the problem.

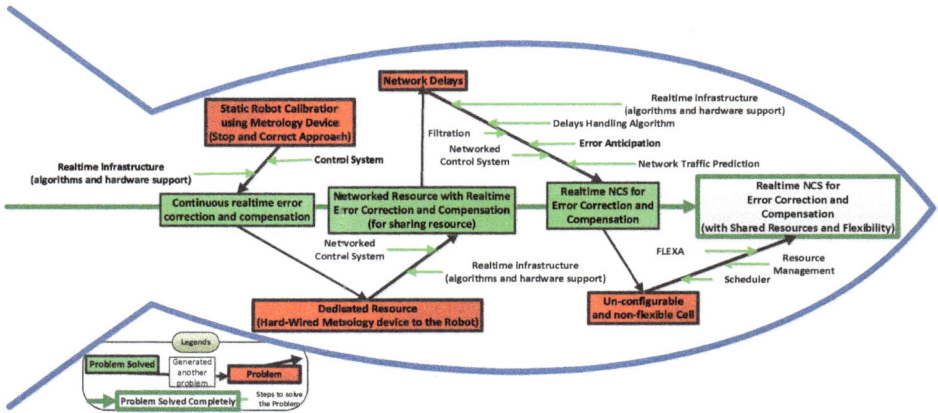

Figure 1. Proposed Solution to the Problem.

The problem depicted in Figure 1 originates from the conventional static approach of Robot Calibration utilizing a metrology device. In this static approach, the error correction occurs through the Stop and Correct Approach, where the metrology device measures the position, and the system performs calculations based on these measurements. However, during this calculation period, both the system and the robot remain idle, leading to suboptimal resource utilisation and increased overall process time.

To address this issue, a feedback control system and realtime infrastructure can be employed. The realtime infrastructure encompasses hardware that supports realtime processing and the implementation of appropriate algorithms. This transforms the system into a continuous realtime error correction and compensation framework. However, this approach introduces the challenge of dedicated resources, particularly the expensive metrology equipment, which may be required by other processes on the factory shop floor.

This challenge can be overcome by incorporating a network and connecting the robot, metrology device, and other equipment to a Networked Control System, integrated with realtime infrastructure. This solution results in a Networked Resource with Realtime Error Correction and Compensation, allowing for resource sharing within this setup. Nonetheless, this approach gives rise to the problem of network delays, as the resources are interconnected over a network shared with other systems that also transmit communication data.

To mitigate network delays, various techniques can be applied, such as Filtration techniques to eliminate noise, Delay Handling Algorithms, Error Prediction (Anticipation), and Network Traffic Prediction utilising realtime infrastructure. The resulting system would be a Realtime Networked Control System for Error Correction and Compensation. However, such a system may suffer from the drawback of un-configurable and non-flexible cells.

To address the management and scheduling of resources when multiple systems and cells require the metrology device simultaneously, the Flexa Control Approach [17] can be applied. This approach integrates the Flexa Cell Controller into the system, ensuring effective resource management and scheduling.

One of the key objectives of the study is to facilitate resource sharing. In the context of automation, resources can be effectively shared through the utilisation of a network. DIA enable users to connect to the network using the same application and seamlessly share the application's state in realtime. This synchronous sharing allows for collaborative activities such as shared whiteboarding, collaborative editing of files, joint editing of files, shared meeting rooms, and multiplayer games. Through DIA, users can interact and collaborate over a computer network, engaging in interactive work on a single application.

A fundamental characteristic of all DIAs is the shared user space and concurrent manipulation of the same data. However, the effective synchronisation and management of resources pose significant challenges for distributed applications. The issue of synchronisation becomes more complex in networked environments due to the presence of other concurrent data communications on the network. In larger networks, the system for sharing the status of each user can become time-consuming, potentially causing delays throughout the entire network. To mitigate this issue, an estimation and filtering approach can be employed to minimise the frequency of updates transmitted over the network. The DRM serves as a foundational protocol within the IEEE Standard for DIAs, particularly in the context of estimation techniques [6].

The Dead Reckoning algorithm is widely utilised in distributed network games, where the delivery of realtime experiences to users is crucial, and factors such as speed and timing play significant roles. This algorithm is derived from deduced reckoning principles and serves as an effective approach in such gaming environments.

The Dead Reckoning Algorithm utilises previous packet information, performs extrapolation to estimate future positions, and employs pre-reckoning to anticipate positions before official reckoning, reducing network packet frequency.

By employing the Dead Reckoning algorithm, fewer packets need to be transmitted over the network, leading to reduced network traffic while simultaneously improving latency. This approach enables efficient utilisation of network resources and enhances the realtime experience for users in distributed network games.

The term "dead reckoning" has a historical origin and was documented in the Oxford Dictionary in 1613. This concept involves the estimation or prediction of a future position based on knowledge of the initial starting position. When considering the concept in terms of speed, it becomes straightforward to illustrate. For instance, let us consider a hypothetical scenario where a ship sets sail at 0900 with a constant speed of 7 mph. The question then arises: Where will the ship be located along its designated course at 1100?

By applying the distance equation, the distance covered by the boat after a duration of 2 h can be calculated as 14 units. This calculation provides an estimation or reckoning of the distance travelled by the object at the specified time.

The concept of DIA has been prevalent for many years, and one of its notable examples is network games, which enable realtime gameplay over a network. DIA aims to achieve realtime interaction while maintaining robustness, reliability, security, scalability, and consistency, among other qualities. However, latency poses a significant challenge for DIA applications. Although increased bandwidth and processor speed have a positive impact on latency control, latencies still exist due to network congestion. In realtime DIAs, these latencies are deemed unacceptable. To address this, the DRM has emerged as a popular technique in positioning systems, aiding in mitigating the impact of latency [3].

The DRM has gained widespread adoption as a predictive contract agreement mechanism within DIA to effectively manage network delays [4]. A study conducted on optimising network performance utilizing DRM provides evidence that this approach can enhance overall network performance and significantly reduce latencies. The findings of the study highlight the effectiveness of DRM in addressing the challenges posed by network delays and its potential to improve the performance of DIA systems [5].

In the current problem scenario, the application of the DRM technique, in conjunction with a smoothing function, can be utilised to extrapolate and predict data and positions. DRM enables the prediction of positions between data updates. However, a potential issue arises when the predicted position does not align with the actual position obtained from the arrived data. To address this problem, a smoothing function, as suggested by [6], is employed. The smoothing function aims to reduce discontinuities by considering time compensation to compensate for packet latency. This approach helps to ensure a more seamless and accurate representation of positions within the system.

2.1. Error Prediction and Estimations for Compensating Network Delays

The fundamental principle of the system is to operate on an open network, allowing for the sharing of resources. In this context, the critical resource to be shared is the Laser Tracker, which carries a significant cost of at least £150 k. However, this requirement of the system to operate on a network introduces the challenge of managing variable network traffic.

Direct connectivity between a resource, such as a Laser Tracker, and a robot in isolation enables fast and efficient communication. However, this approach limits the resource's usage exclusively to that particular system. The objective is to share the resource across the network, allowing other cells or systems to benefit from the error correction capabilities. Introducing the system to the network introduces the challenge of handling delays due to the presence of normal network traffic. While this may not pose a problem for static error correction, the goal is to perform a dynamic error correction [21]. In this dynamic scenario, the error correction system cannot afford to wait for measurements from the laser tracker; instead, it must ensure timely delivery to meet the realtime and dynamic demands of the system. This situation is visually depicted in Figure 2.

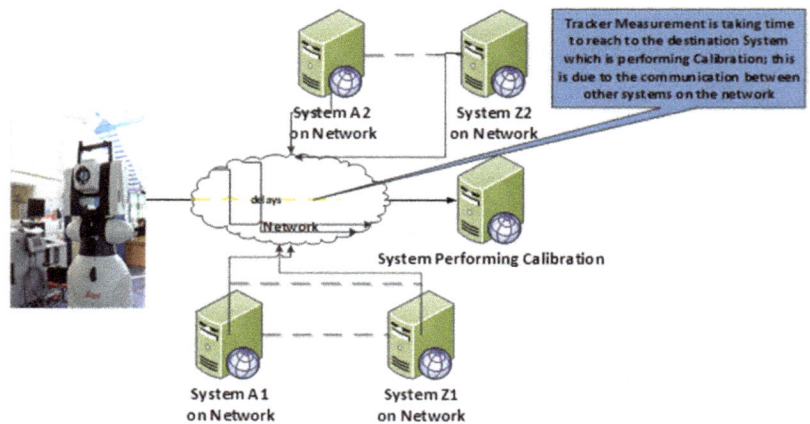

Figure 2. Network Delays for Error Correction System.

2.2. Dead Reckoning—Modelling (DRM)

In order to achieve realtime correction in the calibration system, it is crucial to have regular and uninterrupted feedback from the measurement system. The Dead Reckoning approach, described in previous section, provides a solution by estimating the current error using past observations. By extrapolating or predicting data values between two updates, Dead Reckoning enables the system to accommodate more connected stations and enhance the reliability of receiving measurements from the tracker. This approach improves the overall performance and robustness of the calibration system.

The calibration system is represented as a player within the overall system, operating independently from the tracker system. The term "players" is borrowed from online gaming, where Dead Reckoning is commonly used to address network delays [4,16]. Player 1 represents the tracker system, which sends updates at unpredictable intervals over the network. Player 2 represents the calibration system, which receives measurement information from the laser tracker system. The system model, depicted in Figure 3, illustrates the independent movement of the players and their periodic transmission of robot position updates.

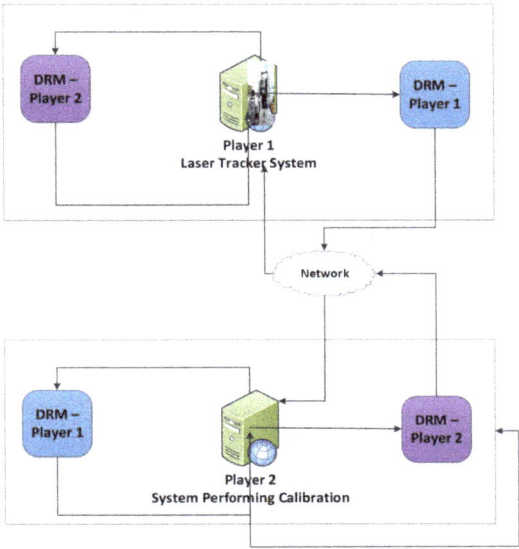

Figure 3. Dead Reckoning Model for NRTEC.

In the system, Player 1 (Laser Tracker System) transmits the measured position of the robot, while Player 2 reports the number of the last received position. DRM is employed by Player 2 (system performing the calibration) to estimate the subsequent measured position provided by Player 1. Additionally, Player 1 utilises Dead Reckoning to track the updates it has sent and confirms that they have been properly incorporated into Player 2's system. The network model accounts for latency, acknowledging that it takes time for updated measurements to be delivered to the other system.

2.3. System Development

System development serves as the final stage of the research, encompassing various aspects such as software modules and kinematics, which are essential for the development of the system [17]. In order to undertake this development, the following equipment was utilised: the Comau NM45 C4G robot, Leica Tracker AT901-MR along with a 3D reflector (SMR) and T-MAC for 6D measurements, the NI CRIO-9024 Realtime Controller, C# for interface development, C++ for realtime error correction and compensation module development, and LabVIEW for running the system on CRIO. These components were instrumental in facilitating the implementation and functionality of the developed system.

The interconnections between the equipment utilised in the research are depicted in Figure 4. The Comau NM45 robot is linked to the Comau C4G Controller, which serves as its controlling unit. The C4G Controller is connected to the Network Switch, with its management being handled by the NRTEC Software 600 v1.20.

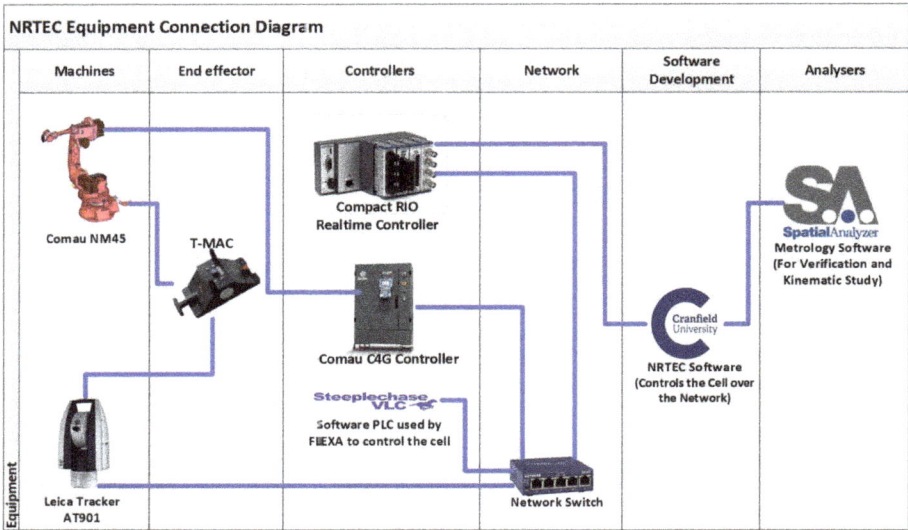

Figure 4. Equipment Connectivity Diagram.

The TMAC, an integral part of the robot's End-Effector, is connected to the Leica Tracker, which tracks its position and performs measurements. The Leica Tracker is also linked to the Network Switch, enabling its management through the NRTEC Software 600 v1.20.

To enable realtime management and execution within the cell, the NRTEC Software 600 v1.20 operates on the Compact RIO realtime controller. The Compact RIO is connected to the Network Switch, facilitating access to the equipment and management by the FLEXA Cell Coordinator.

Steeplechase VLC, a software PLC employed by FLEXA, is responsible for controlling and managing the cells and their resources. It is installed on a PC and connected to the Network Switch.

For analysing the kinematics of the Comau NM45 robot and verifying the results, Spatial Analyzer software was utilised in the research.

2.4. Resource Management System—Flexa Cell Coordinator

The Flexa Cell Coordinator (FCC) is an automated system that efficiently manages the allocation and removal of resources within a Flexa Cell. It coordinates the execution of received programs on the required resources in a non-conflicting manner. Multiple cell coordinators can be run simultaneously by utilizing software Programmable Logic Controller (Soft PLC), eliminating the need for the hardwired binding of resources. The FCC receives programs in the form of recipes through web services, schedules them based on resource availability, and activates sub coordinators equipped with SoftPLC controllers to manage program execution and resource control. Data exchange between the FCC and resources is facilitated, with the FCC acting as the application manager to handle data flow. This bidirectional communication enables the FCC to accept recipe data and send back processed data.

2.5. Integration of NRTEC with FCC

The integration involves several components, including the startup of the FCC and the receipt of recipes, the generation and transmission of status reports, the scheduling and execution of scheduled recipes, and the receipt of recipes from the FCC Database (FDB).

The data flow within the FCC Architecture is depicted in Figure 5, which illustrates the flow of information and interactions between the different components.

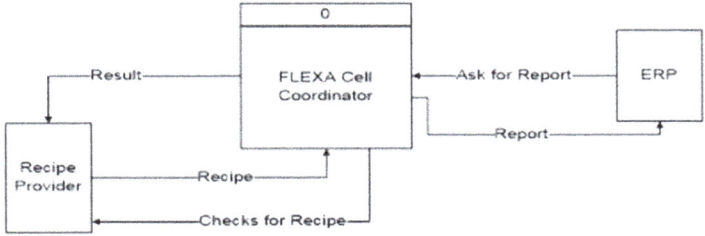

Figure 5. Data Flow—Context Level.

Figure 5 provides a context-level or level 0 diagram of the data flow within the FCC system. It illustrates how the FCC interacts with external entities and manages the flow of data between them. On the other hand, Figure 6 presents a detailed level 1 Data Flow Diagram, showcasing the internal components of the FCC and their respective data flows. Notably, the programs executed on the NRTEC robot are overseen and managed by the FCC Sub-Coordinator within the system [12].

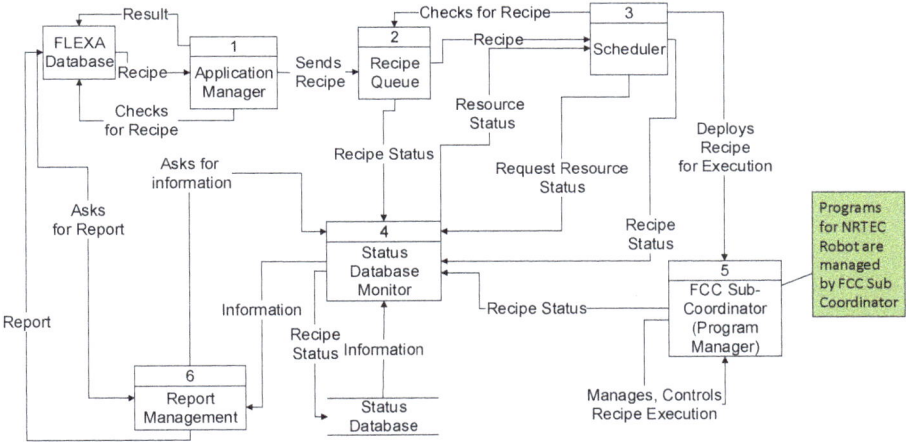

Figure 6. FCC Data Flow—Level 1.

2.6. Dead Reckoning Model

The DRM has been employed in this research to forecast the robot's error when measurement data are unavailable. DRM is a widely utilised technique in various applications such as positioning systems, online gaming, and navigation. It estimates the robot's next position by considering the previous position data. However, it is important to note that DRM is susceptible to errors since it relies on the accuracy of the measurement data obtained from sensors.

Figure 7 illustrates the design and implementation of the DRM, which incorporates a mechanism to trigger a network update if the error exceeds a predefined threshold. This approach ensures that the system responds to significant errors by requesting an update from the network, allowing for more accurate position estimation and error correction.

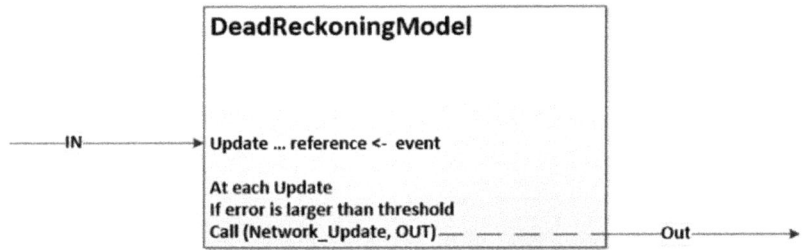

Figure 7. Dead Reckoning Model Object Error Implementation.

The changes in position are determined through the application of dead reckoning, as outlined by the equations provided below. This methodology involves estimating alterations in position based on the robot's prior position and movement data. The equations encapsulate the relationships between the robot's orientation and the alterations in its x, y, and z coordinates, facilitating the prediction of its updated position in response to incremental movements.

$$\Delta x = distance\_per\_step \times cos(robot\_orientation.Rz)$$

$$\Delta y = distance\_per\_step \times sin(robot\_orientation.Rz)$$

$$\Delta z = distance\_per\_step \times sin(robot\_orientation.Rx) \times cos(robot\_orientation.Rz)$$

The predicted position error is initialised randomly and computed through the utilisation of previous data and acquired error insights. Subsequently, the positions are updated by employing the predicted position error, as demonstrated by the equations provided below. This process involves integrating the anticipated error into the calculations, thereby refining the accuracy of the position updates.

$$\Delta x_{with\_error} = \left(1 + predicted_{position_{error}}\right) \times \Delta x$$

$$\Delta y_{with\_error} = \left(1 + predicted_{position_{error}}\right) \times \Delta y$$

$$\Delta z_{with\_error} = \left(1 + predicted_{position_{error}}\right) \times \Delta z$$

Subsequently, the robot's position requires adjustment, accounting for the introduced error. Likewise, the alteration in orientation can be computed employing dead reckoning principles, delineated by the equations presented below:

$$\Delta Rx = angle\_per\_step \times cos(robot\_orientation.Ry) \times cos(robot\_orientation.Rz)$$

$$\Delta Ry = angle\_per\_step \times cos(robot\_orientation.Rz) \times sin(robot\_orientation.Rx)$$

$$\Delta Rz = angle\_per\_step$$

Similar to the positional error, the rotational error is initialised randomly, drawing from previous data and acquired error insights. Subsequently, the rotations are refined through the incorporation of the predicted position error, as depicted by the equations provided below:

$$\Delta Rx_{with\_error} = (1 + predicted\_orientation\_error\_Rx) \times \Delta Rx$$

$$\Delta Ry_{with\_error} = (1 + predicted\_orientation\_error\_Ry) \times \Delta Ry$$

$$\Delta Rz_{with\_error} = (1 + predicted\_orientation\_error\_Rz) \times \Delta Rz$$

Figure 8 presents flowcharts that calculates the disparity between the predicted measurement and the actual measurement by using the set of above equations. The details of the pseudocode are presented in Appendix A. This enables a quantitative evaluation of the variance between the anticipated and observed values. By comparing these values, the system can assess the accuracy and precision of the prediction generated by the DRM. This evaluation is crucial for validating the effectiveness of the error correction and compensation processes within the system.

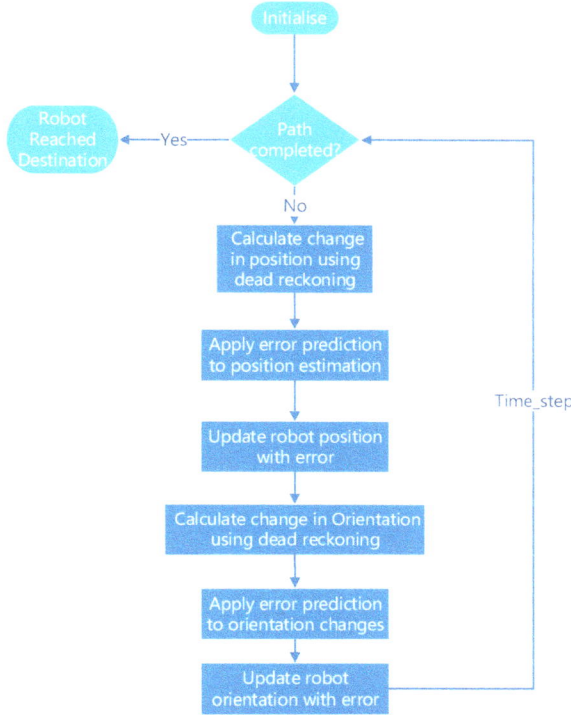

Figure 8. Flowchart for DRM—Estimation of Error in Prediction.

Figure 9 depicts the variation between the projected error and the actual error over the system's progression. The graph illustrates the performance improvement of the system as it advances in time or through successive iterations. As the system evolves, the variance between the anticipated and real errors diminishes, indicating the enhanced accuracy and effectiveness of the error prediction and correction mechanisms. This graph serves as a visual representation of the system's ability to refine its error estimation and compensation capabilities over time.

In the presented graph Figure 9, the red line represents the actual error, while the blue line represents the predicted error. Initially, the real error is observed to be around 0 and −0.1. As the system progresses through approximately 15 cycles, the predicted error gradually converges with the real error. This convergence indicates that the system successfully achieves a close alignment between the predicted error and the actual error.

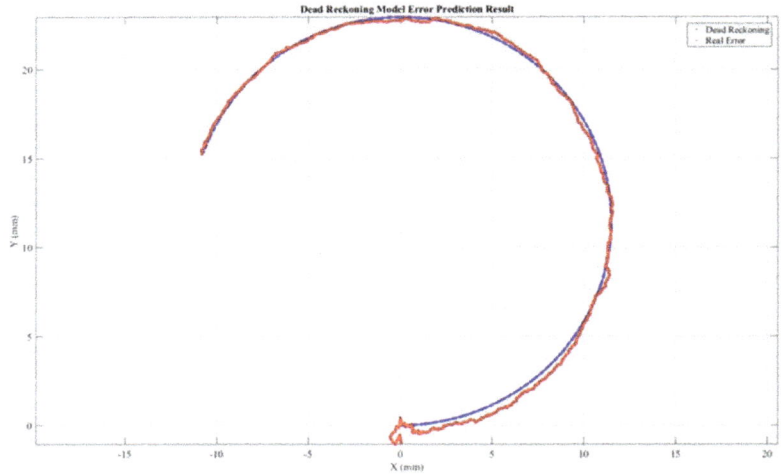

Figure 9. Dead Reckoning Model Error Prediction Result.

2.7. EKF and MPC for the Identification of Errors

To mitigate the impact of noise and for the identification of errors, the system incorporates the use of an MPC in conjunction with an EKF. These control and filtering techniques work together to minimise the influence of noise on the error estimation and correction process. By employing the MPC and EKF, the system enhances its ability to accurately predict and compensate for errors, resulting in an improved overall performance and noise reduction.

Designing an EKF for a serial manipulator (NM45) involves defining the state vector, measurement vector, process models, noise parameters, and the overall estimation process. The state vector 'x' represents the robot's pose in terms of position and orientation. For NM45, the state vector includes:

$$x = [x, \ y, \ z, \ Rx, \ Ry, \ Rz]'$$

The measurement vector 'z' includes the tracker's measurements of the robot's pose. The measurement vector is similar to the state vector:

$$z = [x_{measured}, \ y_{measured}, \ z_{measured}, \ Rx_{measured}, \ Ry_{measured}, \ Rz_{measured}]'$$

The state transition function, '$f(x, u)$', or the process model, illustrates how the robot's condition changes over time with the influence of control inputs u. These inputs are the joint velocities obtained through Jacobians that steer the robot's movements. The specific behaviour of this function is tied to the Comau NM45 robot's forward kinematics [3] and the merging of joint motions to modify its position. The Jacobian matrices 'F' and 'H' represent the partial derivatives of the state transition and measurement functions, respectively. These matrices are crucial for predicting how changes in the state affect the predicted state and the tracker measurements.

The covariance matrices 'Q' and 'R' represent process noise and measurement noise, respectively. These matrices capture the uncertainty and noise associated with the system's dynamics and sensor measurements. Both metrices are tuned and initialised with $10^{-4}I_{6\times6}$. The state is predicted (predicted state vector (x_{pred}) as below:

$$x_{pred} = f(x, u)$$

and the covariance prediction as:

$$P_{pred} = F \cdot P \cdot F^T + Q$$

The equations below describe the essential steps of the Kalman filter update phase, where the predicted state and covariance are adjusted based on measurements and the calculated Kalman Gain. The Kalman gain equation (K) includes the covariance matrix (P_{Pred}), measurement matrix (H) and the noise covariance matrix (R)

$$K = P_{pred} \cdot H^T \cdot \left(H \cdot P_{pred} \cdot H^T + R \right)^{-1}$$

The innovation or residual (y) is determined:

$$y = z - h\left(x_{pred} \right)$$

The update state vector (x) is obtained by using the Kalman Gain (K), residual (y) and predicted state vector (x_{pred})

$$x = x_{pred} + K \cdot y$$

Finally, the updated covariance matrix (P) is determined by subtracting the product of the Kalman Gain matrix (K) and the measurement matrix (H) from the predicted covariance matrix (P_{Pred}):

$$P = (I - K \cdot H) \cdot P_{pred}$$

The DRM and EKF seamlessly combine within the MPC optimisation loop for each time step. The DRM is applied first to predict the state based on the current state and control inputs. Then, the EKF is applied to estimate the state using the predicted state and measurements. The estimated state is used in the MPC optimisation to find optimal control inputs. The control inputs are applied to the robot, and the state is propagated using the DRM. The covariance matrix P is updated based on the prediction error using the Q matrix.

3. Results and Discussion

3.1. Network Load Testing Using iPerf3

During the network load testing, iPerf3 was utilised on two systems to generate network traffic, with a total of 30 packets sent and received within a duration of 90 s. The network performance was monitored using the Capsa Network Analyzer. Despite the visible activity observed in the monitoring windows, the performance of the calibration system was not negatively affected.

Capsa Network Analyzer provided comprehensive insights into the network performance during the iPerf3 test. It confirmed that all network links were properly connected and operational throughout the test. The live communication between the sender and receiver links was highlighted, indicating the successful data transmission. Despite the network being fully occupied (as indicated by the 100% occupancy in the top right pane), no link failures or disruptions were detected, as illustrated by the graph in the middle of the screen illustrated in Figures 10 and 11.

The test was conducted with the robot operating at a speed of 0.1 m/s (100 mm/s). This speed setting was used during the test to assess the system's performance under realistic operational conditions.

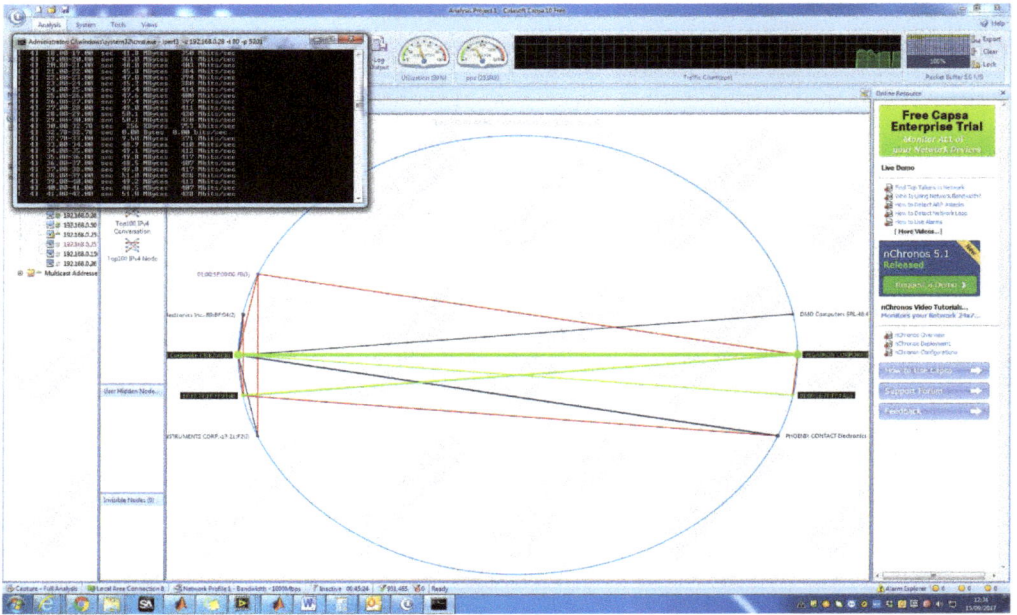

Figure 10. Network Load Testing using iPerf3.

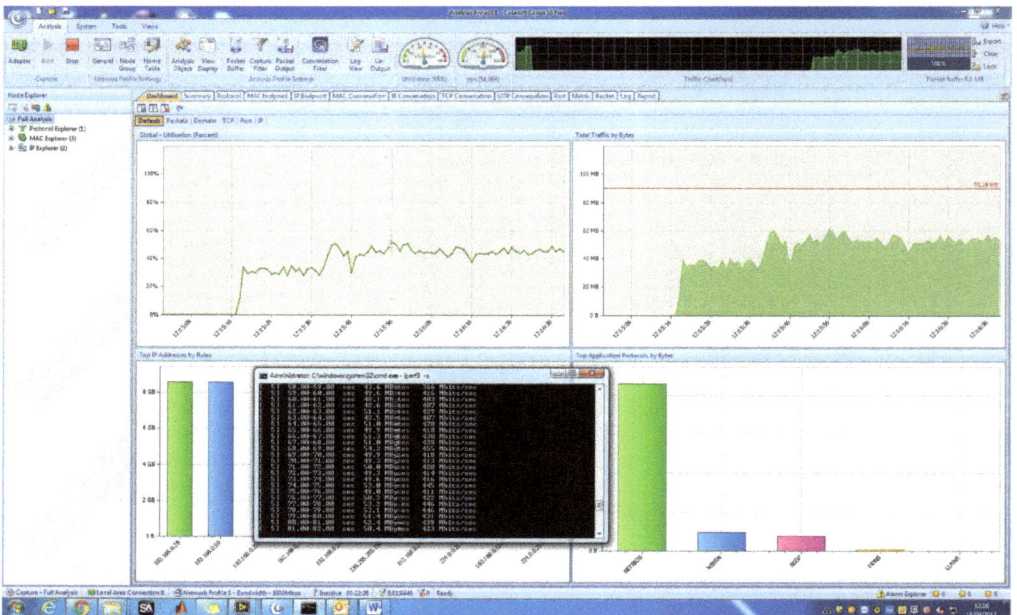

Figure 11. System performance during testing.

3.2. Black Box Testing—Independent System Verification

Independent systems were employed to validate the effectiveness of the NRTEC system and its claimed error correction capabilities. These independent systems were distinct from the equipment and software utilised in developing the NRTEC system. They provided the means to conduct tests independently by leveraging their software development kits (SDKs). These SDKs were further enhanced as part of the testing phase to enable error correction functionalities. Acting as impartial observers, these independent systems served as neutral umpires, facilitating the analysis of the NRTEC system's ability to execute error correction and compensation. Employing these diverse systems, such as the Nikon K-Robot [22], ART Track System [23], and Leica Tracker AT960 [24], ensured a wider range of test results that would not have been attainable solely through the use of the NRTEC Software 600 v1.20 and equipment. The following sections provide comprehensive details of the conducted tests.

3.3. System Verification Using Nikon K-Robot

The K-Robot is a robotic scanning and inspection technology primarily designed for scanning applications. However, for the purpose of testing the NRTEC system's error correction capabilities, the scanning mode of the K-Robot was not utilised. Instead, the K-Robot was employed in a camera capacity to measure the position of the robot. This setup allowed for the collection of relevant data and facilitated the evaluation of the NRTEC system's ability to correct errors. Figure 12 provides a visual representation of the configuration used, showcasing the integration of the K-Robot and the robot for position measurement purposes.

Figure 12. System setup for the verification by Nikon K-Robot.

Due to health and safety considerations, the laser of the Nikon Scanner used in the testing process was masked. Despite this precaution, the positional accuracy of the Nikon Scanner remains at an impressive 5 microns. On the other hand, the Leica Tracker (AT901) offers a positional accuracy of 5 microns plus an additional 5 microns per meter. These accuracy specifications highlight the precision and reliability of the measurement capabilities of both the Nikon Scanner and the Leica Tracker, making them suitable choices for verifying and assessing the performance of the NRTEC system.

Figure 13 provides a visual representation of the verifications conducted using the K-Robot system. The top two graphs illustrate the results of error correction compiled by the NRTEC system and verified by the K-Robot. While the K-Robot graph aligns with the trend of residual error observed in the NRTEC system, there are discrepancies in the actual values. For instance, at point 16, NRTEC claims a residual error of only 0.02 mm, whereas K-Robot verifies it to be up to 0.03 mm. This difference in values can be attributed to the physical location and characteristics of the two systems, as they may provide varying measurements for the same point due to their distinct positions and properties.

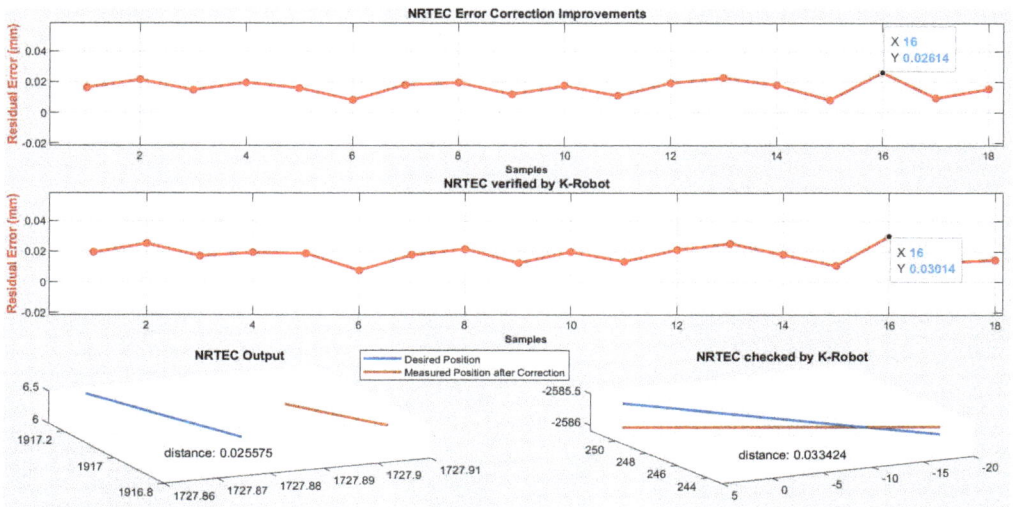

Figure 13. Verification of correction of points by K-Robot.

The bottom graphs display the original 3D points along the X, Y, and Z axes. It is important to note that the measurements of these points differ between the NRTEC and K-Robot systems due to their physical separation. The physical distance between the two systems is depicted in Figure 13. Additionally, the distance between the two lines on the bottom graphs represents the achieved calibration after completing 15 cycles. The red lines indicate the data collected after the calibration process, while the blue lines represent the desired positions for the robot's path.

NRTEC's graph (bottom left) suggests that the error has been reduced to 0.02 mm, supporting the claim of error correction. Conversely, K-Robot's graph (bottom right) verifies the accomplishment of correction up to 0.03 mm. These results demonstrate the effectiveness of the NRTEC system in reducing error and achieving the desired level of accuracy, as validated by the independent verification performed by the K-Robot system.

The verification of the NRTEC system was primarily focused on 3D positional accuracy. It was confirmed by Nikon, the manufacturer of the equipment, that the achieved accuracy of the system was up to 0.03 mm. The verification process specifically utilised the 3D (XYZ) data captured by the Nikon Camera.

In order to perform a comprehensive verification, additional add-ons and the Software Development Kit (SDK) provided by Nikon were employed. These add-ons facilitated the integration of the specific numerical data into the NRTEC system, allowing for a thorough assessment of its performance and accuracy. By utilizing the Nikon SDK, the system was able to process and analyse the 3D positional information obtained from the Nikon Camera, enabling the verification of the NRTEC system's capabilities in relation to the desired accuracy levels.

3.4. System Verification Using Advance Realtime System

Advanced Realtime Tracking (ART) Trackpack, a camera-based measurement system, was employed to perform verifications on the corrections achieved by the NRTEC system. The ART Trackpack utilises four cameras to monitor the movements and positions of three target markers.

To conduct the verifications, a physical setup was arranged within the cell, as illustrated in Figure 14. In this setup, the NRTEC system, equipped with the Leica Tracker AT901, and the test system developed with ART were simultaneously operated. This parallel operation allowed for the collection of data from both systems at the exact same time, enabling a direct and accurate comparison of their performance and correction capabilities.

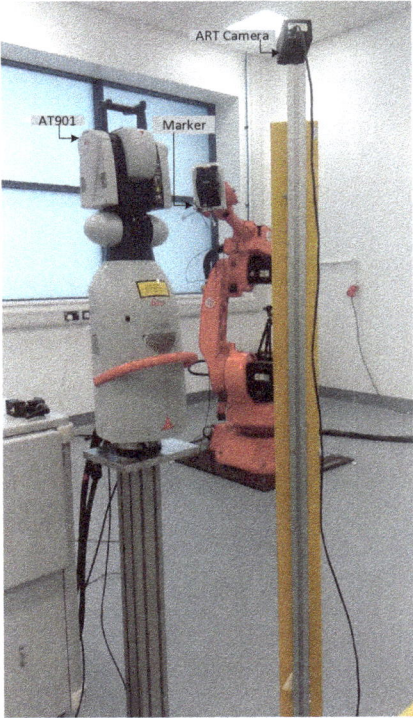

Figure 14. ART setup for NRTEC System Verification.

The following is the result. Figure 15 is the plot of two points of corrected and uncorrected data in the Leica Tracker frame and in the ART Coordinate System. The verifications are performed in 6DoF. The graphs on the left side are the data taken by NRTEC. The graphs on the right side display the data taken by using the ART system. The left and right side has two graphs each. The top one shows the 3D points in the space of 2D. The bottom ones show the original 3D points in an X, Y, Z axis. The points are different for both NRTEC and ART systems, as the systems are physically on different locations and give different value for the same point because of that reason. The physical distance can be observed in Figure 15. The distance between two lines is displayed on the bottom graphs. The blue lines on the graph show the desired position which is planned for the robot path, and the red lines on the graph show the data collected after performing the calibration process. The distance between the desired and measured points' data is displayed on the bottom 3D graphs.

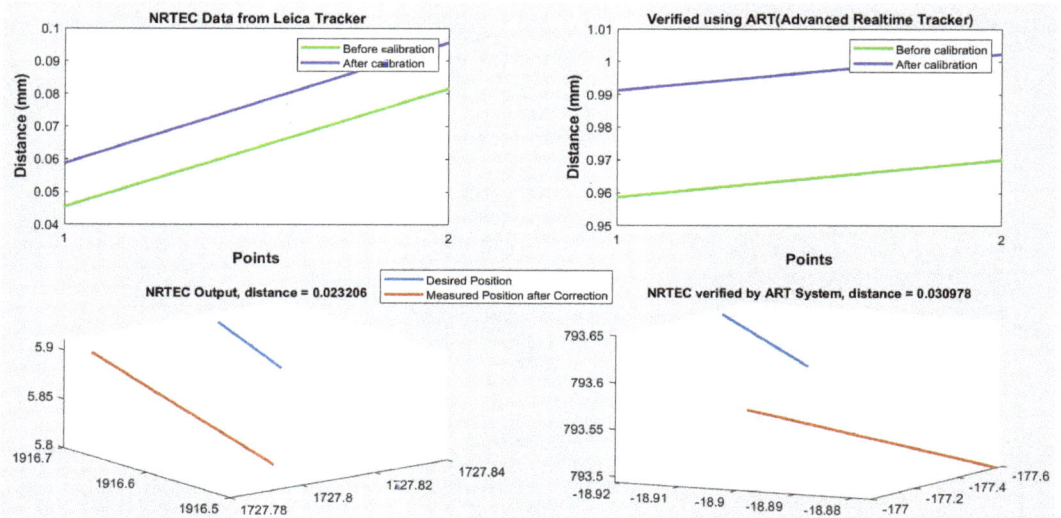

Figure 15. Verification of correction of points by ART.

The results obtained from the analysis of the bottom 3D graphs indicate that there is a difference between the desired position and the measured position after the calibration process. For the NRTEC system (graph on the bottom-left), the distance between these two lines is calculated to be 0.023206, while for the ART system (graph on the bottom-right), the distance is measured to be 0.0030978.

Based on these results, the NRTEC system claims to have achieved a correction of approximately 0 02 mm, whereas the ART system verifies the correction to be around 0.03 mm. Therefore, the ART system's verification supports the claim that the NRTEC system has achieved an accuracy level of up to 0.03 mm.

3.5. System Verification Using AT960

The AT960 Leica Tracker system was utilised to verify the corrections made by the NRTEC system. The AT960 system offers an accuracy of $+/-15$ μm when used in conjunction with the TMAC, and it has a high measurement capacity of 210,000 points per second.

To obtain the 3D points, the AT960 Leica Tracker was used in the experimental setup. The physical configuration of the cell where the testing took place is illustrated in Figure 16. Both the NRTEC system with AT901 and the test system developed with AT960 were simultaneously operated to ensure the collection of data for accurate comparison.

A system was developed to retrieve the points using the AT960 SDK (Software Development Kit). The obtained results are presented below Figure 17. illustrates the plot of two points, depicting both corrected and uncorrected data in the Leica Tracker frame and the AT960 Coordinate System. The verifications were conducted using three degrees of freedom (3DoF) on the AT960 system.

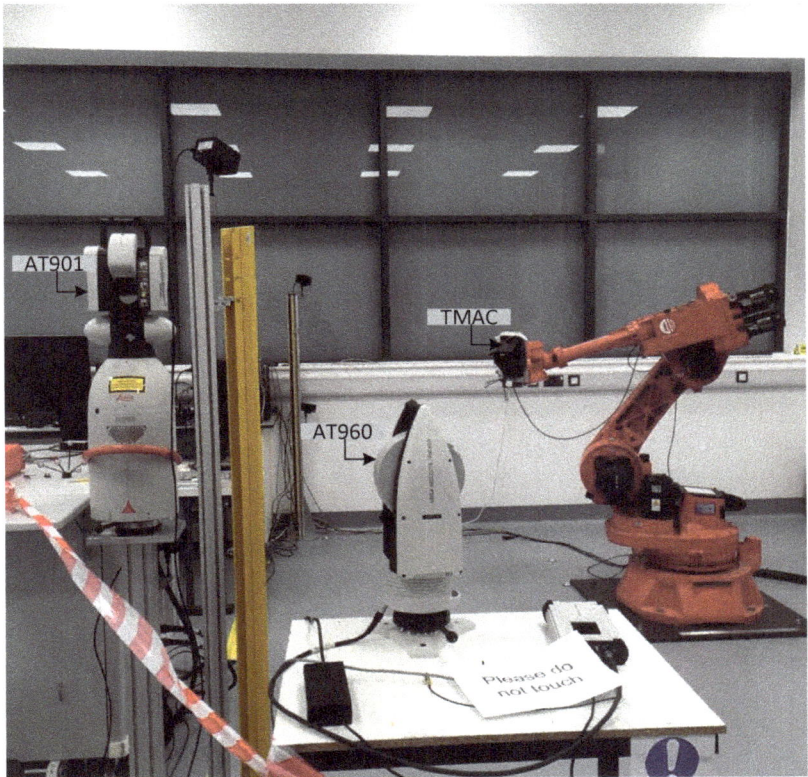

Figure 16. AT960 Leica Tracker Setup for NRTEC Verification.

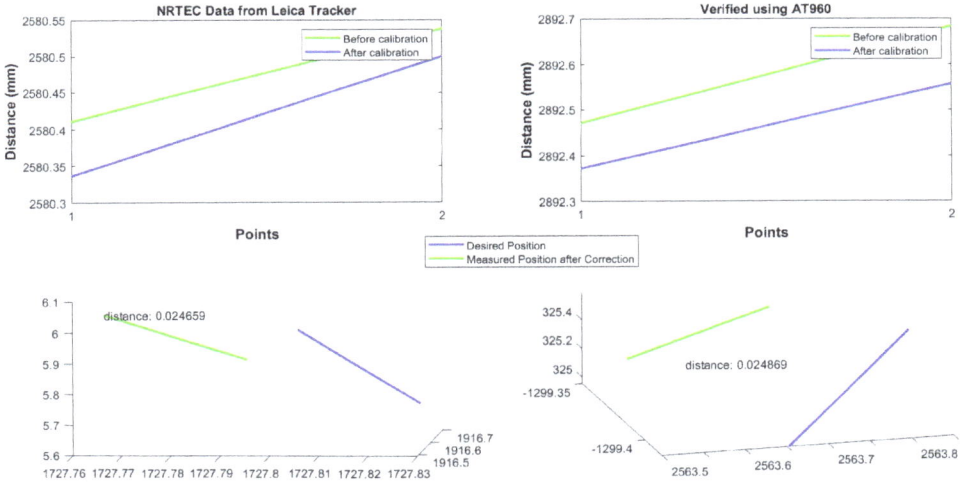

Figure 17. Verification of correction of points by AT960.

Due to the physical attachment of the TMAC to the AT901 using a wire, it acts as a 3D reflector for the AT960 system. Therefore, the data cannot be collected in six degrees of freedom (6DoF) format from the AT960. The left side of the graphs represents the data collected using the Leica Tracker AT901, while the right side displays the data collected using the AT960 system. Each side consists of two graphs: the top graph shows the 3D points in a 2D space, while the bottom graph illustrates the original 3D points in the X, Y, and Z axes.

The points obtained by both the NRTEC and AT960 systems differ since the trackers are physically located in different positions, resulting in varying values for the same point. The physical distance between the two systems can be observed in Figure 17. The distance between the lines depicted in the bottom graphs indicates the difference between the corrected and uncorrected data. The green lines represent the data collected prior to initiating the calibration process, while the blue lines represent the data collected after performing the calibration process. The discrepancy between the corrected and uncorrected data is displayed in the bottom 3D graphs.

The distance between the desired positions and measured positions for NRTEC is 0.024659, while the distance between those lines using the AT960 system is 0.024869. This indicates that the AT960 system aligns with the correction measurements of NRTEC, validating its accuracy.

The graph below in Figure 18 represents the difference between corrected and uncorrected points using both trackers. It illustrates the error correction captured by both the AT960 and NRTEC trackers. The graph displays 6 points in the X, Y, and Z axes in both positive and negative directions. The difference between the corrected and uncorrected points, as measured by the AT960 and NRTEC systems, is indicated by the red error bars.

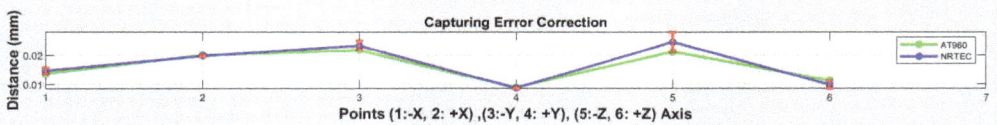

Figure 18. Verification of correction of 6 points in each axis (+, −) by AT960.

The error bars have been drawn to depict the delta range between ±0.000015 and ±0.001029. The error bars are very small, indicating that the difference in the correction measurements performed by both systems is negligible. The data displayed by the error bars demonstrate the reliability of the NRTEC system in comparison to the AT960 system.

Based on the verifications and comparisons with independent systems such as K-Robot, ARTTRACK, and AT960, it has been confirmed that the NRTEC system achieves an accuracy of up to 0.02 mm. The results obtained from these independent systems align with the claimed error correction capabilities of NRTEC, further validating its accuracy and performance.

3.6. Verification Results

The NRTEC system has undergone verification by three independent systems: AT960 Laser Tracker, ART, and K-Robot Camera. Figure 19 showcases the verification of error correction performed by these systems on the NRTEC system. The ART and K-Robot systems validate the correction achieved by NRTEC to be around 0.03 mm. However, the AT960 system, which closely resembles NRTEC's measurement device AT901, confirms the correction claimed by NRTEC to be around 0.02 mm.

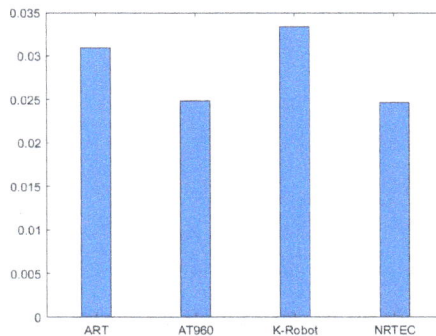

Figure 19. NRTEC Verification by Other Systems.

It should be noted that there are certain limitations associated with K-Robot, as it was used in a camera capacity, and only 3D data could be retrieved from the screen. On the other hand, the ART system utilises four cameras and provides data in 6D. The ART system verifies NRTEC's claim up to 0.030978 mm. While the ART system's accuracy in 2D is reported to be 0.04 pixels, which is equivalent to 0.0105 mm (10 μm) according to UnitConverters.net (2008), there is no available reference or technical information regarding its accuracy in 3D or 6D. In contrast, the AT960 system has an accuracy of 15 μm in 3D according to Metrology (2015).

Based on the available accuracy information for the ART and AT960 systems, it can be concluded that NRTEC is capable of achieving the claimed accuracy of around 0.02 mm.

4. Discussion and Conclusions

The interactions between the robot's position, orientation, control inputs, and measurements involve non-linear relationships due to the complex nature of robotic motion and sensor readings. The incorporation of the DRM further adds to the non-linearity as it involves predicting positions and compensating for errors in a dynamic environment. As a result, the overall system is best described as non-linear due to the complex interactions and relationships between various variables and components.

Testing conducted with 100% network load demonstrated that the system operated without delays. Dead Reckoning addresses network latency by replicating the environment on the other side/node of the network, effectively hiding the latency issue. The system aims to achieve realtime error correction by utilizing a network-based control system, enabling continuous correction without the need for a measure, stop, and correct approach. A Neural Network and Kalman Filter are utilised to improve the correction process, while the Realtime Control System (NI CRIO) provides a suitable platform for network control. Additionally, the cost-effectiveness of the metrology device is achieved by utilizing the Flexa system, allowing multiple machines or robots to use the device through network setup. The implementation of DRM compensates for potential disruptions caused by network traffic, ensuring accurate error correction by predicting and compensating for inaccuracies.

Figure 20 illustrates the data collected at different robot positions during the testing process. Each set of robot positions includes variations in the X, Y, Z axes, as well as rotations in the RX, RY, and RZ axes. The data reveals that after a certain number of correction cycles, the rate of error reduction reaches a plateau. Specific data points have been highlighted to evaluate the differences.

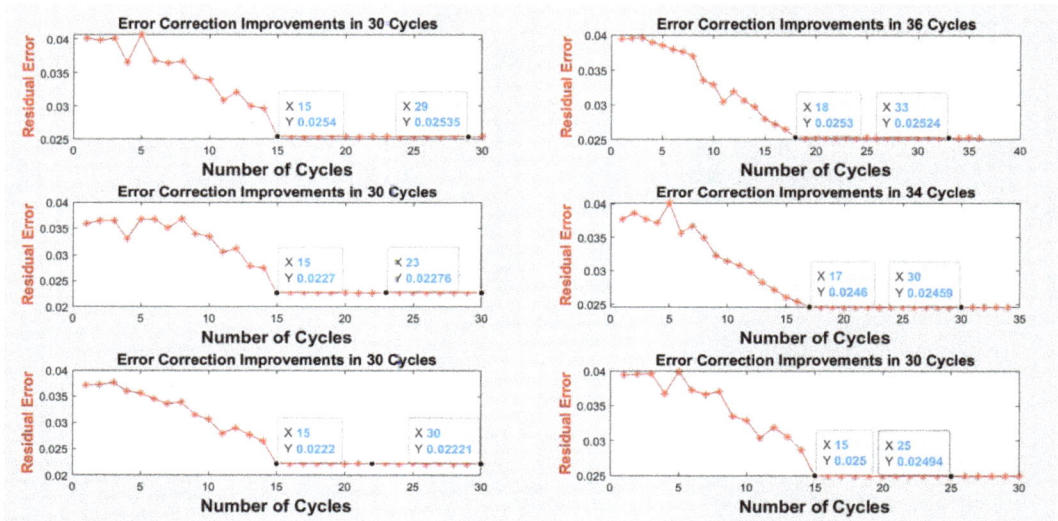

Figure 20. Number of Cycles' Evaluation.

In the first graph (top-left), it can be observed that at cycle 15, the achieved error reduction was 0.0254 mm. Comparatively, at cycle 29, the error reduction was 0.02535 mm, indicating a very minor difference of 0.00005 mm. Continuing the correction cycles beyond cycle 15 only impacts the error reduction in nanometres. Similarly, in the fourth graph (middle-right), where 34 cycles of correction were performed, the difference in error reduction between cycle 17 and cycle 30 is only 10 nanometres. The remaining four data sets exhibit a similar trend. The optimal number of cycles depends on the nature of the application and its accuracy requirements. For this particular study, the cycles were stopped at the point where the improvement in error reduction was in the micrometre range. Based on the graphs below, the optimal number of cycles was determined to be between 15 and 17.

The system achieves the correction threshold of 0.02 mm after 15 cycles, with no further improvement observed. Each cycle takes 225 min, and the overall experimentation and data compilation require approximately 6 weeks. The system demonstrates its learning capability as the accuracy improves with repeated robot path execution. This enhancement is achieved through Neural Network based MPC, enabling the system to learn and correct errors at each path point. After completing the 15 cycles, the Leica Tracker can be disconnected and utilised by other systems as needed.

The system has made significant progress in achieving error correction of 0.02 mm after 15 iterations. This precision has been accomplished by effectively handling network delays through the innovative use of DRM, which leverages gaming technology algorithms. Notably, the system has been rigorously tested and proven to deliver a reliable performance even in busy network environments.

The decision to transition from PID (Proportional-Integral-Derivative) control to MPC was primarily driven by the system's evolving requirements related to multiple inputs and outputs. During the initial stage of the research, PID control sufficed as it provided a simple and straightforward approach for static positional calculations in the three-dimensional space.

However, as the research progressed, the system encountered the need to operate within a normal open network, specifically the university campus network. This posed challenges due to significant packet delays experienced before and after the application of DRM. To address these issues and enhance the system's performance and adaptability, the decision was made to introduce the DRM.

By incorporating MPC into the control architecture, the system gained several advantages. Firstly, MPC enabled simultaneous optimisation of control actions for all inputs and outputs, accommodating the system's multiple requirements more comprehensively. Secondly, MPC's predictive control capabilities allowed for the proactive handling of time-varying dynamics and consideration of network delays, ensuring optimal control decisions based on the anticipated future behaviour. Moreover, MPC's ability to handle constraints facilitated the effective management of system limitations and boundaries on parameters such as position and speed.

The utilisation of MPC reflects the system's commitment to achieving high performance and robustness. It offers the flexibility to integrate advanced algorithms and strategies, enabling adaptation to changing network conditions. Furthermore, the adoption of MPC ensures future adaptability and flexibility, facilitating the incorporation of additional variables, constraints, and objectives as the research progresses.

The adoption of MPC was motivated by the system's evolving needs, including multiple inputs and outputs, time-varying dynamics, and the management of constraints. The system's impressive error correction results, coupled with the ability to handle network delays through DRM, highlight the effectiveness and suitability of MPC in enhancing performance and adaptability within this complex environment.

Figure 21 presents the average completion time in seconds for moves within each cycle. The orange line represents data from the system's stage 1, where PID control was implemented without DRM. The highest average time occurred during the 3rd cycle at stage 1, taking 1498 ms to complete a move. The graph illustrates the performance improvement in terms of time after the implementation of DRM and MPC. Despite some remaining delays, DRM allows for local predicted error calculations using players. The average move time reduced from 1401.467 ms in 15 cycles at stage 1 to an average of 758.8667 ms, which is a 46% decrease after DRM and MPC implementation. The change in robot manipulators between stages 1 and 2 has a minimal impact, since the same equipment is utilised.

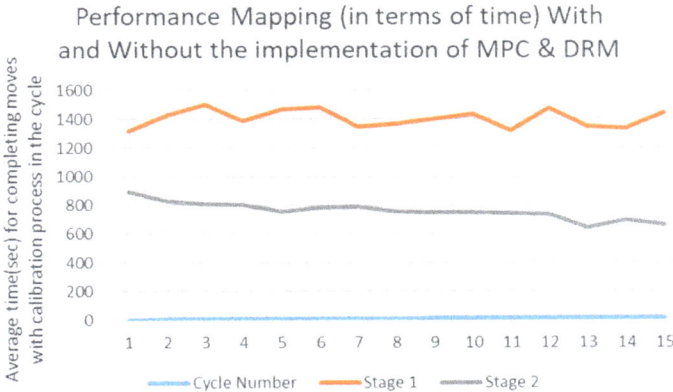

Figure 21. NRTEC Performance with and without the implementation of MPC and DRM.

Notably, the graph reveals the trend of move completion times as cycles progress within each stage. In stage 1, the average time to complete a point remains consistent across cycles. However, in stage 2, there is a significant reduction from 895 s in the first cycle to 663 s, indicating the system's learning capability and improved error prediction with repeated path executions.

In summary, the system successfully handles network delays and achieves realtime error correction. Utilising advanced algorithms and techniques, the system improves its performance and accuracy over multiple cycles. The cost-effectiveness of the metrology device is achieved through a flexible setup that allows its use by multiple machines. By predicting and compensating for inaccuracies, the system ensures accurate error correction even in the presence of network traffic disruptions. The analysis of data collected at different robot positions reveals that the optimal number of correction cycles falls between 15 and 17. The system demonstrates its learning capability, as accuracy improves with repeated execution of the robot's path. The transition from a previous control approach to the current system was driven by the evolving requirements and challenges posed by the network environment. The system's current approach offers improved optimisation, control prediction, and constraint handling, resulting in high performance and adaptability. The system's adoption of these techniques highlights its commitment to achieving accurate and reliable performance, even in challenging network conditions.

Author Contributions: Conceptualization, S.A.; Methodology, S.A.; Software, S.A.; Validation, S.A.; Formal analysis, S.A.; Resources, P.W.; Writing – original draft, S.A.; Writing – review & editing, S.A.; Supervision, P.W.; Project administration, P.W. All authors have read and agreed to the published version of the manuscript.

Funding: This research received no external funding.

Data Availability Statement: Data is available in the body of the paper and in Appendix A.

Conflicts of Interest: The authors declare no conflict of interest.

Appendix A

A1: Pseudocode Details for the Dead Reckoning Model:
Initialise:

```
robot_position = starting_position //Initial position (x, y, z)
robot_orientation = initial_orientation //Initial orientation (Rx, Ry, Rz)
time_step = 0.1 //Time interval between updates
distance_per_step = 0.05 //Distance the robot's end effector moves per time step
angle_per_step = 0.01 //Angle the robot's orientation changes per time step
error_factor_position = 0.02 //Factor to simulate error in position estimation
error_factor_orientation = 0.01 //Factor to simulate error in orientation estimation
While path_not_complete:
    //Calculate change in position using dead reckoning
    delta_x = distance_per_step * cos(robot_orientation.Rz)
    delta_y = distance_per_step * sin(robot_orientation.Rz)
    delta_z = distance_per_step * sin(robot_orientation.Rx) * cos(robot_orientation.Rz)
//Example z update
    //Apply error prediction to position estimation
    predicted_position_error = calculate_predicted_position_error()
    delta_x_with_error = (1 + predicted_position_error) * delta_x
    delta_y_with_error = (1 + predicted_position_error) * delta_y
    delta_z_with_error = (1 + predicted_position_error) * delta_z
    //Update robot position with error
    robot_position.x = robot_position.x + delta_x_with_error
    robot_position.y = robot_position.y + delta_y_with_error
    robot_position.z = robot_position.z + delta_z_with_error
    //Calculate change in orientation using dead reckoning
    delta_orientation_Rx = angle_per_step * cos(robot_orientation.Ry) * cos(robot_orientation.Rz)
    delta_orientation_Ry = angle_per_step * cos(robot_orientation.Rz) * sin(robot_orientation.Rx)
    delta_orientation_Rz = angle_per_step
    //Apply error prediction to orientation changes
```

```
predicted_orientation_error_Rx = calculate_predicted_orientation_error_Rx()
predicted_orientation_error_Ry = calculate_predicted_orientation_error_Ry()
predicted_orientation_error_Rz = calculate_predicted_orientation_error_Rz()
delta_orientation_with_error_Rx = (1 + predicted_orientation_error_Rx) * delta_orientation_Rx
delta_orientation_with_error_Ry = (1 + predicted_orientation_error_Ry) * delta_orientation_Ry
delta_orientation_with_error_Rz = (1 + predicted_orientation_error_Rz) * delta_orientation_Rz
//Update robot orientation with error
robot_orientation.Rx = robot_orientation.Rx + delta_orientation_with_error_Rx
robot_orientation.Ry = robot_orientation.Ry + delta_orientation_with_error_Ry
robot_orientation.Rz = robot_orientation.Rz + delta_orientation_with_error_Rz
//Update time
time_step = time_step + 1
//Check whether the robot has reached the target position
if distance (robot_position, target_position) < threshold_distance:
path_not_complete = false
//Pause for a short time to simulate real-world movement
Wait (time_step)
End While
//Robot has reached the target position
```

References

1. Asif, S.; Webb, P. Realtime Calibration of an Industrial Robot. *MDPI Appl. Syst. Innov.* **2022**, *5*, 96. [CrossRef]
2. Welch, G.; Bishop, G. An Introduction to the Kalman Filter. *Practice* **2006**, *7*, 1–16. Available online: https://www.cs.unc.edu/~welch/media/pdf/kalman_intro.pdf (accessed on 17 June 2023).
3. Mosammam, A.M. Kalman Filter: A Simple Derivation. *Math. Stat.* **2015**, *3*, 41–45. [CrossRef]
4. Kim, W.; Ji, K.; Ambike, A. Networked Real-Time Control Strategy Dealing With Stochastic Time Delays and Packet Losses. *J. Dyn. Syst. Meas. Control.* **2006**, *128*, 681. [CrossRef]
5. Zhang, K.; Huang, H.; Zhang, J. MPC-Based Control Methodology in Networked Control Systems. In Proceedings of the Simulated Evolution and Learning: 6th International Conference, SEAL 2006, Hefei, China, 15–18 October 2006; pp. 814–820.
6. IEEE 1278. *IEEE Standard for Information Tec hnology-Protocols for Distributed Interactive Simulation Applications Entity Information and Interaction*; IEEE Standards: Piscataway, NJ, USA, 1993.
7. Steinhof, U.; Schiele, B. Dead Reckoning from the Pocket—An Experimental Study. In Proceedings of the IEEE International Conference on Pervasive Computing and Communications (PerCom), Mannheim, Germany, 29 March–2 April 2010; pp. 162–170.
8. Delaney, D.; Ward, T.; McLoone, S. On Consistency and Network Latency in Distributed Interactive Applications: A Survey—Part I. *MIT Press J.* **2006**, *15*, 218–234. [CrossRef]
9. Durbach, C.; Fourneau, J.-M. Performance Evaluation of a Dead Reckoning mechanism. In Proceedings of the 2nd International Workshop on Distributed Simulation and Real Time Applications Montreal, Montreal, QC, Canada, 19–20 July 1998; pp. 23–29.
10. Chow, M. Network-Based Control Systems: A Tutorial. In Proceedings of the IECON'01. 27th Annual Conference of the IEEE Industrial Electronics Society (Cat. No.37243), Denver, CO, USA, 29 November–2 December 2001; pp. 4–5.
11. Naghshtabrizi, P.; Hespanha, J.P. Designing an observer-based controller for a network control system. In Proceedings of the the 44th IEEE Conference on Decision and Control Seville, Seville, Spain, 12–15 December 2005; pp. 848–853. [CrossRef]
12. Sinopoli, B.; Schenato, L.; Franceschetti, M.; Poolla, K.; Jordan, M.I.; Sastry, S.S. Kalman Filtering with Intermittent Observations. *IEEE Trans. Autom. Control* **2004**, *49*, 1453–1464. [CrossRef]
13. Pressman, R.; Maxim, B. *Software Engineering: A Practitioner's Approach, 8th ed*; McGraw-Hill Higher Education: Chicago, IL, USA, 2014.
14. IBM Corporation. *SOA in Manufacturing—Guide Book*; A MESA International: Citrus Heights, CA, USA, 2008.
15. Mahmood, Z. *Service Oriented Architecture: A New Paradigm for Enterprise Application Integration*; University of Moratuwa: Moratuwa, Sri Lanka, 2007.
16. Kipper, L.M.; Iepsen, S.; Forno, A.J.D.; Frozza, R.; Furstenau, L.; Agnes, J.; Cossul, D. Scientific mapping to identify competencies required by industry 4.0. *Technol. Soc.* **2021**, *64*, 101454. [CrossRef]
17. Asif, S.; Webb, P. Enhanced Cell Controller for Aerospace Manufacturing. *Aircr. Eng. Aerosp. Technol.* **2015**, *5*, 2.
18. Trunzer, E.; Lötzerich, S.; Vogel-Heuser, B. *Concept and Implementation of a Software Architecture for Unifying Data Transfer in Automated Production Systems*; Institute of Automation and Information Systems Technical University of Munich: Munich, Germany, 2018; pp. 1–17. [CrossRef]

19. Wu, G.; Wang, D.; Dong, H. Off-Line Programmed Error Compensation of an Industrial Robot in Ship Hull Welding. In Proceedings of the International Conference on Intelligent Robotics and Applications (ICIRA), Wuhan, China, 16–18 August 2017. [CrossRef]
20. Jing, W.; Zhou, J.T.; Gao, F.; Liu, Y.; Tao, P.Y.; Yang, G. A Learning-based Approach for Error Compensation of Industrial Manipulator with Hybrid Model. In Proceedings of the 2018 15th International Conference on Control, Automation, Robotics and Vision (ICARCV), Singapore, 18–21 November 2018; pp. 216–221. [CrossRef]
21. Jehaes, T.; De Vleeschauwer, D.; Coppens, T ; Van Doorselaer, B.; Deckers, E.; Naudts, W.; Spruyt, K.; Smets, R. Access network delay in networked games. In Proceedings of the 2nd Workshop on Network and System Support for Games, Redwood City, CA, USA, 22–23 May 2003; pp. 63–71. [CrossRef]
22. Nikon Tokyo Japan. K-Robot. 2019. Available online: http://pdf.directindustry.com/pdf/nikon-metrology/k-robot/21023-1455 56.html (accessed on 25 August 2022).
23. Advanced Realtime Tracking. ART Technical Details. 2017. Available online: https://ar-tracking.com/technology/technical-details/ (accessed on 25 August 2022).
24. Metrology, H. *Leica Absolute Tracker AT960*; Hexagon: Stockholm, Sweden, 2015; p. 9.

machines

MDPI

Article

Process Simulation and Optimization of Arc Welding Robot Workstation Based on Digital Twin

Qinglei Zhang [1], Run Xiao [2,*], Zhen Liu [1], Jianguo Duan [1] and Jiyun Qin [1]

1 China Institute of FTZ Supply Chain, Shanghai Maritime University, Shanghai 201308, China
2 Logistics Engineering College, Shanghai Maritime University, Shanghai 201308, China
* Correspondence: 202030210085@stu.shmtu.edu.cn

Abstract: For the welding cell in the manufacturing process of large excavation motor arm workpieces, a system framework, based on a digital twin welding robot cell, is proposed and constructed in order to optimize the robotic collaboration process of the welding workstation with digital twin technology. For the automated welding cell, combined with the actual robotic welding process, the physical entity was digitally modeled in 3D, and the twin welding robot operating posture process beats and other data were updated in real time, through real-time interactive data drive, to achieve real-time synchronization and faithful mapping of the virtual twin as well as 3D visualization and monitoring of the system. For the robot welding process in the arc welding operation process, a mathematical model of the kinematics of the welding robot was established, and an optimization method for the placement planning of the initial welding position of the robot base was proposed, with the goal of smooth operation of the robot arm joints, which assist in the process simulation verification of the welding process through the virtual twin scenario. The implementation and validation process of welding process optimization, based on this digital twin framework, is introduced with a moving arm robot welding example.

Keywords: arc welding robot; digital twin; welding workstation; placement planning; process optimization

Citation: Zhang, Q.; Xiao, R.; Liu, Z.; Duan, J.; Qin, J. Process Simulation and Optimization of Arc Welding Robot Workstation Based on Digital Twin. *Machines* **2023**, *11*, 53. https://doi.org/10.3390/machines11010053

Academic Editors: Raul D.S.G. Campilho and Francisco J. G. Silva

Received: 5 December 2022
Revised: 25 December 2022
Accepted: 29 December 2022
Published: 2 January 2023

1. Introduction

As the manufacturing industry continues to develop towards intelligence and flexibility, industrial robots are becoming increasingly mature in terms of development and application technologies [1], and are now used as specialized equipment in different specialized fields, such as welding, grinding, handling, and painting, and in specialized application scenarios, such as assembly [2]. As welding is an essential production process, welding robots are widely used in the field of welding. However, traditional welding robots are generally programmed to optimize the path according to manual experience, and it is difficult to ensure the welding quality and welding efficiency. With the new generation of industrial Internet of Things technology [3], ubiquitous perception is becoming a new driver to promote the development of welding processes. The drive for ubiquitous perception is integral in the research and development of new welding systems with digitalization and automation as the main features, so that welding equipment and processes can operate in three-dimensional space and time, providing multi-dimensional ubiquitous perception and transparency [4], which is important to improve product performance quality and production efficiency.

This paper is a study of the arc welding workstation system in the moving arm digital welding system, which consists of a robot system, welding system, welding auxiliary system, sensing system and control system. With the widespread use of welding robots, there are still several problems in the process of welding line applications, such as the following: (1) Programming of complex weldments. Existing robotic welding programming occurs mainly by means of teach-in programming. For the robotic welding of complex weldments, the number of welds is usually large, so it takes a lot of time to teach programming, and it is difficult to achieve the best welding process, due to the lack of holistic grasp of the entire space and welding process operation in manual teaching. Once the relative position of the welded parts and the robot changes, the welding procedure needs to be reprogrammed, which takes longer. (2) The problem of long robot position-seeking time. Since the initial welding of the workpiece by the teaming tool cannot meet the requirements of direct robot welding, it is generally necessary to perform position-seeking operations on the weld seam before robot welding is performed. Usually, a weld requires at least four position finding points, and the actual position of the weld is determined and corrected by the contact between the welding wire and the workpiece weld, on the basis of the shape characteristics of the weld. Once the welded part is relatively complex, the corresponding position-finding work takes longer, thus affecting the overall speed of the robot welding. (3) Long auxiliary time for wire shear gun cleaning. For arc welding robots, a crucial step in the welding operation is the cleaning of the gun nozzle after a certain amount and time of welding. Generally, the gun cleaning device is placed away from the workpiece. The welding robot needs to go back and forth frequently between the welding seam and the gun cleaning station, and if the robot does not choose the gun cleaning position properly, the robot becomes empty, which seriously affects the robot's welding efficiency [5].

In recent years, digital twin (DT) technology has been explored and applied in traditional manufacturing industries [6]. The intersection of multiple disciplines is used to achieve real-time synchronization and faithful mapping of the physical and digital worlds, paving the way for the integration of physical and information spaces [7,8]. Digital twin technology uses data interaction, information fusion, iterative computing, and command optimization to form a real-time intelligent closed-loop system of "data perception—real-time analysis—intelligent decision—precise execution". It uses high-fidelity digital models and sensor data to do the following: reflect the functions of the corresponding physical entities; construct the correlation between physical entities and virtual models; comprehensively interact with, and deeply fuse, physical real-time data, such as operation status, environmental changes and sudden disturbances with information space data, such as simulation prediction, statistical analysis and domain knowledge; enhance the synchronization and consistency between the physical world and the information world; provide more real-time, efficient and intelligent information for product life cycle; provide more real-time, efficient and intelligent services.

The robot's base position is closely related to the robot's structure and the task it is required to perform. Different optimization methods were given by Xue Y et al. [9] and Yang et al. [10] using robot accessibility as the only criterion. Franceschi P et al. proposed a cascade connected two-level optimizer to improve robot maneuverability as a way to optimize robot work cell design [11]. Gadaleta M et al.. in order to reduce the energy consumption of the robot cell, designed a layout optimization tool to optimize the placement of industrial robots relative to a specific task [12]. In addition many existing methods are mainly aimed at improving the performance of the robot, and the relevant criteria include maneuverability or flexibility [13,14], speed performance [15], singularity avoidance [16], collision avoidance [17], etc. Optimization of the base position has now attracted the attention of many researchers, because it can effectively improve the efficiency and quality of task completion.

There are fewer existing studies on the digital twin aspect regarding welding robots. Liu et al. proposed a digital twin-based process knowledge reuse and evaluation method by constructing a digital twin model of process knowledge containing geometric information and real-time states of process equipment, and proposed the use of the model similarity calculation method to filter process knowledge, as well as the real-time processing state and process reusability evaluation method [18]. Li et al. studied the prediction and control of ship assembly and welding process quality using a digital twin approach, established a twin model, connected physical entities to the twin model through twin data, and used artificial neural network algorithms to achieve quality prediction, which, in turn, fed back to physical entities to control welding quality [19]. Wang et al. proposed the use of digital twin technology to analyze welding and welder behavior using human–computer interaction and data-driven methods to improve the operational efficiency and comfort of human users [20].

In response to the above problem analysis, this paper presents a digital simulation-based digital twin framework for welding robot cells to support welding process visualization, welding process optimization and other functions for the low efficiency of welding operations and unreasonable welding process beats in the welding process of large moving arm workpieces. The impact of the robot's initial welding position on the welding process is investigated, and a particle swarm optimization algorithm is used to find the optimal initial welding position under collision-free interference constraints. The integration and mapping of the process and digital simulation is achieved through real-time acquisition and transfer of welding process data. Finally, the feasibility of the method was verified through an example in the digital twin system of a moving arm welding workstation.

The main contributions of this paper are summarized as follows.

1.   By analyzing the process of the arc welding robot workstation, a five-dimensional framework, based on the digital twin of the arc welding robot cell, is proposed, including the physical layer, network layer, data layer, model layer and service layer.
2.   The impact of the initial welding position of the robot on the welding process is investigated, and a particle swarm algorithm is proposed to optimize the initial welding position with the best operational smoothness under the constraints, which improves the overall welding efficiency.
3.   In the service layer of the established digital twin five-dimensional model, real-time data interaction is performed through the data management module for data visualization and optimization of the welding process.

The rest of the paper is organized as follows: Section 2 analyzes the arc welding workstation process architecture and provides a detailed description of the main components of a digital twin system. Section 3 proposes a digital twin-based optimization method for the initial welding position of the welding robot and describes the corresponding methodology. Section 4 uses a case study of an arc welding workstation for research analysis and experiments. Finally, Section 5 presents its conclusions and an outlook for the future.

2. Framework

2.1. Welding Workstation Digital Twin Robot Architecture

2.1.1. Excavator Arm Welding Process Analysis

The structure of the large excavator arm is shown in Figure 1. As shown, the bearing supports at 1, 3, 4, 6, etc. were machined and shaped by turning, and then transferred to the robotic welding workstation after the upper top plate, lower bottom plate, and left and right-side plates were initially spot welded and fixed by the group pairing tool. As the structure of the movable arm was not complicated, but the size was large and the weld seam mostly a space curve, the welding workload was large. When considering the welding process, it is usually necessary to weld the seam between the side plates and the upper and lower panels symmetrically in sections to prevent excessive welding stress and deformation.

The welding process of the excavator arm mainly includes the basic processes of arm loading, displacement machine clamping and positioning, robot welding, and arm discharging. In the welding process, the robot end-effector is welded along the workpiece seam, and the workpiece is fixed and rotated to the corresponding attitude by the positioner so that the welding robot can complete the welding task well. Current industrial production mainly relies on the experience of workers to determine the layout of the welding workstation and robot welding movement. The quality and efficiency of welding are affected by the welding operations, welding process, robot base position, and welding sequence, such as unreasonable situation.

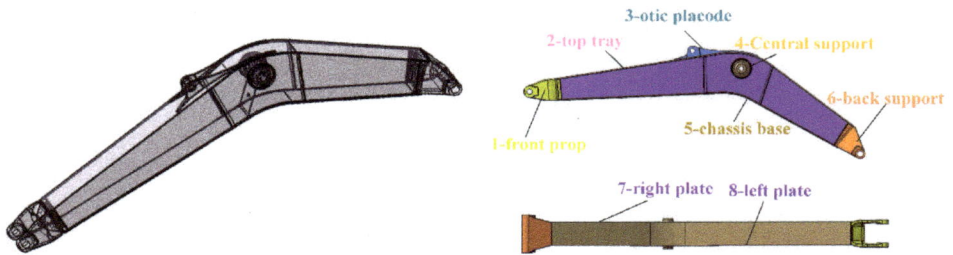

Figure 1. Excavator arm structure diagram.

In summary, the welding robot workstation can be optimized in the following three aspects: (1) through the process of welding parts and position constraints, the establishment of the kinematic model of the welding robot to optimize the trajectory of the welding robot and the displacement machine; (2) through the analysis of the actual production line welding beat to find the unreasonable places for the beat optimization; (3) the welding robot welding process needs to find the position, welding, wire cutting and other operations with the best global operability of the initial welding position, and optimizing the welding beat can also reduce the distance of the robot empty walk during the wire cutting operation, to further improve welding efficiency.

2.1.2. Robot Welding Workstation Cell with Twin Frame

The robot welding cell is the most important component of the welding production line. A typical arc welding robot workstation consists of a robot system, gantry system, positioner system, welding system, etc. According to the process requirements of the welded workpiece, the robot is equipped with the corresponding end-effector welding torch for welding. Figure 2 is a schematic diagram of the structure of a basic welding robot workstation cell, in which the loading table holds the workpiece to be welded, the feed buffer is used to store the finished welded parts, the welding table is used by the displacer to hold the workpiece to be welded, and the robot is used to weld the workpiece. The motion of the welding workstation cell is cyclic, and the welding cycle consists of the delivery of the parts to be welded from the loading buffer to the welding work area, and then the delivery of the finished welded parts to the lower material buffer. After the welding condition is triggered, the workpiece is delivered to the welding operation area, the position changer clamps and fixes it, and the robot motion completes the welding operation through the steps of position finding, welding and gun clearing, etc. After the welding is completed, the finished part flows to the lower material buffer area and the next part to be welded flows to the welding operation area, and the robot returns to the initial position after each weld or welding of the complete workpiece.

Due to the complexity of welding operations and the harsh welding environment, the robot welding workstation in the layout planning, motion control, offline simulation, debugging and verification of the process, there is a long cycle time, and iterative and other characteristics. The use of data interaction, information fusion, iterative computing and decision analysis and optimization methods to build a virtual simulation environment with high-fidelity mapping of the real physical behavior, and to construct a modular, universal, digital twin welding workstation system to realize the integration of the welding process in various aspects, such as model, control and digital services, are key to achieving a breakthrough in the existing technological bottleneck. This paper presents a block diagram of the digital twin welding workstation, as shown in Figure 3, including the physical welding workstation unit, the virtual welding workstation unit, twin data, and service system 4 parts.

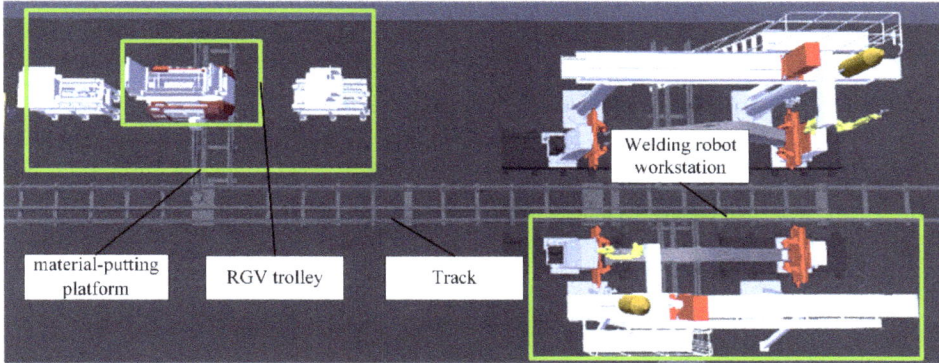

Figure 2. Schematic diagram of the cell structure of the welding robot workstation.

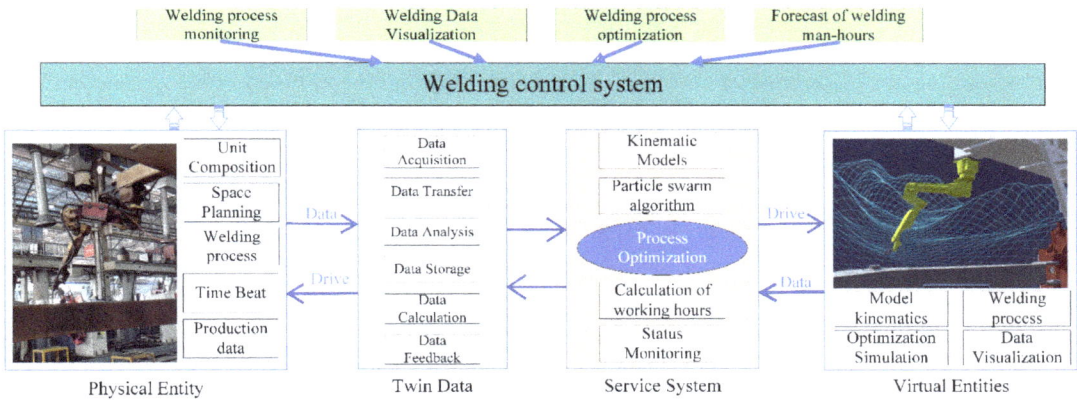

Figure 3. Block diagram of digital twin welding workstation.

(1) The physical unit is an actual welding production system consisting of a welding robot and related production equipment, specifically including physical entities, such as welding robots, welding guns, position changers, controllers, moving arm workpieces, sensors, guide mechanisms, wire shearing and gun clearing devices, etc. It also includes specific welding information associated with robot position changer movement, such as workpiece welding process, welding current, position finding status, welding arc status detection, space planning, size, position, and other data information.

(2) The virtual cell is composed of a virtual digital model of the welding robot workstation, which mainly contains the construction of the model at three levels: elements (such as workstation layout, physical equipment, environment and other production elements), behaviors (such as welding process, linkage and other behavioral characteristics), and rules (such as welding process optimization and other evolutionary rules) to achieve the mapping of digital space to physical space.

(3) The service system is driven by the twin data as the core, providing services such as logic driving and motion control of the digital twin, analyzing and optimizing the welding process, such as welding process, time beat, and robot welding path of the physical entity cell, and mapping it to the virtual entity cell in order to perform motion simulation of the welding process.

(4) The twin data is composed of physical unit data, virtual unit data and service system data, and provides corresponding analysis, verification, and decision information for the service system through data transfer, interaction, and update between each layer.

2.2. Welding Workstation Physical Unit

The physical unit is a collection of all physical entities involved in the welding process, including hardware devices, such as welding robots, shifters, welding guns, welding parts, auxiliary devices, such as controllers, manual demonstrators, gun clearing stations, and different functional areas such as loading and unloading areas. In addition, the physical entity of the welding workstation contains various types of physical sensing equipment (current sensors, displacement sensors) supporting the collection and transmission of information, such as the position of the welding robot and the displacement machine, the position finding signal, displacement signal, and wire shearing and gun clearing. The welding workstation operating system has functions for data and information upload and execution according to data commands, which can provide support for real-time interconnection of virtual and real interactions.

The welding cell for large welded parts mainly includes a six-axis robot, electric control cabinet, positioners, RGV trolley, truss, and other equipment. Each welded part is transferred from the loading area to the welding station, where the robot and the positioner and other equipment work together to complete the corresponding welding task, and the part is then transferred to the lower material area via an RGV trolley.

In this paper, the operational logic of the welding workstation welding operation was clarified to plan the trajectory of the robot operation and realize synergistic cooperation among the devices, so as to avoid problems, such as interference and collision in the welding process of the welding workstation. As the welding process of each weld welding process is the same, one of the selected welding seams (the right-side plate and the top plate between the welding seam) process is shown in Table 1.

Table 1. Workpiece welding process.

Serial No.	Process Content	Rhythm
1	RGV transports the parts to be soldered	5 min
2	Positioner Positioning	50 s
3	Robot Position Finding	1 min 30 s
4	Robot welding seam	7 min
5	Robot wire cutting and clearing gun	1 min
6	Robot returns to initial position	30 s
7	RGV delivery weldments	3 min

In order for the welding robot to complete the welding operation, it is necessary to combine the actual moving arm welding process and path planning. The main purpose of the path planning is to find a reasonable motion path and the best initial welding position under the constraints of environmental and operational tasks, so that the welding robot welding process does not interfere and collide with related equipment in the workstation, and, thereby, improve the efficiency of the welding operation.

2.3. Welding Workstation Virtual Unit

The key to the implementation of a digital twin system for welding workstations lies in the creation of a digital twin model that is equivalent to the actual physical unit. Through digital modeling of virtual scenes, the working environment of the physical welding line can be realistically presented and information decisions can be made for the operation and maintenance of the actual line, etc.

2.3.1. Digital Modeling

The virtual twin model is the core of the realization of the digital twin, and not only requires one-way mapping of physical entities but also needs to have the ability to guide physical entities. The digital modeling object of this paper was the welding robot welding workstation, and digital modeling using a multidimensional model fusion approach, mainly divided into four aspects of geometric modeling, physical modeling, behavioral modeling and rule modeling. The detailed process is shown Figure 4 below.

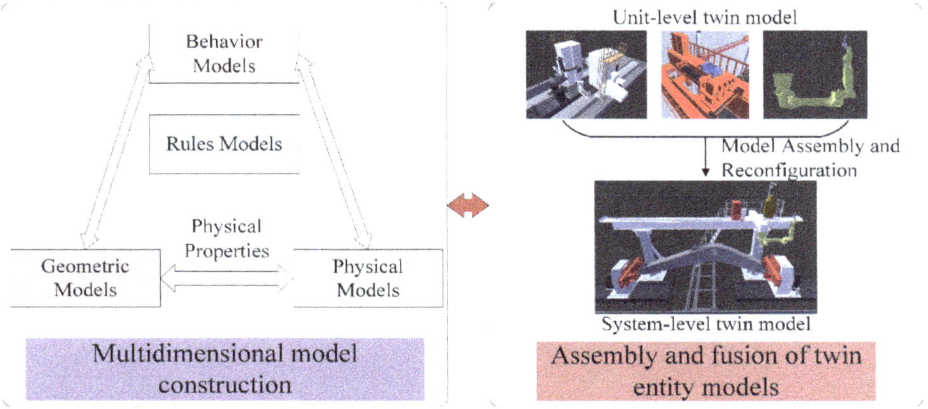

Figure 4. Block diagram of multidimensional model construction.

According to the above multidimensional modeling method, for the geometric dimension, based on the geometric characteristics and parameters of the equipment in the movable arm welding workstation, SolidWorks industrial modeling software was used to construct 3D models of the welding robot, the displacement machine, workpiece, truss and other equipment. For the physical dimensions, based on the material properties and physical parameters of the equipment, 3D Max was used to render the model. For the behavioral dimension, based on the geometric model, the kinematic and kinetic equations of the robot model were defined, the robot joint motion relations were calculated, and the construction was carried out according to the behavioral coupling between the devices. For the rule dimension, XML language was used to describe the operation and evolution rules of the devices, according to the actual welding process and welding procedures. By fusing the 3D visualization model, physical model, behavioral model, and rule model, the virtual model of the welding workstation built in the virtual engine Unity3D software was able to portray the actual operation state and realistically map its physical entities.

2.3.2. Behavior Mapping

The virtual simulation of the twin welding workstation was performed by collecting data from hardware devices, such as on-site robot controllers and sensors, for real-time transmission of the welding process and operational information from the physical shop, using the actual welding data information as the driving source. The virtual twin system has relatively complex model subordination relationships between the welding robot, the welding robot, and the displacement machine, and between the displacement machine and the welded workpiece.

The virtual twin model of the welding workstation can be divided into static and dynamic models, depending on the different states of the equipment in the physical unit. Static models usually consist of different subtypes of models that can be equated as a single whole to establish the hierarchy of models, such as trusses, barrels, storage tables, etc. Dynamic models are divided into two categories according to the operational logic. The first category is for models with the same operational logic, e.g., welding robot, welding gun, etc. and the second category is for models with different operational logic, e.g., workpieces, displacement machines, etc. For the dynamic models, the first category takes the welding robot as an example and establishes the subordination relationship through the parent–child motion relationship between the joints of the robot. The specific structure is shown in Figure 5. For example, the parent–child relationship between the movable arm workpiece and the positioner can be linked by weldment. Transform, parent = positioner. The robot cell in the welding workstation has two states of motion: translation and rotation, and the virtual model is driven by the translate and rotate functions in the virtual scene.

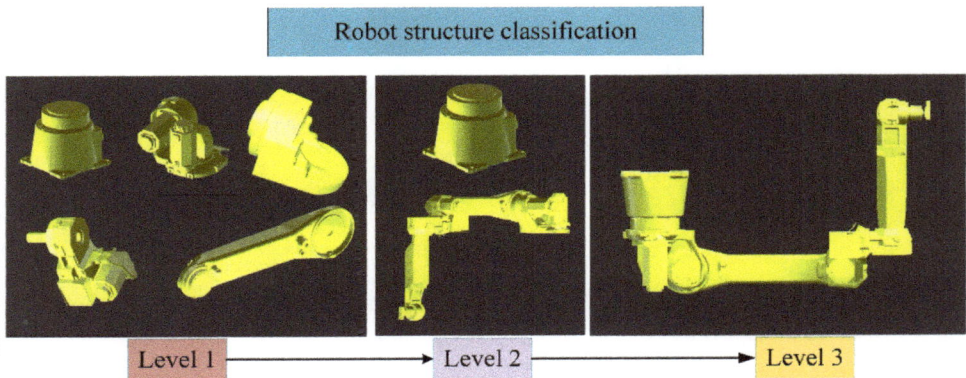

Figure 5. Six-axis robot structure classification.

In the virtual simulation process, the welding robot may collide with itself or interfere with other devices in the scene when performing welding operations, and in this process collision detection can issue an alert signal in time. Collision detection technology determines whether two convex bodies intersect, and the collision detection algorithm for polyhedral intersection detection has high detection accuracy [21]. Unity 3D software supports a variety of collision detection methods. Due to the high accuracy requirements of the virtual welding process, this paper used the orientation bracketing box (OBB) detection method [22] for the collision detection of the model. The OBB wrapping box can wrap the collision body model completely, its direction is determined by the structure of the enclosed model, and collision is detected by the intersection of the wrapping box, which is simple to implement and can effectively reduce the task of collision detection in the virtual system.

2.4. Welding Workstation Twin Data Management System

The twin data is the basic element of the virtual twin system, and the data management system enables data collection, transfer, analysis, and calculation. The physical welding unit and the virtual welding unit interact with each other through the data management system. The welding workstation data management system provides intelligent decision-making and services for the welding operations of the constructed digital twin welding system, based on the integrated management of data and information.

2.4.1. Digital Twin Data for Welding Workstations

The welding robot workstation digital twin system is a dynamically updated and changing virtual system, in which the welding robot and its displacement machine digital twin are process models that continuously generate and update data following the movement of the workpiece welding process. The data of the digital twin is divided into static data and dynamic data. Static data includes data such as geometric characteristics of the equipment, materials, scene location layout, etc., which is used to establish the virtual welding workstation system and realize the transfer mapping of information, such as model data, scene layout, and materials of the real workshop. Dynamic data include data collected by various types of sensors in the robot welding process as well as real-time production data and other data. Dynamic visualization data, as shown in Figure 6, is used to carry out real-time action simulation of the virtual welding robot and to realize the transfer mapping of process data, simulation and service optimization of the workshop. The data of the digital twin of the welding robot workstation mainly includes four kinds of basic data, namely design data, process data, welding process data, and production data, details of which are given below.

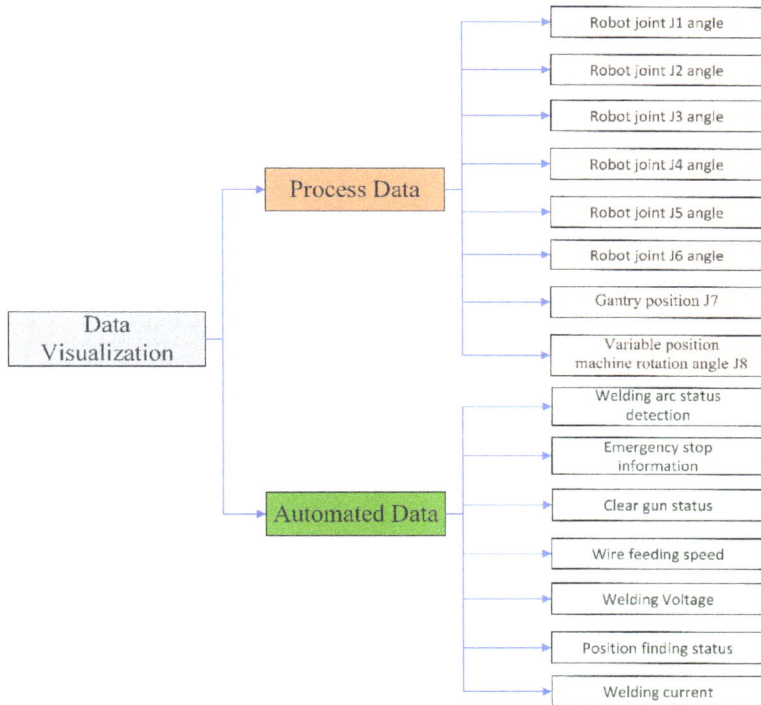

Figure 6. Dynamic data.

(1) Design data is mainly divided into global and local design data. Global design data includes environmental layout, scene rendering, lighting, and other data. Local design data includes unit 3D model data, specifically geometric information, coordinate information, dimensions, materials, parent–child linkage and other data.

(2) Process data mainly includes welding methods for welded parts, specifically process information, welding seam sequence, welding process characteristics, and other parameters.

(3) Welding process data refers mainly to various types of sensor detection data, equipment process information, such as data on the angle of each joint of the welding robot, data on the rotation angle of the translator, and data on the position of the robot base, etc. It also includes automation information, such as data on the position finding status, welding current, wire feeding speed, gun clearing status, running status information, etc. and simulation data, such as virtual welding workstation simulation operation data. Finally, optimization analysis data generated by the service

system data, such as robot kinematic equations, particle swarm optimization (PSO) algorithms, and other data also fall into this category.

(4) Production data includes mainly the total production plan of the workshop, the current number of completed products, product production efficiency and production quality, and other real-time dynamic data.

2.4.2. Data Communication

The robot welding operation process involves the collection of equipment data, resource data, welding material information, and other data, with multiple hardware and software systems, a wide range of data collection methods and interface types, and data showing multiple sources of heterogeneity. The twin system realizes the continuous update and optimization of the unit twin through the interaction of data, and is able to realize the matching articulation of workshop data through the transmission relationship and interface exchange mode between data. The basic flow of data communication in the virtual system of welding workstations is shown in Figure 7.

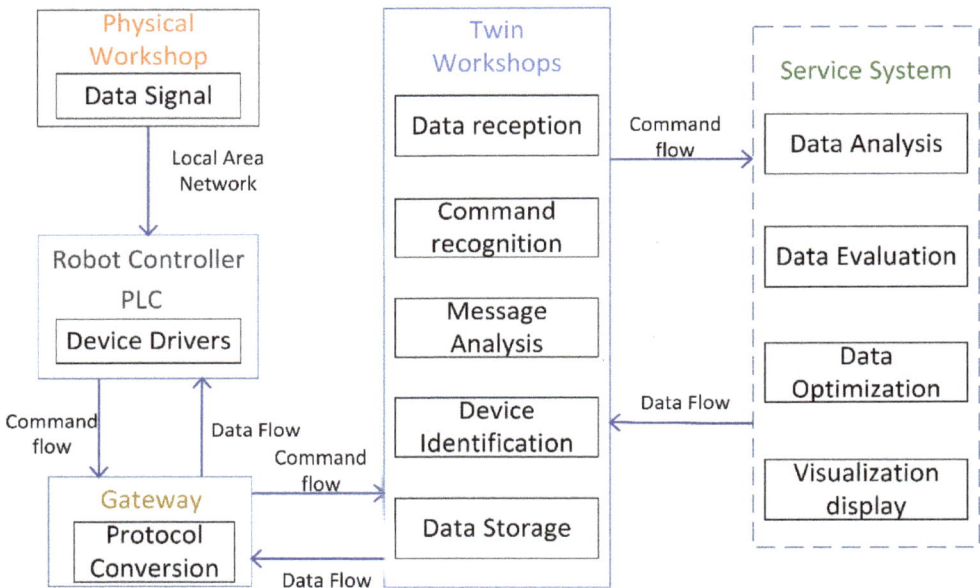

Figure 7. Data communication flow.

The digital twin system for welding workstations built in this paper achieved accurate and complete data collection and integrated management of the welding process through the data communication flow shown above. The collected data was uploaded to the IOT platform for storage via fieldbus, User Datagram Protocol (UDP), Ethernet, Ether CAT and other communication protocols. The API function was used to read PLC information and pass the data to the digital twin workshop system in the form of JSON data strings, and the virtual system achieved virtual real synchronization by sample value and interpolation processing of real-time data. The data processing flow is illustrated in Figure 8.

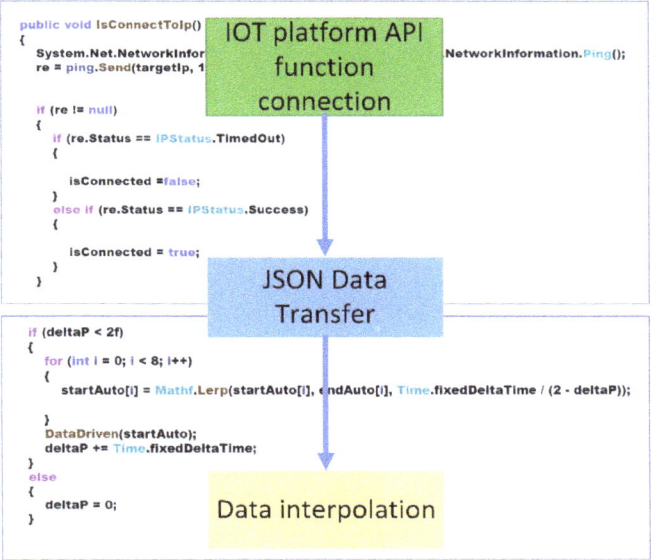

Figure 8. Data processing.

2.5. Welding Workstation Service System

The function of the welding workstation service system is to monitor, predict, and optimize the welding operation process in the workshop. The welding system consists of automated equipment arc welding robots, and shifters working together to complete the welding operation. In order to avoid the impact of the initial welding position set by the robot cell and the welding running trajectory on the overall operational stability of the workstation, the actual welding operation is simulated and optimized in the service system using the twin data transfer between them. Before the actual welding of the physical cell, the service system can use the production line history data to simulate and evaluate the process and production plan through internal algorithms, so that adjustments and optimizations can be made.

The constantly updated physical cell operation status and virtual cell simulation data in the data management system adjusts the production schedule in real time. The welding workstation service system contains mathematical algorithms for robot kinematics, changer kinematics, robot particle swarm path placement optimization algorithms, and man-hour calculations, and the system uses these algorithms for continuous iterative optimization of the digital twin.

3. Methodology

3.1. Robot Kinematics Analysis

The core of the faithful mapping of the moving arm welding workstation is to achieve accurate control of the arc welding robot model. The robot model initially constructed in the twin system is a static model, and the parent–child linkage between the joints was not established, so it was not possible to drive the robot motion through the data. A 6-axis industrial robot has a linkage mechanism with six rotating joints and six links, each joint and corresponding linkage has its corresponding coordinate system, as seen in Figure 9. A simplified diagram of the robot D-H linkage coordinate system is shown. In this paper, we use the D-H parametric method to establish the kinematic model of the robot. Firstly, we use the flush transformation matrix to describe and establish the relative position and orientation relationships between these linkage coordinate systems, and then transform the four matrix elements of linkage length, linkage rotation angle, linkage offset and joint angle to construct the kinematic equations of the robot.

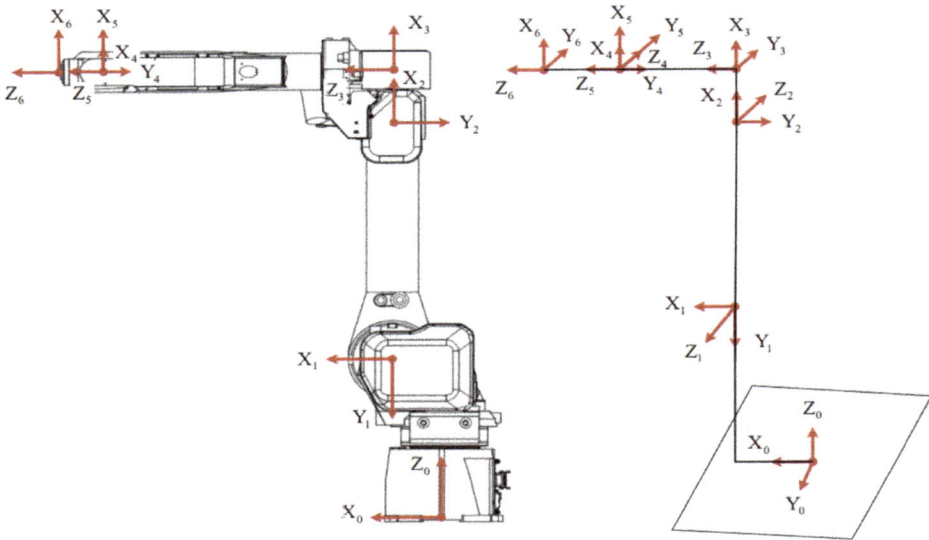

Figure 9. Simplified diagram of the D-H linkage coordinate system of the robot.

In this paper, the FANUC 6-axis robot was used as an example, and the D–H parameters of this robot are provided in Table 2, where a_i is the length of the link i, d_i is the distance between link i and link $i + 1$, α_i is the i the angle of twist of the connecting rod and θ_i is the angle between the connecting rod i and $i + 1$.

Table 2. Six-axis robot D–H parameter table.

Rods i	a_i (mm)	d_i (mm)	a_i (°)	θ_i (°)
1	150	525	$-\pi/2$	0
2	790	0	π	$-\pi/2$
3	150	0	$\pi/2$	0
4	0	860	$-\pi/2$	0
5	0	0	$\pi/2$	0
6	0	100	0	0

The robot kinematic equations are divided into forward kinematics and inverse kinematics. Forward kinematics is obtained from the known joint angle by the flush transformation of the adjacent linkage coordinate system to obtain the robot end position and pose. The inverse kinematics is determined from the known end-effector position and pose to determine the robot joint angle. According to the known D–H parameters, to establish the flush transformation equation between each linkage, the transformation matrix $_{j}^{j-1}T$ is generalized as follows:

$$
_{j}^{j-1}T = \begin{bmatrix} \cos\theta_j & -\sin\theta_j\cos\alpha_j & \sin\theta_j\sin\alpha_j & a_j\cos\theta_j \\ \sin\theta_j & \cos\theta_j\cos\alpha_j & -\cos\theta_j\sin\alpha_j & a_j\sin\theta_j \\ 0 & \sin\alpha_j & \cos\alpha_j & d_j \\ 0 & 0 & 0 & 1 \end{bmatrix}, j = 1, 2, 3, 4, 5, 6 \quad (1)
$$

The transformation matrix of each linkage of the six-axis robot can be obtained according to Equation (1).

$$
{}^0_1T = \begin{bmatrix} c\theta_1 & 0 & -s\theta_1 & a_1c\theta_1 \\ s\theta_1 & 0 & c\theta_1 & a_1s\theta_1 \\ 0 & -1 & 0 & d_1 \\ 0 & 0 & 0 & 1 \end{bmatrix},
$$

$$
{}^1_2T = \begin{bmatrix} s\theta_2 & -c\theta_2 & 0 & a_2s\theta_2 \\ -c\theta_2 & -s\theta_2 & 0 & -a_2c\theta_2 \\ 0 & 0 & -1 & 0 \\ 0 & 0 & 0 & 1 \end{bmatrix},
$$

$$
{}^2_3T = \begin{bmatrix} c\theta_3 & 0 & s\theta_3 & a_3c\theta_3 \\ s\theta_3 & 0 & -c\theta_3 & a_3s\theta_3 \\ 0 & 1 & 0 & 0 \\ 0 & 0 & 0 & 1 \end{bmatrix},
$$

$$
{}^3_4T = \begin{bmatrix} c\theta_4 & 0 & -s\theta_4 & 0 \\ s\theta_4 & 0 & c\theta_4 & 0 \\ 0 & -1 & 0 & d_4 \\ 0 & 0 & 0 & 1 \end{bmatrix},
$$

$$
{}^4_5T = \begin{bmatrix} c\theta_5 & 0 & s\theta_5 & 0 \\ s\theta_5 & 0 & -c\theta_5 & 0 \\ 0 & 1 & 0 & 0 \\ 0 & 0 & 0 & 1 \end{bmatrix},
$$

$$
{}^5_6T = \begin{bmatrix} c\theta_6 & -s\theta_6 & 0 & 0 \\ s\theta_6 & c\theta_6 & 0 & 0 \\ 0 & 0 & 1 & d_6 \\ 0 & 0 & 0 & 1 \end{bmatrix}
$$

$$(2)$$

In Equation (2) above, s denotes $\sin\theta$, and c denotes $\cos\theta$. The positive kinematic equation of the robot is obtained by the concatenated multiplication operation of the linkage transformation matrix.

$$
{}^0_6T = {}^0_1T{}^1_2T{}^2_3T{}^3_4T{}^4_5T{}^5_6T = \begin{bmatrix} n_x & o_x & e_x & p_x \\ n_y & o_y & e_y & p_y \\ n_z & o_z & e_z & p_z \\ 0 & 0 & 0 & 1 \end{bmatrix} \tag{3}
$$

In Equation (3). $\begin{bmatrix} n & o & e \end{bmatrix}$ denotes the end-effector pose vector, and $[p]$ denotes the end-effector position vector. The robot inverse kinematics solution process uses an analytical solution to find the angle of each joint of the robot by stepwise backpropagation of the total transformation matrix of Equation (3) with known end poses.

3.2. Arc Welding Robot Base Position Optimization Problem

Robot base position optimization is a complex nonlinear optimization problem with multiple variables and constraints. The goal is to optimize the process beat during workpiece welding by minimizing the overall robot movement during weld seam welding while maintaining robot kinematic performance. In this section, the problem discussed above is described more clearly in Figure 10. The scenario shown in the following section is illustrated in detail.

The application scenario of a welding robot system with external extension axes in the welding process is shown in the Figure 10. Some of the main coordinate systems in the system are given along with the corresponding coordinate transformation relationships, which are described below.

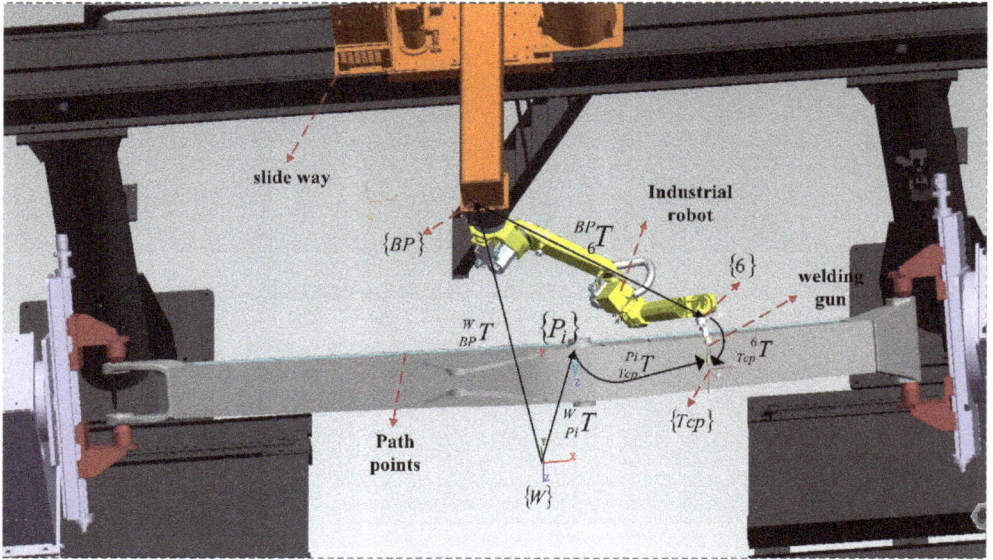

Figure 10. Welding system -related coordinate system.

The value $\{W\}$ is the world coordinate system, also known as the reference coordinate system, $\{Tcp\}$ is the tool TCP coordinate system and $\{6\}$ is the robot end joint coordinate system.

The value $\{P_i\}$ is the coordinate system of weld points on the weld seam, the position vector of discrete points of any weld seam can be obtained by offline programming software $pi = [xi, yi, zi]^T$, and the attitude vector $Ri = [ni, oi, ai]$ and the weld joint coordinate system can be obtained according to the coordinate transformation relationship $\{P_i\}$ transformation matrix with respect to the world coordinate system ${}_{P_i}^{W}T = \begin{bmatrix} ni & oi & ai & pi \\ 0 & 0 & 0 & 1 \end{bmatrix}$. The transformation matrix of the weld point coordinate system with respect to the world coordinate system can be obtained from the coordinate transformation relationship.

The value $\{B\}$ is the robot base coordinate system. Since in practice the first joint rotation axis of the robot is perpendicular to the fixed plane of the robot or the guideway moving plane, and the first three joints of the robot affect the position of the end-effector and the last three joints affect the attitude of the end-effector, it is not necessary to consider the direction of the guideway moving plane in the optimization. This gives the transformation matrix of the robot-based coordinate system with respect to the world coordinate system ${}_{B}^{W}T = \begin{bmatrix} I_{3\times3} & bp \\ 0 & 1 \end{bmatrix}$, where $bp = [x, y, z]^T$ is the number of unknown quantities varies, according to the base placement. For example, the base position is searched in a certain horizontal plane, and z is a known quantity z_0, determined by the height of the robot placement platform. The base position is searched on a horizontal rail, and x, z or y, z are known quantities. In this paper, the robot was suspended on a horizontal rail, so the base position could be expressed as $bp = [x, y_0, z_0]^T$.

Similarly, the transformation matrix of the robot-based coordinate system, with respect to the world coordinate system, can be obtained from the transformation relationship between the coordinate systems in the Figure 10 ${}_{B}^{W}T = {}_{P_i}^{W}T \cdot {}_{T}^{P_i}T \cdot {}_{T}^{6}T^{-1} \cdot {}_{6}^{B}T^{-1}$.

For a given welding task, i.e., the determined welding path, process parameters, robot, and welding gun, the corresponding coordinate transformation matrix are ${}_{P_i}^{W}T$, and ${}_{T}^{P_i}T$, and ${}_{T}^{6}T$. This implies that the robot base position determines the joint angle of the robot. Therefore, considering that the joint angle of the robot determines its kinematics, the problem of optimizing the base position of the mobile robot arm can be defined as follows: under the condition that the welding path, the process parameters, the robot, and the welding gun are determined, an effective optimization method is established to accurately find the most suitable welding position on the rail, i.e., the optimal base position of the robot.

3.3. Optimization of Path Placement Based on Particle Swarm Optimization Algorithm

In this paper, the particle swarm algorithm was used for optimal control of the extended axis joint variables for robot placement. According to the above analysis, each weld is in the optimal welding position during the welding process, the robot end torch is limited to the weld position along the spatial weld movement, and the robot base is placed and moved by satisfying certain constraints and rules, so that the trajectory of each joint of the robot is smoothest during the whole weld, and, finally, the welding operation time is optimized.

The Particle Swarm Optimization (PSO) algorithm was proposed by Kennedy and Eberhart in 1995, inspired by the movement of flocks of birds. The basic concept of the PSO algorithm is to perform a search based on a population of particles, where each particle represents a potential solution and is accelerated towards a better or more optimal solution by constant updating [23]. The original algorithm is described in detail as follows.

In the D-dimensional space, there are N particles, each representing a potential solution to the optimization problem, having both position and velocity properties, and orienting themselves according to their own best solution and with reference to the best solution of the whole population. The current position vector of the first particle X_i as a candidate solution to the optimization problem, i.e., $x_i = \left(x_1^i, x_2^i, \ldots, x_D^i\right)$, and the current velocity is denoted as $V_i = \left(v_1^i, v_2^i, \ldots, v_D^i\right)$. The optimal solution of the current individual search by the first i is denoted as the individual extremum $p_{best}^i = \left(p_1^i, p_2^i, \ldots, p_D^i\right)$, and the optimal solution of current particle group searched by the first particles i is denoted as the global extreme value $g_{best}^i = \left(g_1^i, g_2^i, \ldots, g_D^i\right)$. The velocity and position of the particles are updated by the iterative process, and the kth iteration updates the formula as follows:

$$v_d^i(k+1) = \omega_0 v_d^i(k) + c_1 r_1 \left[p_d^i(k) - x_d^i(k)\right] + c_2 r_2 \left[g_{bestd}(k) - x_d^i(k)\right] \tag{4}$$

$$x_d^i(k+1) = x_d^i(k) + v_d^i(k+1) \tag{5}$$

In the formula, $i = 1, 2, \ldots, N$; $d = 1, 2, \ldots, D$;. ω_0 are the inertia factors, c_1 and c_2 are the learning factors for the individual and global optimal particles, all of which are positive constants and r_1 and r_2 are random numbers uniformly distributed on the interval of $[0, 1]$. The individual extremes and global extremes are obtained by the above particle update formula continuously iterated until the optimal solution, satisfying the termination condition, is obtained.

The optimization of the robot's initial welding position must take into account the speed of each joint, acceleration and other motion attributes, as well as various performance indicators related to robot welding operations. The particle swarm optimization fitness function is shown in the following equation:

$$\min\left\{f = \frac{1}{n} \times \frac{\sum_{i=1}^{n}\left(\sum_{j=1}^{6} k_j \times |\theta_j^i - \theta_j^{i-1}|\right)}{\sum_{j=1}^{6} k_j}\right\} \tag{6}$$

$$s.t.\ \theta_{jmin} < \theta_j < \theta_{jmax}, j = 1, 2, \ldots, 6; d(R, B) > 0 \tag{7}$$

Equation (6) is the objective function, where θ_j^i denotes the angle of the jth joint at the ith weld point, and k_j denotes the weight of the degree of influence of the jth joint on the robot motion performance. Equation (7) is the constraint function, where θ_{jmin} is the lower limit of the jth joint, and θ_{jmax} is upper limit of the jth joint. Furthermore, d is the collision avoidance condition, and R represents the robot and the truss, and B represents the obstacle.

4. Experimental Results and Analysis

To verify the usefulness of the virtual simulation platform for robotic welding workstations, virtual experiments were conducted for robot welding path placement planning and computational analysis of its process beats, followed by digital twin experiments by combining the virtual system with the actual system. The virtual system simulation required interference collision analysis of the robot motion during the welding process, as well as robot reachability testing.

In the arc welding robot path placement planning experiment, the initial robot welding position test and the interference collision test during the welding process were performed. The optimal initial welding position not only reduces the robot's empty walking distance but also optimizes the operating tempo. Collision detection identifies possible collisions and optimizes the robot's operating trajectory and the layout of the corresponding equipment. A collision between the robot and the surrounding equipment can result in downtime or damage to the welding equipment, reducing productivity. Robot accessibility testing is used to test whether the robot can reach the workpiece weld joint at the optimal initial welding position, allowing for maximum motion maneuverability throughout the welding operation, and further determining the optimal placement of the robot with equipment, such as shifters.

4.1. Welding Robot Base Placement Planning and Process Optimization Virtual Experiments

Before the actual welding operation, the welding process was simulated in the virtual simulation environment built using a digital twin, and the initial welding position simulation and trajectory planning of the robot were performed in the virtual environment to verify the rationality of the optimization scheme. As this paper was optimized for welding efficiency, the appropriate welding parameters determined the quality of welding. Ignoring the premise affecting the quality of welding, according to the existing welding process, the appropriate welding parameters were selected, including welding current, welding voltage, welding speed and wire feed speed. The specific parameters are shown in Table 3 below.

Table 3. Welding parameter setting.

Welding Parameters	Welding Current (A)	Welding Voltage (V)	Welding Speed (mm/s)	Wire Feeding Speed (mm/s)
Value	370	32	7.5	108

In this paper, by interpolating the robot trajectory, the initial welding position of the robot was optimized iteratively using an optimization algorithm, taking into account the robot joint angle, process requirements and other constraints, and the movement of the base when the artificial potential field method was used for the welding seam. According to the algorithm results, shown in Figure 11 below, the optimal initial welding position was obtained as 1501.03 mm. By placing the robot, the robot base was controlled to move to the optimized position to complete the welding task.

Based on this method, in this study, to develop the welding robot cell simulation system for multiple path planning, the simulation of the welding process under the original welding process, according to the optimization results, enabled a comparison of the welding work beat results, shown in Table 4.

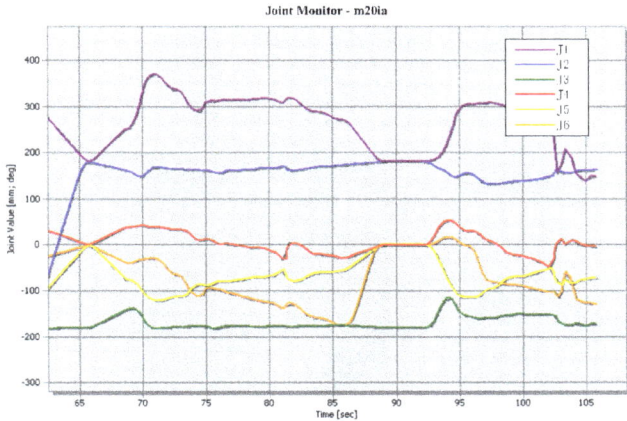

Figure 11. Base joint angle change process.

Table 4. Experimental results of different initial welding positions of the robot.

Initial Welding Position	Welding Operation Time
0 mm	510.44 s
1501.03 mm	489.89 s

As can be seen from the table, after optimizing the initial welding position, the overall welding operation time reduced by 20.55 s, the robot empty walking distance reduced by 15,010.03 mm, and the process optimization efficiency increased by 7.809%. After optimization of the robot base position to obtain the robot joint angle, as shown in Figure 12 below, the joint angle fluctuated slowly during the welding process, which showed that the optimal initial welding position could not only ensure the smooth operation of the robot arm, but also optimized the welding time and achieved the effect of enhancing process optimization.

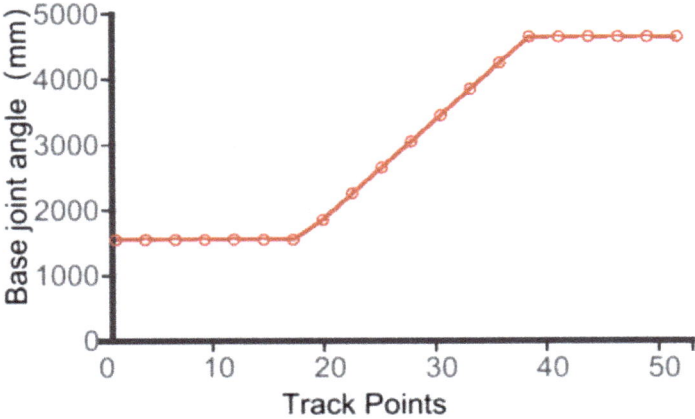

Figure 12. Angle of each joint of the robot.

Figure 13 Virtual welding experiment screenshot shown is a screenshot of the arc welding robot for welding operations and virtual experiments, through the virtual simulation of the welding workstation welding operations, the ability of real-time calculation of the robot motion path, process visualization, and real-time analysis of the weld point accessibility. The experiment proved that the welding robot workstation digital twin system could realize the robot welding operations and process optimization in the virtual system.

Figure 13. Screenshot of virtual welding experiment.

4.2. Digital Twin Experiment

The digital twin system of the robotic welding workstation drives the twin model motion through data scripting and realizes the virtual real mapping between physical space and virtual space through twin data interaction. To further validate the usefulness of the digital twin system, the virtual real system was connected, the initial welding position obtained from the virtual simulation optimization, the path planning transmitted to the physical space, and the data tracking of the welding cell realized through the twin data management system, which provided a comprehensive view of the entire operation. During the creation of the welding workstation arc welding robot twin, the welding process joint data, external axis data, and process data were reflected in the digital twin in real time, thus achieving real-time monitoring of the robot welding process.

Physical space and virtual space reality mapping, as shown in Figure 14, in the welded parts welding solution experiments, first initialized the state of the production line model in the virtual system, so that the virtual cell body of the workpiece, robot, changer and all other components, and the initial state of the physical cell body. This was necessary to ensure that the physical cell and virtual cell data connection achieved virtual–physical space state synchronization, and then the physical and virtual system could be run for welding operations. The experimental results are shown in the table, and the kinematic and path placement optimization algorithms in the welding workstation service system were used to optimize the robot base placement and drive the robot motion through the control system. From Table 5 it can be seen that the global operability of the robot welding operation in the physical welding workstation at the optimal initial welding position, as well as the weld reachability, achieved better production results, along with the normal twin data acquisition and sending.

Table 5. Experimental results.

Unit	Reachability %	Whether Data Is Sent
Physical Robot	100	Yes
Virtual Robot	100	Yes

Figure 15 shows the single-station digital twin system for robotic welding. This paper realized the data interaction between physical space and virtual space through HTTP protocol, completed the real-time synchronization and faithful mapping between the virtual and real systems, and visualized the welding process and process data through the system's

UI interface in real time. Compared with the traditional process beat spreadsheet statistics, the digital twin was real-time and effective in delivering and displaying welding operation process data.

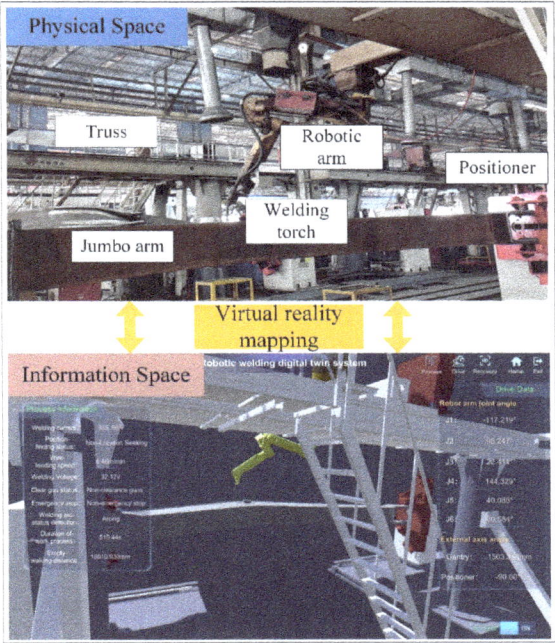

Figure 14. Physical space and virtual space virtual reality mapping.

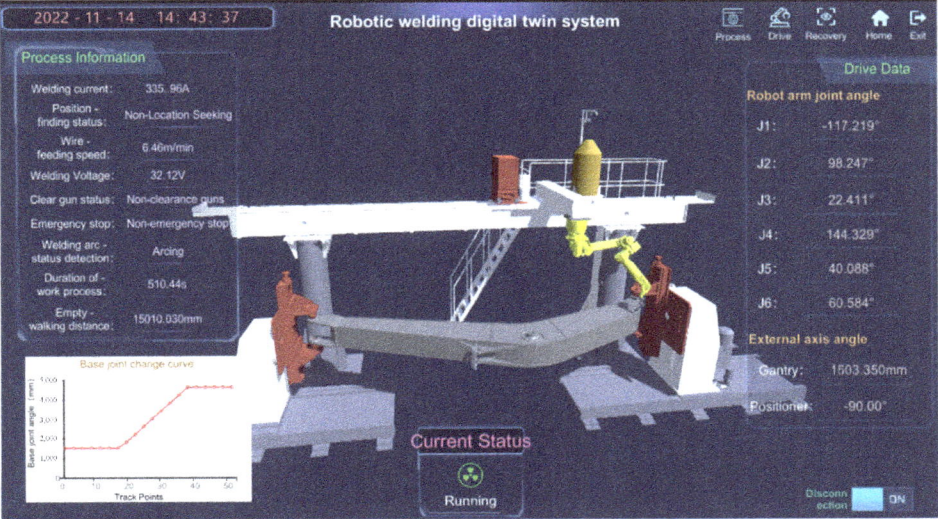

Figure 15. Experimental demonstration of welding twin process.

5. Conclusions

In this paper, an arc welding robot cell process optimization, based on digital twin, was studied. First, a virtual simulation system of the welding robot cell, based on the digital twin, was proposed and then it was constructed. By establishing a digital twin, a digital virtual simulation platform of the welding process and welding environment of the welding robot was created. Secondly, the optimal initial welding position was optimized, based on the welding process, using the particle swarm seeking algorithm to find the optimal initial welding position, and the kinematic analysis calculation and path optimization of the robot. Finally, the mapping relationship between the virtual system and the physical system model was established by using various types of sensor data transmission, and data communication was carried out using a socket network to achieve a faithful mapping of the virtual welding workstation to the actual welding workstation and to realize real-time monitoring of the welding process.

In this paper, we completed the simulation of welding motion, visualization of real-time operation data, monitoring of the welding process and optimization of the robot welding process in the digital environment for the large movable arm welding process, realized efficient collaboration between the robot and the positioner during the welding process, optimized the robot idle time, improved the efficiency of the welding operation, and laid the data foundation for subsequent robot research and automation of the arc welding process.

At present, the research is still in its infancy and a lot of work needs to be done. Subsequent research will focus on the following aspects in greater depth: (1) the use of a digital twin service system, improving the dynamic arm welding process database, from the welding process of welded parts (weld quality, weld welding sequence) and other aspects of process optimization; (2) for the real-time welding process, network communication caused by real-time data delays needs to be further explored; (3) the use of data visualization to achieve health management of production equipment, fault warning and continuous iterative optimization of the operating state.

Author Contributions: Conceptualization, R.X. and Q.Z.; methodology, R.X.; software, R.X.; validation, R.X., Z.L. and Q.Z.; formal analysis, R.X.; investigation, R.X.; resources, R.X.; data curation, R.X.; writing—original draft preparation, R.X.; writing—review and editing, R.X. and Z.L. and J.D. and J.Q.; visualization, R.X. and Z.L.; supervision, J.D. and J.Q. All authors have read and agreed to the published version of the manuscript.

Funding: This research received no external funding.

Institutional Review Board Statement: Not applicable.

Informed Consent Statement: Not applicable.

Data Availability Statement: Not applicable.

Conflicts of Interest: The authors declare no conflict of interest.

References

1. Hoebert, T.; Lepuschitz, W.; Vincze, M.; Merdan, M. Knowledge-driven framework for industrial robotic systems. *J. Intell. Manuf.* **2021**, 1–18. [CrossRef]
2. Wang, T. Research Status and Industrialization Development Strategy of Chinese Industrial Robot. *Chin. J. Mech. Eng.* **2014**, *50*, 1–13. [CrossRef]
3. Bai, Y. Industrial Internet of things over tactile Internet in the context of intelligent manufacturing. *Clust. Comput.* **2017**, *21*, 869–877. [CrossRef]
4. Qu, Y.J.; Ming, X.G.; Liu, Z.W.; Zhang, X.Y.; Hou, Z.T. Smart manufacturing systems: State of the art and future trends. *Int. J. Adv. Manuf. Technol.* **2019**, *103*, 3751–3768. [CrossRef]
5. Luo, Y.; Fang, Z.; Guo, J.; Lu, H.; Li, J. Research on the virtual reality technology of a pipeline welding robot. *Ind. Robot. Int. J. Robot. Res. Appl.* **2020**, *48*, 84–94. [CrossRef]
6. Cimino, C.; Negri, E.; Fumagalli, L. Review of digital twin applications in manufacturing. *Comput. Ind.* **2019**, *113*, 103130. [CrossRef]
7. Tao, F.; Qi, Q.; Liu, A.; Kusiak, A. Data-driven smart manufacturing. *J. Manuf. Syst.* **2018**, *48*, 157–169. [CrossRef]

8. Lu, Y.; Liu, C.; Kevin, I.; Wang, K.; Huang, H.; Xu, X. Digital Twin-driven smart manufacturing: Connotation, reference model, applications and research issues. *Robot. Comput. Integr. Manuf.* **2019**, *61*, 101837. [CrossRef]
9. Xue, Y.; Sun, Z.; Liu, S.; Gao, D.; Xu, Z. Stiffness-Oriented Placement Optimization of Machining Robots for Large Component Flexible Manufacturing System. *Machines* **2022**, *10*, 389. [CrossRef]
10. Yang, J.; Yu, W.; Kim, J.; Abdel-Malek, K. On the placement of open-loop robotic manipulators for reachability. *Mech. Mach. Theory* **2009**, *44*, 671–684. [CrossRef]
11. Franceschi, P.; Mutti, S.; Pedrocchi, N. Optimal design of robotic work-cell through hierarchical manipulability maximization. *Robot. Comput. Manuf.* **2022**, *78*, 102401. [CrossRef]
12. Gadaleta, M.; Berselli, G.; Pellicciari, M. Energy-optimal layout design of robotic work cells: Potential assessment on an industrial case study. *Robot. Comput. Manuf.* **2017**, *47*, 102–111. [CrossRef]
13. Di, J.; Xu, M.; Das, N.; Yip, M.C. Optimal multi-manipulator arm placement for maximal dexterity during robotics surgery. *arXiv* **2021**, arXiv:2104.06348.
14. Ren, S.; Xie, Y.; Yang, X.; Xu, J.; Wang, G.; Chen, K. A Method for Optimizing the Base Position of Mobile Painting Manipulators. *IEEE Trans. Autom. Sci. Eng.* **2017**, *14*, 370–375. [CrossRef]
15. Aspragathos, N.A.; Foussias, S. Optimal location of a robot path when considering velocity performance. *Robotica* **2002**, *20*, 139–147. [CrossRef]
16. Tao, L.; Quan, Q.; Chunjiang, Z.; Feng, X. Task planning of multi-arm harvesting robots for high-density dwarf orchards. Nongye Gongcheng Xuebao/Trans. Chin. Soc. Agric. Eng. **2021**, *37*, 1–10. [CrossRef]
17. Constantinescu, D.; Croft, E.A. Smooth and time-optimal trajectory planning for industrial manipulators along specified path. *J. Robot. Syst.* **2000**, *17*, 233–249. [CrossRef]
18. Liu, J.; Zhou, H.; Tian, G.; Liu, X.; Jing, X. Digital twin-based process reuse and evaluation approach for smart process planning. *Int. J. Adv. Manuf. Technol.* **2019**, *100*, 1619–1634. [CrossRef]
19. Li, L.; Liu, D.; Liu, J.; Zhou, H.-G.; Zhou, J. Quality Prediction and Control of Assembly and Welding Process for Ship Group Product Based on Digital Twin. *Scanning* **2020**, *2020*, 1–13. [CrossRef]
20. Wang, Q.; Jiao, W.; Wang, P.; Zhang, Y. Digital Twin for Human-Robot Interactive Welding and Welder Behavior Analysis. *IEEE/CAA J. Autom. Sin.* **2021**, *8*, 334–343. [CrossRef]
21. Kim, M.; Sung, N.-J.; Kim, S.-J.; Choi, Y.-J.; Hong, M. Parallel cloth simulation with effective collision detection for interactive AR application. *Multimedia Tools Appl.* **2019**, *78*, 4851–4868. [CrossRef]
22. Chang, J.-W.; Kim, M.-S. Efficient triangle–triangle intersection test for OBB-based collision detection. *Comput. Graph.* **2009**, *33*, 235–240. [CrossRef]
23. Coello, C.A.C.; Toscano-Pulido, G.T.; Lechuga, M.S. Handling multiple objectives with particle swarm optimization. *IEEE Trans. Evol. Comput.* **2004**, *8*, 256–279. [CrossRef]

machines

MDPI

Article

Production Planning Process Based on the Work Psychology of a Collaborative Workplace with Humans and Robots

Felicita Chromjakova

Department of Industrial Engineering and Information Systems, Faculty of Management and Economics, Tomas Bata University in Zlín, Mostní 5139, 76001 Zlín, Czech Republic; chromjakova@utb.cz

Abstract: This study focuses on discerning how economics, as it pertains to work psychology, is lent a new perspective by the compatibility of humans and robots cooperating in the manufacturing sector. The stability of production plans, flexibility of the organizations, and the management of production constitute the basis for such analysis. In this context, initial findings revealed that steady performance by an individual was significantly influenced by a production plan, while the cycle and lead times in place fundamentally affected the behaviour of employees. Observations were made over five years of 200 workers at 100 manufacturers. Times given over to operations and cycles, and throughput, were primarily defined by the technical cycle of the robot. The secondary element of production planning was the employee, whose operator cycle time was informed by that of the robot. The authors set out to deduce which key factors altered the work psychology in situ in manufacturing environments where collaboration occurred between humans and robots. Prerequisites for optimal psychological conditions were identified (the cooperating human, production planner, collaborative workplace, standardized durations of complete tasks, distance between the worker and robot, and data analytics of production flow). Ensuring circumstances are optimal in terms of work psychology is essential to raising productivity and employee performance. Results showed that the operator was directly dependent on the robot in relation to mutual, continuous production flow. A model of production plan stability was devised, informed by the dependence of specific parameters of the planning model. Research was conducted on the reliance of selected parameters, leading to establishment of prerequisites for an optimal work psychology setting in enterprises with such a collaborative structure.

Keywords: Industry 4.0; human–cobot interaction; job; psychology; process

check for updates

Citation: Chromjakova, F. Production Planning Process Based on the Work Psychology of a Collaborative Workplace with Humans and Robots. *Machines* 2023, 11, 160. https://doi.org/10.3390/machines11020160

Academic Editors: Raul D.S.G. Campilho and Francisco J. G. Silva

Received: 10 December 2022
Revised: 17 January 2023
Accepted: 18 January 2023
Published: 23 January 2023

1. Introduction

Fundamental to the matter is ensuring the psychological balance of a person in this kind of collaborative workplace. Production planners make a series of trade-offs in schedules every day in the pursuit of maximum production capacity. A robot intended for such an environment must be highly efficient in regards to productivity with minimal risk of failure. In this context, determining an optimal production schedule requires taking the physical and psychological capabilities of the worker collaborating with the robot into consideration.

A key prerequisite for optimal manufacturing performance is having a stable production process, integrated with Operator 4.0. Operator 4.0 is a smart and skilled operator who performs not only 'cooperative work' with robots, but also 'work aided' by machines as and if needed, using human cyber physical systems, advanced human–machine interaction technologies, and adaptive automation towards 'human-automation symbiosis work systems'. Within the setting of a human and a robot being directly part of the collaborative workplace, a paradigm shift occurs in how such a collaborative environment is set up. A traditional approach functions in line with the derived arrangement of machine and human operations, while a digitally controlled system is scheduled according to the cycle time of the robot and the managed technology. However, intelligent production demands a significant alteration be made in terms of the psychology of the worker, since human–human and robot–human relationships change when a robot possesses an equal collegial position as a worker.

Thorvald et al. [1] identify a 'collaborative operator' as a human cooperating with collaborative robots (cobots). This cooperation is based on coexistence in a shared physical space between two agents: human and cobot. The Cognitive Operator 4.0 represents knowledge and cognitive interactions in which it exchanges process information with the help of technologies to symbiotically perform a task. It is an important part of process management by collaborative workplace the relationship between workload and workers' situation awareness [2]. Artkin [3] defines situation awareness as the ability of smart agents (both humans and artificial) to understand what is happening in the surrounding environment. Based on the Endsley definition [4], the human-decision-making process depends on comprehension of a human meaning and the projection of their status in the near future. Our research, conducted in 100 manufacturing companies, confirmed that as long as a person has experience in a given workplace, it is important to transform this into the operational standards of the workplace, and, at the same time, share his experience in the team of colleagues (cobot is the colleague, robot process, and operation standard improvement). This significantly helps reduce stress before and during the implementation of collaborative cooperation. We called this moment 'predictive response ability' and the core content will be analysed in the upcoming months.

The process of production planning and scheduling requires flexibility from a human perspective, in addition to which adhering to a plan must constitute a priority and the general approach remain flexible [5]. When production flows smoothly and in accordance with the plan, the psychological burden on the employee is normal. This balance is upset by any outages that affect a work shift. A serious issue arises if a robot has to wait for the input or output of a human, heightening the psychological demands placed on the latter. A number of system-oriented connections pertain to maximal cooperation between a human and cobot (a "collaborative robot") within a collaborative work environment [6], as currently being investigated by more than 100 SMEs (in the automotive sector); further description of the matter is given in this paper.

Production planning and scheduling based on digitally controlled processes have affected the psychology of human labour. Procedures governed electronically are managed by software ("SW", by a "digital manager", as defined later); hence, people react exclusively to the prompts provided by the controlling SW. Industrial enterprises have been endeavouring to find a suitable method of combining such technology with lean production activities, and the introduction of Industry 4.0 has innovated manufacturing processes and the use of equipment [7]. In fact, smart technologies have brought about improvement in production-related performance by affording real-time assessment of it [8]. The basis for electronic management of planning and scheduling in manufacturing is knowledge of interrelationships between the parameters of procedures. Each activity has its own specific variable and interval range, and it is important to collect data on ongoing operations, with subsequent analysis revealing the dependencies of such variables in processes. Performance in this context directly depends on the potential to achieve full capacity by the human, robot, and given technology. Assessing the effectiveness of a person is now possible in real time, as is the evaluation of that of the robot and, maybe, the advanced production technology in place. Though it has been observed that a robot performs work consistently at a planned rate, performance in this regard by an employee will occasionally diminish for various reasons. The requirement to provide comfortable conditions for them is crucial in relation to the aspect of work psychology, since workers frequently experience stress and frustration stemming from their performance of tasks. From this point of view, it is necessary to concentrate on planned and online data gathered from the workplace, with analysis making it possible to predict how the worker shall behave as per the manufacturing schedule. Statistical analysis of production processes shows that the ability to manage data from smart processes permits transformation of them into directives for correct management of employees.

A workplace with collaborative production is typically one where workers are forced to concentrate greatly in cooperation with robots. Practical experience and knowledge of various operational situations that take place routinely are critical, because employees undertake duties which they actively control and conduct repeatedly. In contrast, the robot is subordinate to the lead time designated to it with precisely set cycle durations. The psychological difficulty of human–robot cooperation could eventually result in frustration amongst workers; for example, in the event of an involuntary delay in entering a production-related request, or should the robot be ahead and receive work late, leading to it outpacing the I/O of its human co-worker [9]. Disruptions in production flow on the part of the worker might also occur. A way of avoiding issues like these would be to schedule defined downtimes in the activity of workers, then adapt the cycle time of the robot accordingly.

2. Theoretical Review

Recently, there has been an increased interest in conducting research on work psychology. Factors, such as frequent turnover of staff, instability in work positions, a rise in the coefficient of absenteeism, and an increase in occupational diseases, are behind the trend. One tool that facilitates data analysis of cooperation in a collaborative human–robot environment is dubbed "process mapping", which involves obtaining continuous information on arguments affecting the psychological well-being of workers at a production facility [10]. Investigation has focused on defining the performance of staff, i.e., parameters for quantifying and qualifying human performance, though work psychology was considered by the given author as largely ignored [11]; results therein showed that efficient work systems directly correlated with the psychological state of staff. Arefin [12] stated that a psychologically positive manufacturing environment directly influenced the performance of employees. The literature also reports that organization of work practices through targeted psychological measures affects such performance, while Amor [13] wrote that the latter could be managed by applying work psychology as a basic mechanism. Elsewhere, proper orientation of work psychology brought about benefits in psychological capital and job performance, with an associated reduction in reports of burnout [14].

The current recession, economic fluctuations and circumstances relating to the COVID-19 pandemic have highlighted how important it is to consider the psychological aspects of collaborative work, a constituent part of the rapidly growing trend of Industry 4.0 [15]. Two key concepts exist in psychology, those of the "digital worker", and the "digitally controlled worker" (hereinafter referred to as "DW" and "DCW", respectively). A digital worker is a software-controlled robot or machine directed in its operations by elements of artificial intelligence, whereas a digitally controlled worker is a person who does tasks strictly in accordance with prompts given by software. Such an employee in an environment of collaborative production becomes both an initiator of ethics and work standards, and a user of them [16]. The digitization of work brings people and technology ever closer, enabling staff to implement related operations and voice their opinions in the enterprise. Moreover, it facilitates the management and control of duties performed remotely (e.g., at home), and aids employees in pursuit of their jobs and responsibilities. In essence, the DW acts as the primary support mechanism for the DCW. This advancement has encouraged investigation of psychology of work with respect to improving manufacturing processes and innovating associated technologies [17]. A stable process methodology is required to enable this, and the ratio of the DW to DCW determines the foundation upon which the consequent work psychology is based. An analysis of 100 industrial companies in the automotive sector revealed that the more the DW prevails over the DCW, the greater the amount of attention that has to be paid to the manual labour of staff and organization of their duties; thus, consideration is necessary as to how compatible the competence of a person (know-how and skills) relating to digital technologies is with the given production process [18].

The analysis and monitoring of humans integrated into collaborative workshop operations brought information about the changes in workers that occur in a regular work shift in a collaborative workplace (human–robot combination). Before the definition of the parameters was investigated, the analysis of the production shift plan by 200 selected operators was performed (Figure 1).

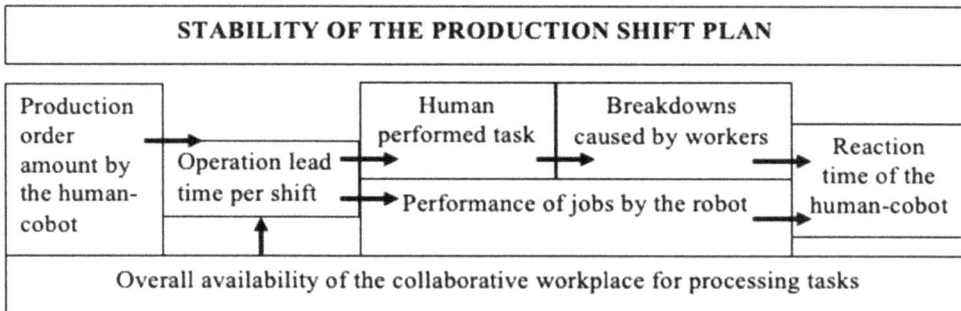

Figure 1. Mutual relationships in the collaborative workplace "human-cobot" (source: author).

The parameters of 'human–cobot collaboration' were identified in collaborative assembly workplaces. Based on the process and ergonomic analysis, the selected workplaces were identified. The selection process was determined by the frequency of the operation, the temporal stability of the operation in the given workload, and the workplace process quality. Continuity with the robot was also important. Herein, information was gained from conducting analysis and field research of people performing operations within a collaborative environment, specifically those on a regular shift at a workplace shared by humans and robots. The following parameters were under study: The following parameters were investigated here:

- the physical load on a staff;
- the psychological stress of a person;
- the stability of a person's workflow;
- the rotation of employees across various workplaces;
- technologically (digitized) assisted monitoring of the performance of individuals;
- the flexibility of workers with regard to downtime in production;
- the proportion of operations undertaken by digital technologies and employees.

The following process-oriented assumptions for optimal work psychology by collaborative workplace were identified:

- flexibility of the employee (settings to ensure a suitable workplace and production flow in real time);
- productivity of the employee (conflict-free, continuous performance of operations with regard to the capacity of the robot);
- set, standardized inputs for processes and manufacturing operations to permit fulfilment of duties by the employee;
- online monitoring of data to aid the organization and management of production;
- horizontal and vertical integration of process data, setting IDs for coded process actions linked to the instigator of an action (employee, machine, robot, information, etc.).

Subsequently, the mutual correlation of the parameters 'human–cobot collaboration' and assumptions oriented to the collaborative process was investigated (Table 1).

Table 1. Mutual correlation of human–cobot collaboration parameters and collaborative process-oriented assumptions (source: own research).

Human–Cobot Collaboration Parameters	Process Oriented Assumptions of Work Psychology				
	Man Flexibility	Man Productivity	Stable Standardisation	Online Data Monitoring	Process Data Integration
physical load on a person at work	27 M	120 M	56 M	137 M	7 M
	12 W	12 W	9 W	63 W	12 W
psychological stress of a person	97 M	121 M	95 M	127 M	78 M
	43 W	56 W	8 W	45 W	53 W
the stability of a person's workflow	132 M	**137 M**	137 M	127 M	112 M
	63 W	**64 W**	63 W	61 W	63 W
worker rotation across multiple workplaces	109 M	127 M	37 M	**137 M**	127 M
	57 W	48 W	24 W	**62 W**	54 W
digitally set monitoring of a person's work performance	126 M	**137 M**	45 M	135 M	111 M
	58 W	**660 W**	27 W	58 W	52 W
worker flexibility due to production process downtime	**137 M**	110 M	127 M	135 M	97 M
	58 W	57 W	56 W	**57 W**	46 W
share of digital and human-driven production activities	116 M	87 M	12 M	35 M	7 M
	35 W	54 W	16 W	23 W	13 W

Research was conducted on a sample of 200 operators. Of these, 137 were men (M) and 63 women (W). Each operator was monitored 30 times to objectively assess the static or dynamic load during the job execution. Each row contains the resulting value of the number of operators who had a psychological/stress problem during dynamic changes that were related to random situations at the given workplace and at the given moment of the operation.

Overall, the employees of manufacturing companies became experts at new production technologies and gained experience from digitally controlled production. Should technical conflicts arise, they have the option to discuss and resolve the situation with a colleague or a digital manager. Numerous companies are training employees to adopt a suitable digital and technological perspective.

3. Materials and Methods—Research Limitations

Based on the stability model for the work shift schedule, identification was made as to prerequisites for work psychology and standardized definitions of operations at the collaborative workplace were determined.

With a digitized process at its core, the concept described herein encompasses the aspects of the employee, robot, and digital technology; the flow of data is controlled by a SW algorithm. Key support variables pertain to the extent of digital literacy on the part of the employee and robot, although they obviously differ widely in actual content and procedure, and in relation to applicable work psychology. Cobots are designed to be more akin to co-workers than mere tools, and fluent interaction between operators and their robotic counterparts is critical to the "task and high performance" of the former [19]. Important facets include technological factors (e.g., occupational safety), appropriate configuration of the cobot, employee-centred matters (e.g., fear of redundancy), and gaining an appropriate level of trust with the robot [20]. In the context of work psychology, the link between efficiency and employee satisfaction is crucial, since it represents a function of how much control the worker has over their role in the human–cobot manufacturing team [21].

Based on rules for production planning and scheduling, the following pillars of a collaborative environment were identified:

- human–cobot cooperation is directly dependent on there being sufficient standardized processes in place;
- a production planner is able to identify key parameters for ethical human–cobot cooperation;
- a standardized workplace ensures optimal ethical conditions to achieve the desired level of performance from a human–cobot workforce;

- compliance assessments adhere to the same regulations for humans and cobots, with respect to the duties conducted;
- workplace processes and environment standards are governed by principles of eliminating waste and benefiting from efficacy on the part of the robot (minimization of e-waste);
- all crucial responsibilities of the employee and competences of the cobot are identified and applied as a basis for standardization.

The author conducted interviews with production managers at selected industrial enterprises, the primary topic being work psychology in collaborative workplaces. The following parameters were specified in the production operator's research: the operator controls the communication options with the robot, the robot gives feedback to the human about the performance at the workplace, and the human–cobot conflict resolution process is standardized. As a part of empirical research, we focused on in-depth interviews with assembly operators. Each operator had at least one year of job experience in the given collaborative workplace. The age structure of the operators was between 35 and 52 years old. Empirical research was conducted during 2021–2022. To evaluate the results, we used qualitative research, where we identified and investigated the effects on work psychology at the operator's workplace. Subsequently, we defined significant parameters for the human–cobot collaborative workplace through a comparative analysis from 2 operators, at minimum, by workplace (shift change view on problem). Analysis revealed that up to 86% of the respective employees were affected by psychological stress stemming from the production process; a crucial finding in the opinion of the authors (Table 2). The robot, acting as a colleague in the collaborative process, has the capacity to recognize when an issue arises that impacts the psychological state of the worker during human–cobot cooperation, and respond to it accordingly. The efficiency with which this kind of relationship functions is reliant on the mental well-being of the human and their degree of trust in cooperative endeavours with the robot. These two aspects were seen to be decisive in the research conducted at industrial companies. Manufacturing productivity and ergonomics issues brought extensive opportunities for cobot safety, workplace design, and task scheduling [22].

Table 2. Analysis of the human–cobot collaborative workplace (source: own research).

Human–Cobot Workplace Analysis: Factors Affecting Work Psychology Relating to Planning and Scheduling Practices					
Factor of Work Psychology	Aspect of Human–cobot Cooperation (Share in % Labour Undertaken by Human and Robot for a Planned Task)	Number of Collaborative Workspaces Monitored at 200 Selected SMEs	Stability of the Production Plan in Collaborative work (% of Mutual, Planned Tasks Conducted)	Number of Psychological Conflicts That Arose between the Human and Robot	Acceptable Lead Time of Collaborative Workplace (% of Plan Completed during Actual Lead Time)
Job definition	70/30	134	80	(1–3)	(90–95)
	40/60	27	60	(6–8)	(80–85)
	30/70	18	70	(2–3)	(90–95)
	20/80	17	70	(3–5)	(80–85)
	10/90	4	90	(2–6)	(90–95)
Production inputs (availability)	90/10	176	90	(8–10)	(80–85)
	80/20	23	75	(7–12)	(85–90)
	75/25	1	80	(0–1)	(80–85)
Continuous production flow before and after collaborative endeavours	80/20	147	80	(5–7)	(80–85)
	85/15	34	85	(4–6)	(90–95)
	90/10	17	85	(2–3)	(80–85)
	95/5	2	90	(0–1)	(90–95)
Technological support provided by staff	YES	170	95	(2–8)	(85–90)
	NO	27	80	(15–20)	(75–80)
	PARTIALL	3	85	(14–17)	(80–85)

Comment: Total monitored, 200 SMEs—collaborative workplace; of this, 130 SMEs—standardised rules by hybrid workplaces (balanced cooperation human-cobot); 180 SMEs—standardised human performance in a flexible production time; 114 SMEs—production planning and scheduling of digitised processes in cooperation with humans, and according to the reflection on the ethical rules based on actual shop floor conditions.

Descriptive statistics were generated to determine correlations between the given factors for work psychology. Analysis of the data showed that the primary issue affecting production planning and scheduling in a collaborative environment was the availability (i.e., readiness) of the workplace to perform tasks. When such circumstances are favourable, a positive physiological state (satisfaction) prevails amongst staff. The lead time of the employee is dependent on the frequency of continuous production flow, which corresponds with performance per shift (in relation to lead time).

The analysis of the collaborative environment was conducted to verify the scientific hypotheses below:

Hypothesis 1 (H1). *The psychological well-being of a person directly relates to a defined work task.*

Hypothesis 2 (H2). *The production plan and associated schedule are conditioned by availability at the collaborative workplace (in the context of achieving the desired level of performance).*

Hypothesis 3 (H3). *The time taken to react to an issue at a collaborative workplace is a controlled process.*

The following findings were arrived at from investigation of the 200 industrial enterprises:
H1: 183 SMEs ("yes"), 10 SMEs ("no"), and 7 SMEs ("partially"—conflicts arose through changes in production priority).
H2: 165 SMEs ("yes"), 29 SMEs ("no"), and 6 SMEs ("partially"—conflicts arose from production standards).
H3: 120 SMEs ("yes"), 22 SMEs ("no"), and 58 SMEs ("partially"—conflicts arose based on the technologically related cause of unavailability at the collaborative workplace).

The robot performs and undertakes tasks according to precisely defined instructions. It can continue in its duties with adherence to the given production schedule, and has exactly determined functionality for jobs, in addition to which it is able to operate in flexible modes to react to changes in arrangements. The employee represents the immediate co-worker of the robot. Unforeseen circumstances may invoke stress in the worker through having to adapt to alteration in production flow; hence, possessing a state mental readiness is crucial in a collaborative environment. Since the concentration of the worker is affected by the factor of availability (in connection with tasks), attention should be paid to how such tasks are prioritized. The associated learning curve plays an important role, too, especially regarding the application of skill sets of the worker for the performance of operations.

4. Results

Research was conducted in cohort comprising 100 industrial companies and 200 operators. Of these, 130 were men (65%) and 70 women (35%). Almost 85% of the operators had a graduated from secondary school, while 15% possessed a university qualification. The collaborative robots were primarily products by KUKA, FANUC, and ABB.

The correlation of human–cobot collaboration parameters and collaborative process oriented assumptions (Table 1) showed that the key factors of work psychology are connected with the stability of a person's workflow/man productivity, worker rotation across multiple workplaces/online data monitoring, digitally set monitoring of a person's work performance/man productivity, worker flexibility due to production process downtime/man flexibility, and worker flexibility due to production process downtime/online data monitoring. A critical parameter of work psychology is the stability of the work process, which is why the research of collaborative workplaces was first focused on the stability of the production shift plan (Figure 1). Similar arguments are presented in various research papers, such as [23], fluent interaction between the operators and their robotic counterparts; by [24], "resiliency" and "flexibility" capabilities of humans; or by [25], shared workspace for human and robots. Flexibility, cognitive skills, and dynamic thinking by humans are most important parameters of effective work psychology by collaborative workplaces.

The main objective of research was to obtain data on work psychology in a collaborative environment, with investigation of staff employed in such a facility in relation to the following parameters (Figure 1):

1. Amount of production orders for the human–cobot collaborators per shift (POA_{hc}).
2. Operation lead time per shift (OLT_s).
3. Cycle time of the human (CT_{man}).
4. Human-originated breakdowns (BD_{man}).
5. Reaction time of the human–cobot arrangement (RT_{hc}).
6. Stability of the plan for modifying production (S_{sp}).
7. Overall availability (capacity) to process production tasks (OAC).

The structure of the proposed indicators corresponds to the production planning indicators. Workload efficiency is based primarily on a stable production plan and a technologically acceptable work schedule. The objective of the empirical research was to determine the stability of the work process in the collaborative workplace. These parameters were chosen as they are a constituent part of planning and scheduling. From a human point of view, a technological acceptable work schedule dictates the stability of a production plan, and such a schedule is a precondition for realizing optimal work psychology in a collaborative environment. A total of 20 repeated observations were made for each employee, the priority being to discern parameters crucial to their cooperation with the robots, since data analytics revealed that the latter did not deviate from the given schedule within the period of a month. Only two companies experienced variance in performance by the robot through unexpected downtime, i.e., 3% of their planned lead time (Table 3). On the contrary, the humans showed several deviations from the planned production time. Therefore, we investigate deviations in work performance by humans, which are related to work psychology.

Table 3. Production flow analysis in a collaborative workplace: sample analysis and calculation of basic variables (selection) (source: author).

Amount of Production Orders Fulfilled Collaboratively per Shift	Lead Time of Operations per Shift	Cycle Time-Human	Human-Originated Breakdowns	Reaction Times	Overall Availability for Realization of Production Tasks	Stability of Shift Plans
POA_{hc} [min]	OLT_s [min]	CT_{ma} [min]	BD_{man} [min]	RT_{hc} [min]	OAC [min]	S_{sp}
420	420	12	21	3		1.00
	$OAC = OLT_s - CT_{ma} - BD_{hum} - RT_{hc}$				384	0.91
410	420	9	10	6		0.98
	$OAC = OLT_s - CT_{ma} - BD_{hum} - RT_{hc}$				395	0.96
400	420	4	12	2		0.95
	$OAC = OLT_s - CT_{ma} - BD_{hum} - RT_{hc}$				402	1.01
420	420	3	5	5		1.00
	$OAC = OLT_s - CT_{ma} - BD_{hum} - RT_{hc}$				407	0.97
395	420	7	9	6		0.94
	$OAC = OLT_s - CT_{ma} - BD_{hum} - RT_{hc}$					1.01

The setting for volume of production crucially affects work psychology, as a prerequisite for this to be optimal is workplace stability from the perspective of staff; after all, they have all the inputs necessary for continuous performance during a shift. It was seen that work psychology benefited from greater stability (consistency) in performance as per workflow, and a higher level of quality in this respect was observed. Table 1 illustrates how such stability in a collaborative environment was dependent on reactions given to instances of downtime.

POA_{hc} = 420 min	OAC 384 min	S_{sp} = 0.91
POA_{hc} = 410 min	OAC 395 min	S_{sp} = 0.96
POA_{hc} = 400 min	OAC 402 min	S_{sp} = 1.01
POA_{hc} = 420 min	OAC 407 min	S_{sp} = 0.97
POA_{hc} = 395 min	OAC 398 min	S_{sp} = 1.01

Subsequent attention was paid to intervals of permitted limits for the parameter S_{sp} for stability. Data on 200 collaborative employees revealed a dependence in the connection between productivity (OAC) and available performance (S_{sp}). The causes of instances of downtime were also analysed, highlighting a correlation between the human and robot as per the times required to fulfil tasks. The instant a mutual task commences is key, and work psychology influences matters at this point within a sequence of collaborative, connected stages (time for initiating a task + time taken to do a task + time for finalizing a task). It is important for a person in such a collaborative team to be situated in front of the robot or perform duties behind it within a clearly defined time frame.

The collaborative robot has the ability to spot when a problem will occur, and send a pre-emptive signal to the operator, followed by a recommendation on rectifying it, thereby lending psychological support to the person. Work psychology is affected by various situations involving the operator:

- staff may intuitively react and decide to tackle an issue with regard to their physical and psychological safety, potentially conflicting with process management;
- the operator follows recommendations given by the robot, diminishing their emotional state, and increasing the chance of errors being made in decision-making and duties;
- the employee chooses to solve a matter in their way, based on personal experience, rather than taking the advice of the cobot, and the human could feel a sense of accomplishment or disappointment in subsequent performance.

In addition to planning and scheduling, achieving a state of work psychology is reliant on solid preparation of all attributes involved in managerial processes (Figure 2). Important elements relating to production flow that affect the psychological well-being of staff include the types of components incorporated in products, assembly tolerances, and time constraints for resetting machinery, and preparatory or final operations. These fundamentally impact the productivity of staff and place emphasis on performance. In the context of an assembly line, for example, the employee is expected to function with a high degree of flexibility, prioritizing it over the aspect of time.

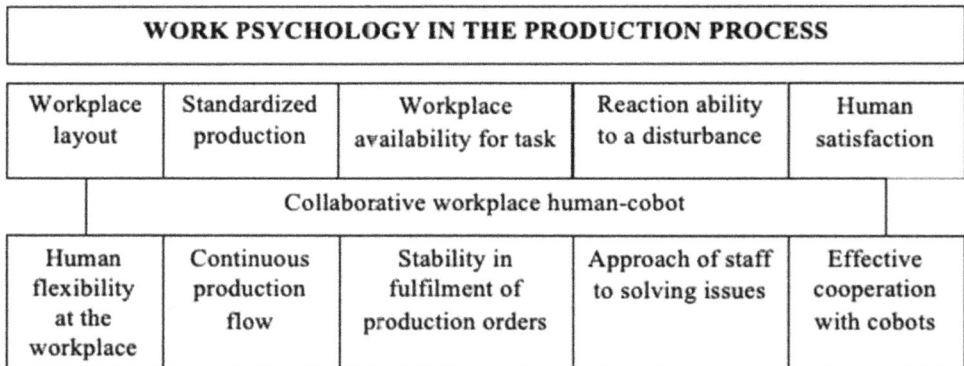

WORK PSYCHOLOGY IN THE PRODUCTION PROCESS				
Workplace layout	Standardized production	Workplace availability for task	Reaction ability to a disturbance	Human satisfaction
Collaborative workplace human-cobot				
Human flexibility at the workplace	Continuous production flow	Stability in fulfilment of production orders	Approach of staff to solving issues	Effective cooperation with cobots

Figure 2. Work psychology in the production process—basic pillars (source: author).

Optimizing working conditions in a collaborative environment for employees necessitates that the latter are suitably qualified for their positions. Stabilizing the production schedule during a shift (to fulfil quotas) around specific lead times is currently a major issue in the industrial sector. In relation to this, significant stressors in a collaborative unit comprise the mismanagement of material and insufficiency in the skill sets of workers. Staff need to feel a sense of confidence about conducting duties and giving good performance, and their mental state can be impacted in the event an issue arises and complicates matters. Continuous control procedures determine the relevant actions to deal with incidents.

The digital literacy that workers possess plays an important role in work psychology. In the given context, it relates to the capacity to collaborate with the digitally controlled machine and presupposes certain cognitive skills, abilities, and attitudes [26]. Reports in the literature have highlighted that digitization exerts a knock-on effect on psychosocial risk at work [27]. Tasks that require physical or manual skills and basic cognitive awareness shall be automated first, whereas more demanding jobs necessitate social, emotional, and technological skill sets instead [28].

5. Discussions and Social Implications

Working in front of a robot leads to a rise in the level of stress experienced by the human, making the situation less stable and altering readiness for performance. In effect, the person is psychologically drawn in to align with the actions of the robot, hence their immediate surroundings are crucial, and it is necessary to ensure all inputs are present and correct prior to the commencement of work. The stress level was investigated by empirical research. The core inputs were human–cobot cooperation and process collaboration. According to the results presented in Table 1 and the results achieved by the research, we distinguish by human factors, such as stability, rotation, and flexibility, as a principal stress factors. The level of risk associated with these factors increases proportionally with the requirements on the digital process monitoring and the productivity index by human working in a collaborative workplace. Therefore, the intentions (diagnostics) in the field of the work psychology should lead to the optimisation of workload in given lead time. The production plan for the work shift determines the optimal production schedule. This fact is essential for a stable work environment and, at the same time, for an optimal psychological adjustment of the human being. De Longis [29] refers to the fact that burnout syndrome may affect the temporal dynamics of negative emotions at work, making them more persistent over time and resistant to change. Due to the use of ICT, workers can be accessible, even after standard working hours, preventing adequate recovery and psychological detachment [30].

Collaborative workplace inputs are technological production sources (material, machine standards, job instructions, etc.), they are necessary components for the production performance as an output. Eliminating stress by human and the improvement of the work psychology setting by collaborative workplace is necessary before the performance. It is conditioned by optimal setting of the parameter "Reaction time of the human–cobot arrangement (RT_{hc})". This variable additionally encompasses the aspect of mental concentration applied by the operator to their labours. The less disorder they have around them before undertaking duties, the more stable their resultant performance. The authors intend to conduct further research on this matter. In our research, we investigated the stress level using an indicator 'Stability of the production change plan' (see Table 3). The presented stability value below 1.0 means an increased level in human stress. The value under 1.0 represents the stability of the production planned tasks completion. In our next empirical research, we want to give attention to the effective psychological collaboration between artificial intelligence and human stress by process task realization in planned throughput time in a collaborative workplace. Artkin [3] stats that the failure detection, quality inspection, enhancing workplace safety, and people concentration on performance are important elements of workplace optimization by a human. Therefore, our further research in this area will focus on the analysis of the interruption parameter by the worker before the robot (Table 4).

Table 4. Work psychology process approach schema (source: author).

Work Psychology at the Collaborative Workplace: Key Processes for Future Research	Specification for Research Purposes
Structure of the planning and scheduling process	Detailed structure of production flow; controlled process for enumerating customer orders; digitization of procedural steps in the cycle of planning, scheduling, and production flow.
Automatization of the production process	Lead-time coordination between production planning and logistics processes, and actual workplace performance and lead times.
Maintenance	Actual, relevant knowledge and skills pertaining to processes and related issues involving technology and workplace practices.
Organization of tasks and their realization	Competence and skill sets of collaborative staff pertaining to actions for actual production and potential disruption, application of KPIs for monitoring performance and eliminating performance wastes.

The robot has the ability to identify technical circumstances that could give rise to disruptions or other issues and respond immediately should the matter have been correctly diagnosed by it. Nevertheless, staff single out processes and deal with problems by drawing on their personal knowledge and experience. The decision-making process they apply transpires differently from that of the robot, since the person finds a solution on the basis of whether they had caused the disruption or not. Past research conducted revealed that people usually relied on the robot to provide a suitable reaction, only seeking the best solution afterwards. This aspect contributes to how arrangements are set for the worker, primarily since the collaborative environment is managed by technology. The person becomes, in essence, a component within the digitally controlled workplace, subordinate to how processes are managed, and the responses of the robot.

In contrast, it is more comfortable to work behind the robot. Staff there are not "feeding" the robot but take items from it. The associated work psychology differs from having to feed it with input, as the primary concern is to keep up with the output rate of the robot. The employee to the front of the robot is under pressure at the start of the operation, whereas the emphasis for the one situated behind it is the duration of its output. This means two different variables exist in relation to the setting of work psychology, and these "push" and "pull" actions are crucial for an optimal work psychology setting; therefore, an enumerated specified order is needed for proper scheduling of production.

In our research, we dealt with the concept of 'push' production (data analysis in Table 3), which is connected with the production planning and scheduling system in the investigated industrial companies. The reason of the push system preference was the human experience in a mass repetitive production. The data from the workplaces provided us with relevant outputs to understand the essence of the psychology of work in a collaborative workplace. The reason for the reliability of the data was the volume and the repeatability of the production. Thus, workers gain relevant experience and knowledge to manage and improve work psychology. Based on this experience, we will start similar research in 2023 in companies oriented to the pull production concept. It is characterized by the piece or small series production. We will try to generalize the results of our research so far, and develop a methodology for managing work psychology in pull production companies.

The goal of future research is to set parameters for "reaction times for productive workers", where the position the operator holds (in front of or behind the robot) is important to analysis. The results of such research shall subsequently be linked with a methodology for planning and scheduling production at the enterprises under study. Distinguishing between planned and actual production capacity is crucial to implementing correct planning and management. As can displayed in Table 1, safeguarding processes are in place to counteract as a stabilizing element of work psychology. Though stability is reduced, and the psychological well-being of the operator is negatively affected, and the level of work psychology diminished. Results prove that correctly enumerating orders for manufacturing goods is crucial to projecting a sense of work psychology, since it represents the impulse for setting the latter. The role of the planner in this is to make sure the production schedule of allotted time slots takes into consideration the capacity of staff for output. Any such plan is informed by the activity of the robot, so it follows that the duties of the employee are scheduled to align with the robot.

6. Conclusions and Recommendations for Future Study

The psychology of a workplace closely relates to a sense of having ownership over human labour. Advances in collaborative environments and related procedures have emphasized the need to consider the staff employed in such a digitally controlled domain [9], which presupposes aspects of hybrid research are applied in evaluation of such a site. Problems stemming from reduced performance by a worker that collaborates with a robot could give rise to psychological disorders [6].

Items for future research have been identified in relation to the limits of this study. The first shall be to analyse to what extent agreement prevails in human–cobot feedback in the event of an unforeseen issue in production. It will then be necessary to discern the dependence of the consequent revision process in real time and psychological readiness of staff for performance. Identifying psychological factors affecting employees in terms of motivation for performance, or barriers that diminish it, would be the overarching purpose of the investigation. Findings shall inform how a stable production plan is set up in a collaborative environment. It is known that a correlation exists between production planning and the effectiveness of human performance in this context. Hence, emphasis shall be placed on the actual content of processes governing production planning and scheduling. The following research questions have been formulated to this end:

- RQ 1: Tasks undertaken at a collaborative workplace are clearly defined in structure and content at a system-wide level.
- RQ 2: The lead times in a collaborative workplace adhere to a flexible schedule and stable plan for production.
- RQ 3: Work psychology is crucial to achieving the desired level of performance in a collaborative environment.

A collaborative work environment creates prerequisites for integrating the employee into a manufacturing line with digitized technology, within which they have the opportunity to improve and refine the digitally controlled process, promoting a sense of positive work psychology. Fruitful human–robot cooperation at sections of the manufacturing facility benefits production, and settings of planning and scheduling processes [31].

Data analysis herein focussed on gaining a comprehensive understanding of the human–cobot relationship, and results were applied to define parameters for optimizing operations in hybrid workplaces. Standardization of the manufacturing environment, as laid out in a production plan, and realized on human–cobot assembly lines, necessitates relevant scheduling of procedures based on the capacities of the human and cobot, in turn influencing the real-time schedule, which is primarily informed by the durations of cobot activities [32].

The authors consider one of the greatest challenges of current research on work psychology in a collaborative environment to be ensuring a process for stable production, the flexibility, skill sets and rotation of workers, and the technical and technological compatibility of the operator and robot. It is a realistic to assume that another year of research shall increase comprehension of the tolerance of reaction time on the part of the operator to restarting an interrupted process. Work psychology is primarily about the satisfaction of employees, and, herein, about a sense of satisfaction stemming from good cooperation with a cobot. In the future, knowledge of the requirements of human and robot colleagues will constitute an important field of research for industrial engineers, especially with regard to managing efficient and effective processes.

Funding: This research did not receive external funding.

Institutional Review Board Statement: The study was conducted according to the content of the research projects conducted by the Department of Industrial Engineering and Information Systems, Faculty of Management and Economics, Tomas Bata University in Zlin (VaV-IP-RO/2022/01).

Data Availability Statement: The author does not have permission to share data.

Conflicts of Interest: The author declares no conflict of interest.

References

1. Thorvald, P.; Berglund, A.F.; Romero, D. The Cognitive Operator 4.0. In *International Conference on Manufacturing Research*; Advances in Manufacturing Technology XXXIV; IOS Press: Amsterdam, The Netherlands, 2021; Volume 15, pp. 3–8. [CrossRef]
2. Chromjakova, F. Process Stabilisation–Key Assumption for Implementation of Industry 4.0 Concept in Industrial Company. *J. Syst. Integr.* **2017**, *8*, 3.
3. Artkin, F. Applications of Artificial Intelligence in Mechanical Engineering. *Eur. J. Sci. Technol.* **2022**, *45*, 159–163. [CrossRef]
4. Endsley, M.R. Toward a Theory of Situation Awareness in Dynamic Systems. *Hum. Factor* **1995**, *37*, 32–64. [CrossRef]
5. Abd Rahman, M.S.B.; Mohamad, E.; Abdul Rahman, A.A.B. Development of IoT—Enabled data analytics enhance decision support system for lean manufacturing process improvement. *Concurr. Eng.* **2021**, *29*, 208–220. [CrossRef]
6. Cohen, Y.; Shoval, S.; Faccio, M. Strategic View on Cobot Deployment in Assembly 4.0 Systems. *IFAC-PapersOnLine* **2019**, *52*, 1519–1524. [CrossRef]
7. Delgosha, M.S.; Hajiheydari, N. How human users engage with consumer robots? A dual model of psychological ownership and trust to explain post-adoption behaviours. *Comput. Hum. Behav.* **2021**, *117*, 106660. [CrossRef]
8. Obrenovic, B.; Du, J.; Godinic, D.; Baslom, M.M.M.; Tsoy, D. The Threat of COVID-19 and Job Insecurity Impact on Depression and Anxiety: An Empirical Study in the USA *Front. Psychol.* **2021**, *12*, 3162. [CrossRef]
9. Chromjakova, F.; Bobák, R.; Tuček, D.; Hrušecká, D.; Hrbáčková, L. Predictive Modelling and Diagnostics of Production Planning and Scheduling Processes. Research grant VaV-IP-RO/2022/; Tomas Bata University in Zlín: Zlín, Czechia, 2022.
10. Judge, T.A.; Weiss, H.M.; Kammeyer-Mueller, J.D.; Hulin, C.L. Job attitudes, job satisfaction, and job affect: A century of continuity and of change. *J. Appl. Psychol.* **2017**, *102*, 356–374. [CrossRef]
11. Fogaça, N.; Rego, M.C.B.; Melo, M.C.C.; Armond, L.P.; Coelho, F.A. Job Performance Analysis: Scientific Studies in the Main Journals of Management and Psychology from 2006 to 2015. *Perform. Improv. Q.* **2018**, *30*, 231–247. [CrossRef]
12. Arefin, S.; Alam, S.; Islam, R.; Rahaman, M. High-performance work systems and job engagement: The mediating role of psychological empowerment. *Cogent Bus. Manag.* **2019**, *6*, 1664204. [CrossRef]
13. Amor, A.M.; Xanthopoulou, D.; Calvo, N.; Vázquez, J.P.A. Structural empowerment, psychological empowerment, and work engagement: A cross-country study. *Eur. Manag. J.* **2021**, *39*, 779–789. [CrossRef]
14. Gong, Z.; Chen, Y.; Wang, Y. The Influence of Emotional Intelligence on Job Burnout and Job Performance: Mediating Effect of Psychological Capital. *Front. Psychol.* **2019**, *10*, 2707. [CrossRef] [PubMed]
15. Romero, D.; Stahre, J.; Wuest, T.; Noran, O.; Bernus, P.; Fast-Berglund, Å.; Gorecky, D. Towards an Operator 4.0 Typology: A Human-Centric Perspective on the Fourth Industrial Revolution Technologies. In Proceedings of the CIE 2016: 46th International Conferences on Computers and Industrial Engineering, Tianjin, China, 29–31 October 2016; pp. 1–11.
16. Mathieu, C. Defining knowledge workers' creation, description, and storage practices as impact on enterprise content management strategy. *J. Assoc. Inf. Sci. Technol.* **2021**, *73*, 472–484. [CrossRef]
17. Krzywdzinski, M.; Pfeiffer, S.; Evers, M.; Gerber, C. *Measuring Work and Workers: Wearables and Digital Assistance Systems in Manufacturing and Logistics*; No. SP III 2022-301; Wissenschaftszentrum Berlin für Sozialforschung (WZB): Berlin, Germany, 2022.
18. Kim, S.; Lee, H.; Hwang, S.; Yi, J.-S.; Son, J. Construction workers' awareness of safety information depending on physical and mental load. *J. Asian Arch. Build. Eng.* **2021**, *21*, 1067–1077. [CrossRef]
19. Paliga, M.; Pollak, A. Development and validation of the fluency in human-robot interaction scale. A two-wave study on three perspectives of fluency. *Int. J. Hum. Comput. Stud.* **2021**, *155*, 102698. [CrossRef]
20. Kopp, T.; Baumgartner, M.; Kinkel, S. Success factors for introducing industrial human-robot interaction in practice: An empirically driven framework. *Int. J. Adv. Manuf. Technol.* **2020**, *112*, 685–704. [CrossRef]
21. Paliga, M. Human–cobot interaction fluency and cobot operators' job performance. The mediating role of work engagement: A survey. *Robot. Auton. Syst.* **2022**, *155*, 104191. [CrossRef]
22. Liu, L.; Guo, F.; Zou, Z.; Duffy, V.G. Application, Development and Future Opportunities of Collaborative Robots (Cobots) in Manufacturing: A Literature Review. *Int. J. Hum. Comput. Interact.* **2022**, 1–18. [CrossRef]
23. Kolbeinsson, A.; Lagerstedt, E.; Lindblom, J. Classification of Collaboration Levels for Human-Robot Cooperation in Manufacturing. In *Advances in Manufacturing Technology XXXII*; IOS Press: Amsterdam, The Netherlands, 2018; pp. 151–156.
24. Simões, A.C.; Pinto, A.; Santos, J.; Pinheiro, S.; Romero, D. Designing human-robot collaboration (HRC) workspaces in industrial settings: A systematic literature review. *J. Manuf. Syst.* **2022**, *62*, 28–43. [CrossRef]
25. Mukherjee, D.; Gupta, K.; Chang, L.H.; Najjaran, H. A Survey of Robot Learning Strategies for Human-Robot Collaboration in Industrial Settings. *Robot. Comput. Manuf.* **2022**, *73*, 102231. [CrossRef]
26. Bejaković, P.; Mrnjavac, Ž. The importance of digital literacy on the labour market. *Empl. Relat. Int. J.* **2020**, *42*, 921–932. [CrossRef]

27. Palumbo, R.; Casprini, E.; Montera, R. Making digitalization work: Unveiling digitalization's implications on psycho-social risks at work. *Total. Qual. Manag. Bus. Excel.* **2022**, 1–22. [CrossRef]
28. Lopez-Sanchez, J.I.; Arroyo-Barriguete, J.L. Robot and Automation: Which are the Impacts on the Productivity, Jobs, and Inequality of the Countries. In *INBOTS 2021 Conference, Biosystems and Biorobotics*; Grau Ruiz, M.A., Ed.; Springer: Cham, Switzerland, 2022; Volume 30. [CrossRef]
29. De Longis, E.; Alessandri, G.; Sonnentag, S.; Kuppens, P. Inertia of negative emotions at work: Correlates of inflexible emotion dynamics in the workplace. *Appl. Psychol.* **2021**, *71*, 380–406. [CrossRef]
30. Giorgi, G.; Ariza-Montes, A.; Mucci, N.; Leal-Rodríguez, A.L. The Dark Side and the Light Side of Technology-Related Stress and Stress Related to Workplace Innovations: From Artificial Intelligence to Business Transformations. *Int. J. Environ. Res. Public Health* **2022**, *19*, 1248. [CrossRef]
31. Fletcher, S.R.; Webb, P. Industrial Robot Ethics: The Challenges of Closer Human Collaboration in Future Manufacturing Systems. In *A World with Robots*; Springer: Cham, Switzerland, 2017; pp. 159–169. [CrossRef]
32. Goleman, D.P. *Emotional Intelligence: Why it Can Matter More than IQ for Character, Health and Lifelong Achievement*; Bantam Books: New York, NY, USA, 1995.

MDPI

Article

Brain–Computer Interface and Hand-Guiding Control in a Human–Robot Collaborative Assembly Task

Yevheniy Dmytriyev ©, Federico Insero ©, Marco Carnevale © and Hermes Giberti *©

Dipartimento di Ingegneria Industriale e dell'Informazione, Università degli Studi di Pavia, Via Ferrata 5, 27100 Pavia, Italy
* Correspondence: hermes.giberti@unipv.it

Abstract: Collaborative robots (Cobots) are compact machines programmable for a wide variety of tasks and able to ease operators' working conditions. They can be therefore adopted in small and medium enterprises, characterized by small production batches and a multitude of different and complex tasks. To develop an actual collaborative application, a suitable task design and a suitable interaction strategy between human and cobot are required. The achievement of an effective and efficient communication strategy between human and cobot is one of the milestones of collaborative approaches, which can be based on several communication technologies, possibly in a multimodal way. In this work, we focus on a cooperative assembly task. A brain–computer interface (BCI) is exploited to supply commands to the cobot, to allow the operator the possibility to switch, with the desired timing, between independent and cooperative modality of assistance. The two kinds of control can be activated based on the brain commands gathered when the operator looks at two blinking screens corresponding to different commands, so that the operator does not need to have his hands free to give command messages to the cobot, and the assembly process can be sped up. The feasibility of the proposed approach is validated by developing and testing the interaction in an assembly application. Cycle times for the same assembling task, carried out with and without the cobot support, are compared in terms of average times, variability and learning trends. The usability and effectiveness of the proposed interaction strategy are therefore evaluated, to assess the advantages of the proposed solution in an actual industrial environment.

Keywords: collaborative robotics; brain–computer interface; hand guiding; haptic interface; multimodal communication; assembly task

check for updates

Citation: Dmytriyev, Y.; Insero, F.; Carnevale, M.; Giberti, H. Brain–Computer Interface and Hand-Guiding Control in a Human–Robot Collaborative Assembly Task. *Machines* **2022**, *10*, 654. https://doi.org/10.3390/machines10080654

Academic Editor: Raul Campilho

Received: 5 July 2022
Accepted: 31 July 2022
Published: 5 August 2022

Publisher's Note: MDPI stays neutral with regard to jurisdictional claims in published maps and institutional affiliations.

1. Introduction

Production in small and medium enterprises (SMEs) is usually characterized by high variability and low volumes of batches, which makes flexibility and highly adaptable production systems necessary requirements. Robotic installations easily fulfill these needs, so that their usage is quickly increasing in industry [1,2].

In SMEs, where manufacturing often relies on manual production processes, collaborative robots (cobots) are potentially easy-to-use tools that can improve productivity without loss of the advantages provided by a human-centered system [3]. Human workers indeed have an innate ability to adapt to unexpected events and to maintain strong decision-making skills, even in a dynamic and complex environment. They have flexibility and adaptability for learning new tasks and intelligence, while the robots can guarantee physical strength, repeatability, and accuracy. Human-Robot Collaboration (HRC) aims at being complementary to conventional robotics, increasing the human participation in terms of shared time and space [4]. With HRC, humans and robots can share their best skills, provided that the involved devices are designed for both safety and interaction.

A safe, close interaction with human operators is only part of the technical issues to be faced to exploit the advantages of a collaborative working station. Much effort is dedicated, still mainly at a research level, to enhance the design of human-aware robots, able to cope with uncertainties due to human presence [5–7] and to get, at the same time, command and guidance messages from them [8]. Several communication technologies are being developed to reach a real Human-Robot Collaboration (HRC). Interaction modes, such as gesture recognition [9], vocal commands [10], haptic controls [11], and brain-computer interfaces [12] are some of the technologies exploitable to provide inputs to the robot, allowing the operators to dynamically alter the robotic behavior during its functioning or taking control of the ongoing task [8,13,14]. These technologies can also be grouped into a multimodal communication strategy, enabling the possibility of programming a robot in a more adaptive and unstructured way.

All of the above-mentioned techniques have been extensively discussed in [8], where a metric for evaluating the performances and the type of information exchanged between the human and the robot is proposed. The type of messages are classified as Command and Guidance messages, achievable through different interfaces. The former (i.e., command messages) communicates to the robot simple commands such as "next" or "stop", not requiring any kind of parameter to be exchanged. Guidance messages, on the other hand, provide instructions regarding how the robot should move, thus requiring a continuous flow of positional or force information. For this reason, command messages can be considered the simplest way of interaction, nevertheless requiring a suitable design of the programmed task.

According to [7,15], assembly tasks are typical tasks, which well suit collaborative robotic applications since the operator's action influences the behavior of the robot and vice versa. However, the most common use of collaborative robots for assembly tasks in manufacturing lines consists of workstations suited for sequential assembly [13], in which the robot performs the simpler operations and the most complex or variable ones are left to the human. Whenever possible, the worker carries out the last manipulations on the assembled product at the end of the assembly line so as to limit the need for interaction with the robot. A more complex cooperative assembly, i.e., parallel assembly, is characterized by human intervention taking place in parallel to robot activities, so that the last assembly steps can also be carried out by the robot. In this second scenario, timing and coordination between humans and robots are critical factors, which might severely affect the collaboration, which therefore strengthens the need for a suitable interaction strategy.

Different collaborative strategies can be exploited according to the degree of task interconnection and dependency [16,17]. When the possibility of a direct command message from the operator to the robot is enabled through a suitable user-interface, human's intentions can be communicated to the robot. It is then possible to alternate between Independent strategies– in which the operator and the robot work simultaneously and independently on their own tasks– and Supportive strategy– in which the operator receives assistance from the robot. The switch between Independent and Supportive phases can be enabled based on the timing given by the operator, who works independently on his own task until he gives a triggering signal to the robot [18]. This strategy can go under the name of Collaboration On-Demand since the possibility of switching from independent to supportive phase is left to the operator. Once a consent exchange is detected by the machine, the operator can be assisted by the robot in a supportive manner.

Under specific circumstances, e.g., when the presence of environmental noise or poor illumination occur, or when the operator's hands must be engaged in the assembly process and cannot be used to give commands, technologies like voice or gesture recognition might result unsuitable, and therefore the possibility to exploit techniques based on Brain–Computer Interface (BCI) is conveniently investigated. The potentiality of this techniques in applications for Industry 4.0 has been pointed out [19], but, to the authors' knowledge, few industrial applications have been actually developed. BCI is extensively studied in the neuroscience field [20–22]. Applications to robotics have been developed in different contexts [23,24], other than industrial applications.

In this work, we test the concept of collaboration on demand by setting up a Brain–Computer Interface (BCI) to transfer command messages from the operator to the cobot, thus enabling a supportive behavior in which the operator can use hand guiding control to be assisted by the robot. The communication strategy and the collaboration on-demand are deployed in a proof-of-concept assembly task developed on a TM5-700 cobot.

The BCI allows keeping the operators' hands free for the assembling task and can be adopted in noisy environments. Moreover, the possibility to insert few BCI sensors in the personal protective equipment (i.e., helmets) would pave the way for application in industrial environments.

The paper is organized as follows: Section 2 describes the overall framework of the collaboration on-demand, which adopts the BCI and the hand guiding control. In Section 3, the proof-of-concept assembly task is presented to validate the proposed solution. Results are presented in Section 4 and conclusions are drawn in Section 5.

2. Collaboration On-Demand Strategy Exploiting BCI-SSVEP and Hand Guiding

The different phases of an assembly process can be classified based on their repetitivity and complexity. The robot can take charge of the most repetitive and unskilled operations, while the human operator can execute complex activities. Collaborative robots in assembly tasks can also provide benefits in handling large and heavy objects [25], both thanks to hand-guided operational mode and to the capability of reducing the apparent mass of heavy work pieces by a factor of ten or more [26]. This means that the physical strain of the workers is significantly reduced.

The cooperation between human and cobot relies on the synchronization between the different phases and on a proper task assignment. In a collaboration on-demand strategy, the workcell is human-centered, and the robot is intended as a supportive tool for the worker. Figure 1 schematically represents the process by highlighting the alternating switch between two different phases, namely independent and supportive phases. During the independent phases, the robot and the operator work in a coexistence scenario, operating on different workpieces and processes. As soon as the operator needs the robot's assistance, the supportive phase can be enabled: the operator and robot work on the same workpiece interactively, and the robot must behave according to human intentions. This requires the robot to be warned of human intention in advance through a command message [8], which can be sent by a worker through a proper user-interface with the desired timing. The operator can therefore use the robot as a flexible tool whose supportive behavior is triggered at will, according to their needs and timing.

When the supportive function ends, the independent behavior can start again, and robot and operator keep on carrying out their planned activities. Furthermore, the reverse transition from supportive to independent phase can be controlled by the human through another command message so that only two different messages are needed to deploy the described interaction scheme.

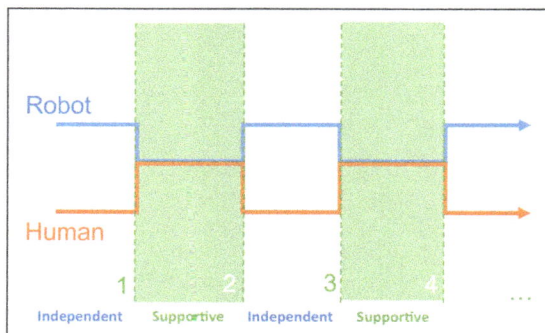

Figure 1. Collaboration on-demand.

In the particular industrial case proposed in this paper, the switch between independent and supportive phases is controlled through a BCI, which allows sending a command message without the use of hands. The BCI is used in a reactive mode with the Steady State Visually Evoked Potentials (SSVEP) method [19]: the operator looks at an external monitor with images blinking at two different frequencies corresponding to the two command messages needed to switch between the two operating modes. During the supportive phase, hand guiding control is exploited to position the objects in a co-manipulation mode. Hand guiding control has been realized by means of a six-component load cell mounted on the robot wrist, as described more in detail in the following paragraphs.

An exemplary assembly task has been considered as a test case consisting of pick and place operations, relative positioning of objects, and joint connections with bolts and nuts. The task can be therefore divided into simple, repetitive operations, such as pick and place, and more complex ones, such as joint connections. The former are assigned to the robot, whereas the more complex manipulations subtask can be left to the operator – as soon as the cobot has picked up the assigned component, it positions itself in a stand-by pose, waiting for the operator to take control of the process. As soon as it is ready, the operator switches to supportive operational mode so that the intermediate step of positioning the object relative to the previous one can be conducted in a cooperation mode through hand-guiding control. The robot can then assist the operator by holding and moving large and heavy objects according to its payload, leaving the operator the flexibility to properly position the component with the desired timing and then join it to the other parts of the structure being assembled.

Figure 2 shows the framework proposed, in which the operator can interact with the robot manipulator through two different interfaces. The activities related to the Brain–Computer Interface are described on the left side of the figure, whereas those related to load cell and haptic control are reported on the right side of the scheme. Looking at the left-hand side of the scheme in Figure 2, SSVEP signals are collected by electrodes, processed and then referenced to suitable Command Messages to be sent to the robot controller. On the right side of the scheme, the load cell provides a haptic interface featuring hand guiding control, which introduces Guiding Messages to properly position the robot. The task flow depends at the same time on the pre-programmed robotic subtasks and on the real-time commands given by the human operator. The latter makes the decision about when the robot must move independently and when to switch to a supportive phase.

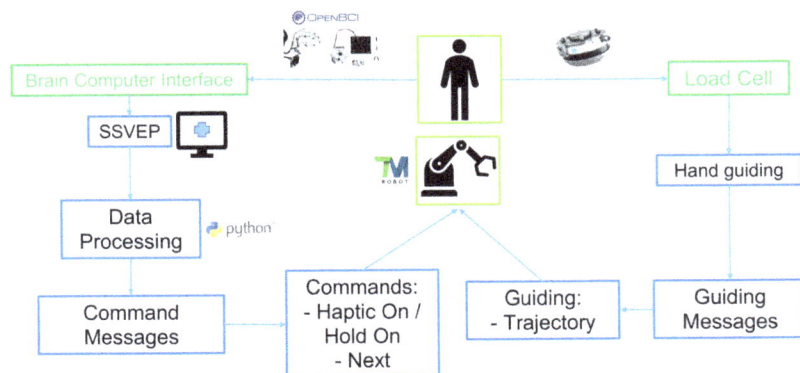

Figure 2. BCI-SSVEP with Load Cell sensor Framework.

Each BCI-SSVEP signal is assigned to a pre-defined specific robotic function. When the operator and the robot end their independent phases, the operator can activate the hand guiding mode by providing the command "Haptic On". The "Hold On" command is given when the operator needs the robot to keep precisely the same position in which it has been placed through hand guiding control so that the assembly operation can take place. The same SSVEP signal can be used for the two cases. When the operator finishes connecting the two parts by screws and bolts, the "Next" command allows the robot to get back to its independent phase. The logic applied therefore requires only two signals, hence, two different stimuli: one associated with the "H" command message (which can have two different states: "Haptic On" or "Hold On"), and another for the "N" ("Next") command.

Figure 3 represents the same collaboration on-demand framework from the perspective of the human operator and the robot, which are seen as parallel players in the task. Command and guidance messages are in charge of timing and controlling the task – when the operator sends the "Haptic On" command the two branches of the independent phases merge, starting the supportive phase (collaborative) [16], in which operations are headed by the operator and supported by the robot. After the hand-guided phase, in which the operator can place the component on hold by the robot relative to the structure being assembled, the "Hold On" command allows keeping the component in place so that bolting operations can take place. Afterward, the "Next" command makes the operator and robot get back to their independent phases. The former can continue to perform assembly operations for which he does not need any support, the latter can pick the next component to be handed to the operator. He will pick it up just when he needs it, and he will take charge of the process through another "Haptic On" command.

The following subsections detail how BCI data are treated to manage the switching between independent and supportive collaborative phases and how the hand guiding control is deployed on the adopted Techman TM5-700 cobot.

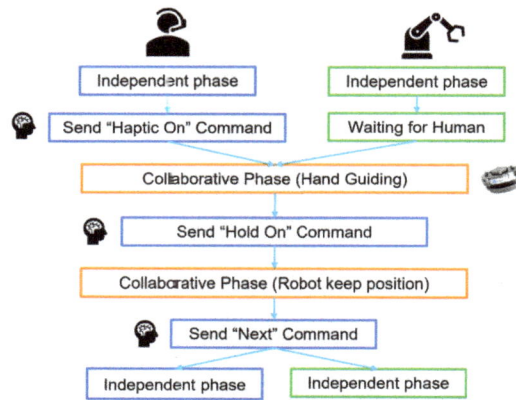

Figure 3. Collaboration on-demand with BCI and hand guiding.

2.1. Brain–Computer Interface Based on SSVEP

Steady State Visually Evoked Potentials (SSVEP) response is a BCI reactive technique based on Electroencephalogram (EEG) signals, having the highest potentiality for industrial applications [19]. It allows interacting with the robot without the use of hands [27,28], gestures, or voice commands, and it does not require much training sessions. When the operator, wearing an headset, is subjected to a visual stimulus, the signals gathered from the response of the visual cortex area, suitably processed and analyzed, allow to provide inputs to the robot controller.

In the proposed work, EEG signals are acquired using a cap with electrodes located in accordance with the international Ten-Twenty system [29], the representation of which is reported in Figure 4. SSVEP signals are recognized when the retina is subjected to visual stimuli blinking in the band of 3.5–75 Hz [22] even if, in common practice, the upper bound of the frequency range of the stimuli can be limited to 20 Hz [20]. SSVEP signals are characterized by a high Signal to Noise Ratio (SNR) [20,21]; hence, it is possible to recognize different stimuli with low processing effort and good precision. The visual cortex area reflects the same frequency of the stimulus.

Two kinds of commercial electrodes can be exploited for the purpose: wet, with an appropriate electrolyte gel applied on sensors, or dry [30]. Practical use of dry electrodes would be easier in industrial environments, as it is not necessary to apply gel to allow electrical contact.

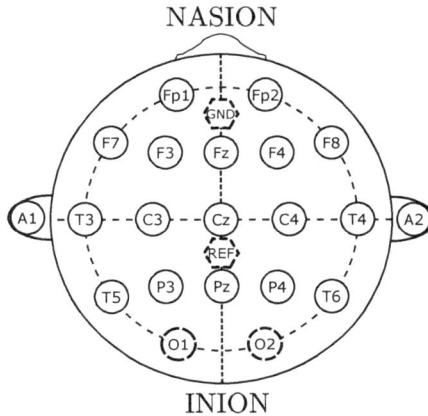

Figure 4. International ten-twenty system. The subset of exploited electrodes O1, O2, REF, and GND is highlighted with dashed bold line.

In order to deploy a methodology suitable for industrial applications, the lowest number of sensors has to be exploited; according to the ten-twenty system, the response of the visual cortex area can be acquired through the electrodes O1 and O2 highlighted in Figure 4. Electrodes O1 and O2 are placed on the occipital area, which is the one responsible for visual processing [31]. Ground (GND) and reference (REF) electrodes placed on the midline sagittal plane of the skull must be used for referencing and denoising the signal.

When the operator is subjected to a visual stimulus blinking at a given frequency, the same frequency and its multiples can be identified in the signal spectra. The generable visual stimuli depend on the type of devices adopted. In this work, an LCD monitor with a refresh rate of 60 Hz is used, in which two blinking windows are displayed [32,33], corresponding to the two command messages "H" and "N" needed for the robotic application. An OpenBCI headset and a Cython-Daisy board have been used in the setup, operating at a sampling rate of 125 Hz. The OpenBCI GUI software receives signals from the Cython-Daisy board over the LSL communication protocol.

The SSVEP frequency recognition software, and its integration with the robotic workflow described in the previous section, have been specifically developed. The algorithm flow is shown in Figure 5. When the robotic task starts, the first step of the SSVEP frequency recognition program is to open the communication with the OpenBCI software and flush the data buffer. Signals O1 and O2 of the visual cortex area are then gathered with a time window of 1 s. They are averaged and filtered using a digital, fourth order, Butterworth band-pass filter in the range 4–45 Hz. Finally, a Hamming window is applied to the signal, and a Fast Fourier Transform (FFT) is performed to identify the frequencies related to the visual stimulus. The FFT main peaks are analyzed to seek peaks that exceed a certain threshold at pre-set frequencies (the stimulus frequency and its first two multiples), which correspond to an actual will of the operator to give a command message.

These operations are performed in a continuous loop, and the corresponding command messages are sent to the robot only if a "cooperative flag" has been activated, meaning that the robot has completed its independent phase and is available to shift to cooperative operational mode. This prevents false positive commands from being transmitted to the robot before its independent phase has been completed.

Robotic commands corresponding to each command message can be general subtasks programmed in the robot controller [18,23,34] or a sequence of robot movements controlled by the PC. In the present work, robotic subtasks are programmed in the robot controller and triggered by the PC when the corresponding command message is detected. The mentioned signal processing software, as well as the one in charge of sending commands to the robot, have been realized in Python 3.8 with SciPy *v1.8.1* and Pymodbus *v2.5.3* libraries.

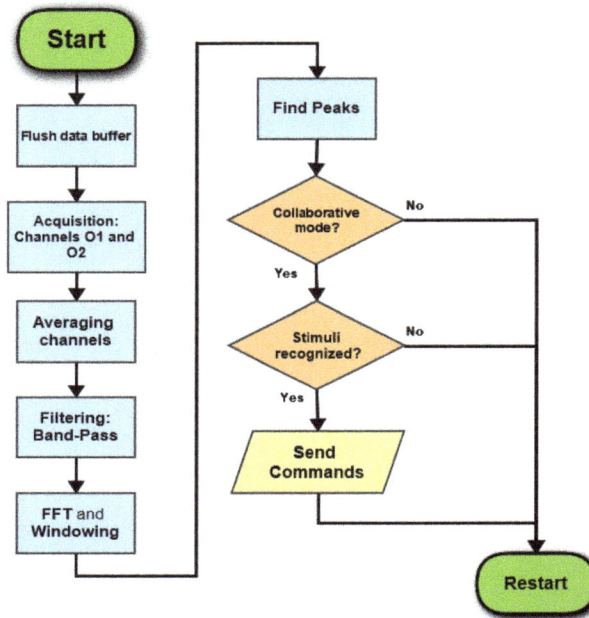

Figure 5. SSVEP frequency recognition software flowchart.

The issue of having a quick reactive response is essential for the fluidity of the robotic task. For this reason, to improve the system promptness, a time window length of 1 s was tested for frequency recognition in SSVEP and compared with a 2 s one [18]. The poorer spectral resolution of 1 Hz obviously affects the set of exploitable stimulus frequencies and worsens somehow the signal to noise ratio [35]. Figures 6 and 7 report the SSVEP responses to a visual stimulus blinking at 8 Hz and 10 Hz, respectively. For each frequency, the results corresponding to the time windows of 2 s and 1 s are reported in Figures 6a,7a and Figures 6b,7b , respectively. For all the considered cases, it is possible to detect the main frequency of the stimulus, with the first multiple frequencies also visible in the case of higher spectral resolution.

Since, in the present application, the operator is moving during the assembly task, the issue of motion artifacts had to be faced. Motion artifacts result in a signal spectrum with several peaks spread in the analyzed frequency range [36], which might generate false and unintended command messages. This issue had to be solved in real time during the signal processing phase by discarding all the occurrences presenting any peak at frequencies not corresponding to the expected response.

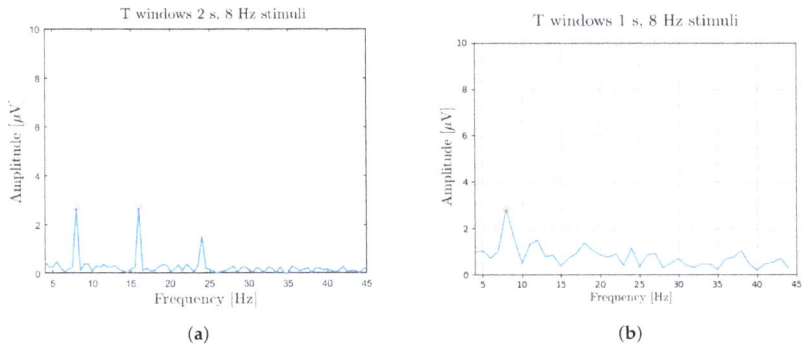

Figure 6. FFT analysis of SSVEP response for a 8 Hz visual stimulus. (**a**) Time window 2 s. (**b**) Time window 1 s.

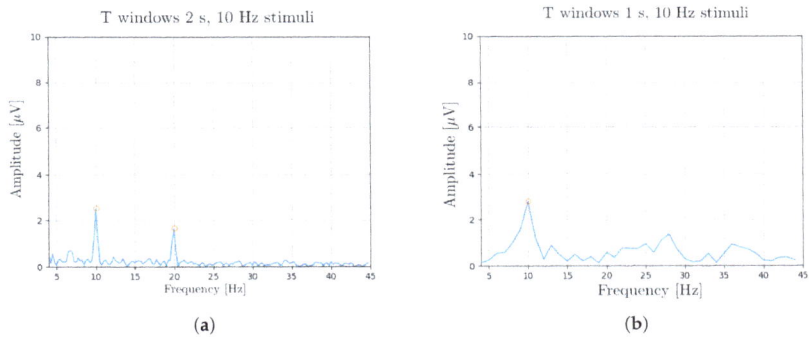

Figure 7. FFT analysis of SSVEP response for a 10 Hz visual stimulus. (**a**) Time window 2 s. (**b**) Time window 1 s.

2.2. Haptic Control Based on End Effector: Hand Guiding Control

In the proposed assembly task, positioning and co-manipulation operations require guidance messages for motion control and, in particular, for the proper positioning of the workpiece to be joined to the structure being assembled. Hand-guiding control is suitable for this purpose, assisting the operator in positioning and sustaining the weight of the parts to be assembled.

In the exemplary test bench used for demonstration, hand-guiding control has been developed by exploiting an OnRobot HEX-E v2 force sensor, measuring six axis components, mounted on the robot wrist. The system is able to sense the gripped workpiece and compensate for its weight. The arm is guided according to the forces sensed and imposed by the operator's hand, based on the proposal in [25,37,38], with a control scheme adjusted to fit the programming features of the Techman TM5-700 cobot.

A *force control* block in TMFlow software, activated during the cooperative phases, allows following any driving force provided by the operator (i.e., force in *x*, *y* and *z* directions). According to the ongoing phase in the task and, in particular, to the piece to be assembled in the pre-defined sequence, different directions for control can be enabled or disabled to have either a planar or 3D motion. For the first part to be assembled, positioning requires only a Cartesian linear motion in *x*, *y* and *z* coordinates. For the second positioning phase, the angular orientation along the Z-axis of the wrist reference frame is involved. Since only the rotation along the Z-axis is enabled, the gravity force always acts along the Z-axis of the robot's wrist.

The force control block integrates a PID control on the sensed forces and torque applied to the end-effector. To compensate for the gravity effects on the raised objects, measurements are reset at the beginning of each force control routine, and then stored to be used as reference force/torque for the PID controller.

PID gains used for the Cartesian linear motion are: $k_p = 0.15$, $k_d = 0$, and $k_i = 0.00001$; for the rotation along the Z-axis: $k_p = 0.012$ and $k_i = k_d = 0$.

3. Application of the Proposed Framework in an Exemplary Assembly Task

The discussed control structure has been validated by means of an exemplary assembly task developed with a TM5-700, 6 kg payload. The same control structure can, however, be applied to a bigger robot, in which the higher admitted payload allows assembling heavier components, thus enhancing the effectiveness of the proposed solution.

The task is divided into the following steps:

1. The operator is preparing a long aluminum profile and assembling a corner joint with bolts and nuts, while the robot is picking a short aluminum profile and positioning it in front of the operator, ready to be hand-guided. During this phase, the operator and the cobot are working with independent collaboration strategies.

2. In order to activate the cooperative mode through the BCI signal ("Haptic On" command), the operator looks at the monitor window with the letter "H" blinking at a frequency $f_1 = 8$ Hz. The person can then exploit the hand-guiding control and place the beam in the final position where it is to be assembled. After a second command message is given through BCI ("Hold On" command, blinking letter H), the robot keeps the component in position, and the operator can connect the two aluminum profiles with a corner joint and nuts. In this step, all activities are carried out with supportive strategies.

3. After finishing the assembly of the two parts, the operator gives the "Next" command by looking at the "N" letter blinking on the monitor at a frequency $f_2 = 10$ Hz. The operator prepares independently a second corner joint and gathers bolts and nuts, and the cobot executes the next routine, picking another long aluminum profile, handing it to the operator, and waiting motionless for the next BCI message. In this phase, the operator and robot are working independently.

4. When ready, the operator once again activates the cooperative mode through the BCI signal ("Haptic On" command) and hand-guides the beam into the proper position to be assembled. After a second command message through BCI ("Hold On"), the robot keep the component in position, and the operator can join the two aluminum profiles with a corner joint and nuts. Once assembly is complete, the person communicate to the robot the end of the task, through the last "Next" command.

The following Figures 8–10 highlight the main steps of the assembly task: Figure 8 shows an independent phase, representative of phase numbers 1 and 3 in the numbered list above, where the operator is mounting an angular joint while the robot is bringing the beam into place. It is possible to notice in the background the monitor with the blinking windows exploited as visual stimuli for BCI.

In Figure 9, the switch from independent to supportive phase is enabled by the operator who looks at the "H" letter on the monitor, to activate the hand-guiding mode. This switch only takes 1 s, thanks to the FFT analysis discussed in the previous Section 2.1.

Finally, Figure 10 shows the supportive phase in which the operator exploits hand-guiding control to position the beam to be assembled. When a proper positioning has been achieved, the hand guiding control can be deactivated by looking at the "H" window again, so that the robot holds the component steady in the desired position. Then, when the operator finishes the assembly, the robot can be moved by looking to the "N" window. In this way, the proper task timing is given by the operator, so that the robotic code is able to obviate any difficulties or setbacks in the assembly phase.

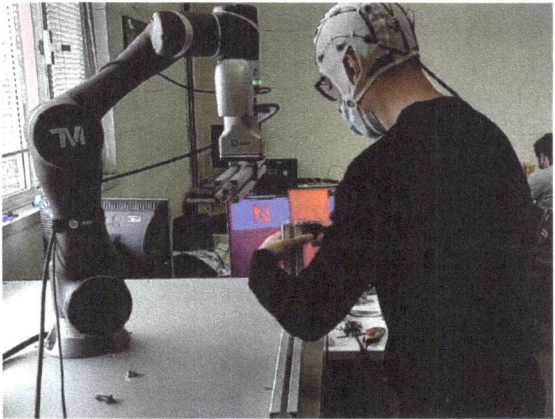

Figure 8. Independent phase. Step 3 of the assembly task.

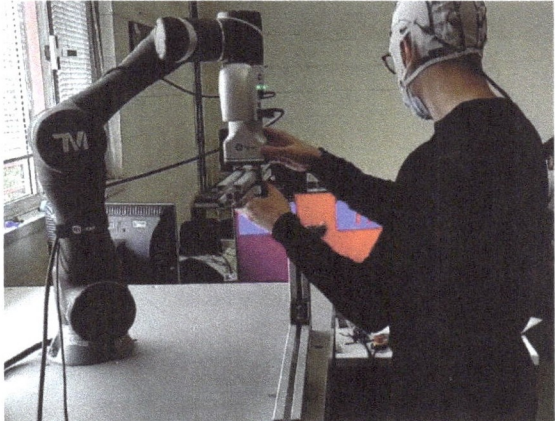

Figure 9. The operator looking at the "H" window is activating the hand guiding.

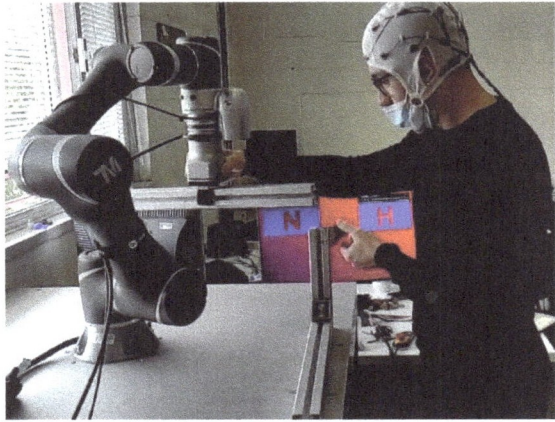

Figure 10. Step 4 of the assembly task. Supportive phase with hand guiding.

Figure 11 shows the flowchart of the robotic program. Independent phases are represented with yellow blocks (continuous line contour), decision phases in which the robot is waiting for BCI inputs in blue (rhomboidal shape, dashed line contour), and supportive phases in green (dashed line contour).

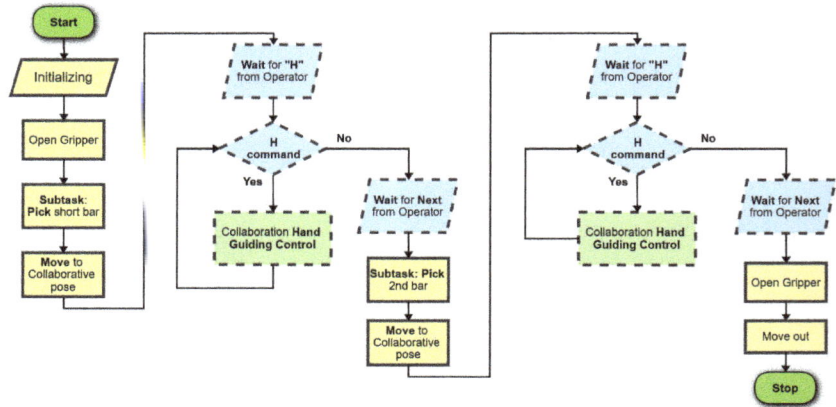

Figure 11. The robotic task flowchart.

Right after the start block, initialization actions and variable setting take place. The subtasks to be executed in independent mode are then carried out (i.e., Open Gripper, Pick bar Subtask, Move to handing pose where to wait for the operator). When the independent phase is over, the cobot is ready for entering the cooperative mode, it activates the "cooperative flag" enabling the possibility to receive the BCI command "H" from the operator. Then it waits for the operator's command: measurements from the BCI device and results from data analysis start being taken into account by the robot controller. SSVEP frequencies gathered from the BCI headset are compared with the pre-set frequencies to recognize the operator's choice. The program makes a step forward when the frequency corresponding to the H signal is detected.

The H command can have two binary statuses (i.e., 1, yes; 0, no). In the former status, the hand-guiding control is activated. In the latter, the hand-guiding is deactivated and the system goes into a new decisional block, waiting for the BCI next (N) command. At this point, the second independent subtask starts and the described cycle is repeated.

4. Results

In order to assess the potential improvements achievable through the described collaborative strategy, two series of tests have been carried out by repeating the assembly twenty times firstly purely manually and then with robotic assistance. Both assembly series were repeated by two different operators (named in the following S and Y), to also take into account variability due to the human factor. The cycle times were measured for each repetition.

Figure 12a,b show the distributions of the cycle times corresponding to the two operators, S and Y, respectively. In each figure, the cycle times related to the manual assembly (labeled with *Not assisted*), and to the human/robotic assembly (labeled with *Assisted*) are reported, together with the values of average μ and standard deviation σ and the representation of the corresponding Gaussian distribution.

Both S and Y operators experienced a clear reduction of the average cycle times in the assisted case compared to the not assisted one (71.3 s against 101.05 s for operator S and 86.25 s against 102.3 s for operator Y, corresponding to a reduction of 29.44% and 15.69% respectively), mainly thanks to the assistance of the robot in properly positioning the component to be assembled by means of the hand guiding control and holding it in the proper position during the bolting. Moreover, the cases in which the operator is assisted by the cobot show a lower standard deviation of cycle times, meaning that the presence of a predetermined workflow mitigates the variability of the cycle times, helping the operator be more regular in operations.

The standard deviation of the cycle times is equal to 6.19 s against 10.47 s for operator S, and 8.28 s against 12.49 s for operator Y.

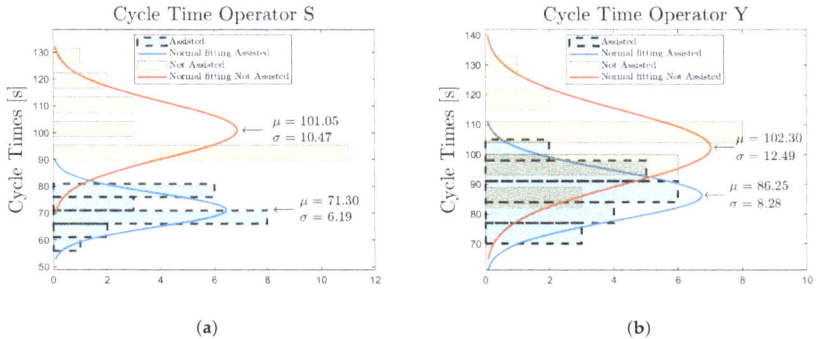

Figure 12. Assembly cycle times distributions for Assisted and Not Assisted cases with mean and standard deviation. (**a**) Operator S. (**b**) Operator Y.

Figure 13 shows the same cycle times as in Figure 12, reported this time in sequential order, as gathered during the execution of the test series. A learning effect is visible in all the series of tests, with a reduction of the cycle time occurring when the number of repetitions of the task increases, due to the fact that the operators improve their skills during the activity. The linear regression lines plotted in the figures can be used to estimate the learning rate of each operator. The linear regression slopes in the assisted cases show a reduction of the cycle time for both the operators (0.78 s/iteration for operator S, 0.51 s/iteration for operator Y), highlighting a beneficial effect of the robotic system. On the other hand, in the not assisted case, a more erratic behavior is observed, with operator S showing almost no learning (0.15 s/iteration), and operator Y showing a relevant learning rate (0.74 s/iteration). The results are summarized in Table 1.

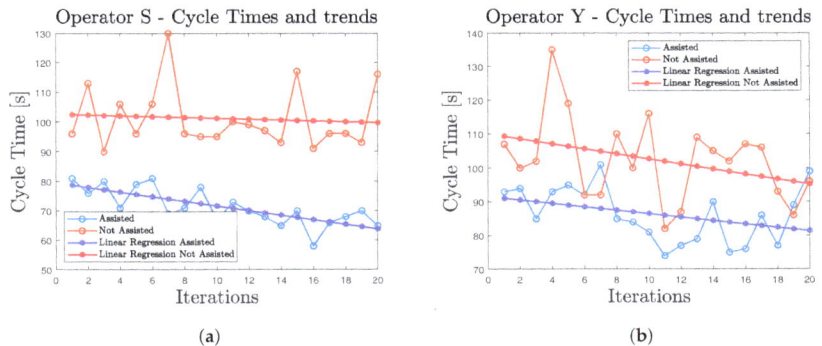

Figure 13. Learning rates. (**a**) Operator S. (**b**) Operator Y.

Table 1. Results summary table: Assisted vs Not Assisted.

Assisted vs. Not Assisted Cases Summary Table				
Operator		**Avg. Cycle Time** μ	**Cycle Time** σ_{STD}	**Learning Rate**
Operator S	Not assisted	101.05 s	10.47 s^2	0.15 s/iteration
	Assisted	71.30 s	6.19 s^2	0.78 s/iteration
	Percentage difference	−29.44%	−40.88%	80.77%
Operator Y	Not assisted	102.30 s	12.49 s^2	0.74 s/iteration
	Assisted	86.25 s	8.28 s^2	0.51 s/iteration
	Percentage difference	−15.69%	−33.71%	−31.08%

The following paragraph points out all possible issues related to wrong or poor signals from the interfaces which might arise during the development of the task.

Regarding command messages (i.e., through BCI), the following cases can be highlighted:

- A low signal-to-noise ratio, caused by a poor quality of the electrode/gel and skin contact, can cause a non-acquisition of the frequencies associated with command messages, leading to a slowdown of the task, as the robot will keep on waiting for a command through BCI.
- In the presence of motion artefacts, the commands are ignored. This is a positive feature if the motion artefact is to be discarded, but, if motion artefacts occur right when the operator is turning their head to look at the blinking signal, the desired command message would not be delivered. A new signal acquisition would be required, causing a delay of a few seconds in the overall task.
- Peripheral vision of operators might lead to the acquisition of incorrect command during the task, even if the operator is not purposely looking at the blinking signals. The influence of peripheral vision can be reduced by locating the blinking stimuli far from the peripheral field of view.

As for the guidance messages (i.e., through hand-guiding control):

- A wrong guidance message can be generated only if the operator guides the robot in a singularity position. Since the hand guiding control uses the motion control operating in the working space, in singularity positions, the motion control fails to compute the inverse kinematic, leading to a failure of the collaborative mode. This possibility can be completely avoided, and can be considered as human error.

5. Conclusions

A collaborative on-demand strategy is implemented in the present work, with the aim of demonstrating the possibility of exploiting the Brain–Computer Interface to give command messages to the robot controller in an industrial assembly task. The BCI provides the operator with the chance for controlling and giving proper timing to the robotic task during the assembly operations, without the need to use hands to push physical buttons or interact with a gesture recognition system.

A proper BCI command can switch the robot operational mode from independent to cooperative, thus activating a hand-guiding control by which the cobot and the human operator can interact in a supportive manner during the assembly task. The performance of this approach has been experimentally validated with two different operators, resulting in a significant reduction not only of the average cycle time (−29.44% and −15.69%), but also of its variability (i.e., standard deviation of the cycle times), thus leading to a more predictable productivity.

Nowadays, the BCI technology is not commonly applied in robotics for industrial applications, for example in assembly tasks. The proposed work demonstrates that if BCI electrodes could be fitted in the helmets provided as personal protective equipment in an industrial environment, this technology could be a considerable option for Industry 4.0 applications.

However, some limitations are still present. The human involvement in the completion of the shared task requires the overall assembly process to be clearly structured in advance, having clear in mind the alternative phases to be assigned to the human operator and to the robot. This introduces strict constraints in the design phase of the task.

Compared to standard robotic applications, the fact that the timing control of the assembly task is entrusted to the human operator may introduce delays due to the operator's behavior. Some might be due to human errors, but others are related to technological limitations of the selected interface: the signal to noise ratio in data generated by BCI, normally rather high, might show relevant variations depending on the subject wearing the BCI helmet. In such circumstances, the peak of the expected frequency may require more than one time-window to be detected, negatively affecting the overall execution time.

Motion artifacts, faced in this work by discarding all the occurrences presenting any peak at frequencies not corresponding to the expected response, might also occur when the operator looks at the blinking signal to generate the command message. In such a circumstance, the corresponding command would not be given, implying a delay due to the need of a new signal acquisition.

For the explained reason, the issue of signal-to-noise ratio, depending on the subject wearing the helmet, and motion artifact are still issues to be more comprehensively investigated in a further development of the work.

Author Contributions: Conceptualization, H.G. and M.C.; methodology, M.C., Y.D. and F.I.; software, Y.D. and F.I.; validation, M.C., F.I. and Y.D.; formal analysis, H.G. and M.C.; investigation, M.C., F.I. and Y.D.; resources, H.G.; data curation, M.C. and F.I.; writing—original draft preparation, F.I.; writing—review and editing, M.C.; visualization, Y.D., F.I. and M.C.; supervision, H.G. and M.C. All authors have read and agreed to the published version of the manuscript.

Funding: This research received no external funding.

Institutional Review Board Statement: Ethical review and approval were waived for this study, since the manuscript only involves frequency data in reaction to a blinking screen frequency.

Informed Consent Statement: Informed consent was obtained from all subjects involved in the study. Written informed consent has been obtained from the involved subjects to publish this paper.

Data Availability Statement: The data presented in this study are available on request from the corresponding author.

Acknowledgments: The authors would like to thank Stefano Ciardiello for carrying out part of the experimental tests during his BSc thesis.

Conflicts of Interest: The authors declare no conflict of interest.

Abbreviations

The following abbreviations are used in this manuscript:

BCI	Brain–Computer Interface
SSVEP	Steady State Visually Evoked Potentials
SNR	Signal to Noise Ratio
EEG	Electroencephalography
HRC	Human Robot Collaboration
FFT	Fast Fourier Transform

References

1. Executive Summary World Robotics 2021 Industrial Robots. 2021. Available online: https://ifr.org/img/worldrobotics/Executive_Summary_WR_Industrial_Robots_2021.pdf (accessed on 1 January 2022).
2. Giberti, H.; Abbattista, T.; Carnevale, M.; Giagu, L.; Cristini, F. A Methodology for Flexible Implementation of Collaborative Robots in Smart Manufacturing Systems. *Robotics* **2022**, *11*, 9. [CrossRef]
3. Castelli, K.; Zaki, A.M.A.; Dmytriyev, Y.; Carnevale, M.; Giberti, H. A Feasibility Study of a Robotic Approach for the Gluing Process in the Footwear Industry. *Robotics* **2021**, *10*, 6. [CrossRef]
4. Vicentini, F. Collaborative Robotics: A Survey. *J. Mech. Des.* **2020**, *143*, 040802. [CrossRef]

5. Casalino, A.; Mazzocca, E.; Di Giorgio, M.G.; Maria Zanchettin, A.; Rocco, P. Task scheduling for human-robot collaboration with uncertain duration of tasks: A fuzzy approach. In Proceedings of the 7th International Conference on Control, Mechatronics and Automation (ICCMA), Delft, The Netherlands, 6–8 November 2019; pp. 90–97. [CrossRef]
6. Chen, F.; Sekiyama, K.; Cannella, F.; Fukuda, T. Optimal Subtask Allocation for Human and Robot Collaboration Within Hybrid Assembly System. *IEEE Trans. Autom. Sci. Eng.* **2014**, *11*, 1065–1075. [CrossRef]
7. Casalino, A.; Geraci, A. Allowing a Real Collaboration Between Humans and Robots. In *Special Topics in Information Technology*; Springer: Cham, Switzerland, 2021; pp. 139–148. [CrossRef]
8. Gustavsson, P.; Holm, M.; Syberfeldt, A.; Wang, L. Human-robot collaboration—Towards new metrics for selection of communication technologies. *Procedia CIRP* **2018**, *72*, 123–128. [CrossRef]
9. Liu, H.; Wang, L. Gesture recognition for human-robot collaboration: A review. *Int. J. Ind. Ergon.* **2018**, *68*, 355–367. [CrossRef]
10. Kaczmarek, W.; Panasiuk, J.; Borys, S.; Barach, P. Industrial Robot Control by Means of Gestures and Voice Commands in Off-Line and On-Line Mode. *Sensors* **2020**, *20*, 6358. [CrossRef]
11. Gustavsson, P.; Syberfeldt, A.; Brewster, R.; Wang, L. Human-robot Collaboration Demonstrator Combining Speech Recognition and Haptic Control. *Procedia CIRP* **2017**, *63*, 396–401. [CrossRef]
12. McFarland, D.J.; Wolpaw, J.R. Brain-Computer Interface Operation of Robotic and Prosthetic Devices. *Computer* **2008**, *41*, 52–56. [CrossRef]
13. Villani, V.; Pini, F.; Leali, F.; Secchi, C. Survey on human–robot collaboration in industrial settings: Safety, intuitive interfaces and applications. *Mechatronics* **2018**, *55*, 248–266. [CrossRef]
14. Wang, L.; Gao, R.; Váncza, J.; Krüger, J.; Wang, X.; Makris, S.; Chryssolouris, G. Symbiotic human-robot collaborative assembly. *CIRP Ann.* **2019**, *68*, 701–726. [CrossRef]
15. Matheson, E.; Minto, R.; Zampieri, E.G.G.; Faccio, M.; Rosati, G. Human–Robot Collaboration in Manufacturing Applications: A Review. *Robotics* **2019**, *8*, 100. [CrossRef]
16. El Zaatari, S.; Marei, M.; Li, W.; Usman, Z. Cobot programming for collaborative industrial tasks: An overview. *Robot. Auton. Syst.* **2019**, *116*, 162–180. [CrossRef]
17. Scoccia, C.; Ciccarelli, M.; Palmieri, G.; Callegari, M. Design of a Human-Robot Collaborative System: Methodology and Case Study. In *International Design Engineering Technical Conferences and Computers and Information in Engineering Conference*; American Society of Mechanical Engineers: New York, NY, USA, 2021; Volume 7.
18. Dmytriyev, Y.; Zaki, A.M.A.; Carnevale, M.; Insero, F.; Giberti, H. Brain computer interface for human-cobot interaction in industrial applications. In Proceedings of the 3rd International Congress on Human-Computer Interaction, Optimization and Robotic Applications (HORA), Ankara, Turkey, 9–11 June 2021; pp. 1–6. [CrossRef]
19. Douibi, K.; Le Bars, S.; Lemontey, A.; Nag, L.; Balp, R.; Breda, G. Toward EEG-Based BCI Applications for Industry 4.0: Challenges and Possible Applications. *Front. Hum. Neurosci.* **2021**, *15*, 705064. [CrossRef] [PubMed]
20. Norcia, A.M.; Appelbaum, L.G.; Ales, J.M.; Cottereau, B.R.; Rossion, B. The steady-state visual evoked potential in vision research: A review. *J. Vis.* **2015**, *15*, 4. [CrossRef]
21. Ravi, A.; Heydari, N.; Jiang, N. User-Independent SSVEP BCI Using Complex FFT Features and CNN Classification. In Proceedings of the IEEE International Conference on Systems, Man and Cybernetics (SMC), Bari, Italy, 6–9 October 2019; pp. 4175–4180. [CrossRef]
22. Beverina, F.; Palmas, G.; Silvoni, S.; Piccione, F.; Giove, S. User adaptive BCIs: SSVEP and P300 based interfaces. *PsychNology J.* **2003**, *1*, 331–354.
23. Perera, C.J.; Naotunna, I.; Sadaruwan, C.; Gopura, R.; Lalitharatne, T.D. SSVEP based BMI for a meal assistance robot. In Proceedings of the IEEE International Conference on Systems, Man, and Cybernetics (SMC), Budapest, Hungary, 9–12 October 2016; pp. 002295–002300. [CrossRef]
24. Ortner, R.; Guger, C.; Prueckl, R.; Grünbacher, E.; Edlinger, G. SSVEP Based Brain-Computer Interface for Robot Control. In *Computers Helping People with Special Needs*; Miesenberger, K., Klaus, J., Zagler, W., Karshmer, A., Eds.; Springer: Berlin/Heidelberg, Germany, 2010; pp. 85–90. [CrossRef]
25. Ogura, Y.; Fujii, M.; Nishijima, K.; Murakami, H.; Sonehara, M. Applicability of Hand-Guided Robot for Assembly-Line Work. *J. Robot. Mechatron.* **2012**, *24*, 547–552. [CrossRef]
26. Krüger, J.; Lien, T.; Verl, A. Cooperation of human and machines in assembly lines. *CIRP Ann.* **2009**, *58*, 628–646. [CrossRef]
27. Angrisani, L.; Arpaia, P.; Esposito, A.; Moccaldi, N. A Wearable Brain–Computer Interface Instrument for Augmented Reality-Based Inspection in Industry 4.0. *IEEE Trans. Instrum. Meas.* **2020**, *69*, 1530–1539. [CrossRef]
28. Angrisani, L.; Arpaia, P.; Moccaldi, N.; Esposito, A. Wearable Augmented Reality and Brain Computer Interface to Improve Human-Robot Interactions in Smart Industry: A Feasibility Study for SSVEP Signals. In Proceedings of the IEEE 4th International Forum on Research and Technology for Society and Industry (RTSI), Palermo, Italy, 10–13 September 2018; pp. 1–5. [CrossRef]
29. Klem, G.H.; Lüders, H.O.; Jasper, H.H.; Elger, C. The ten-twenty electrode system of the International Federation. The International Federation of Clinical Neurophysiology. *Electroencephalogr. Clin. Neurophysiol. Suppl.* **1999**, *52*, 3–6.
30. Hinrichs, H.; Scholz, M.; Baum, A.K.; Kam, J.W.Y.; Knight, R.T.; Heinze, H.J. Comparison between a wireless dry electrode EEG system with a conventional wired wet electrode EEG system for clinical applications. *Sci. Rep.* **2020**, *10*, 5218. [CrossRef] [PubMed]

31. Xing, X.; Wang, Y.; Pei, W.; Guo, X.; Liu, Z.; Wang, F.; Ming, G.; Zhao, H.; Gui, Q.; Chen, H. A High-Speed SSVEP-Based BCI Using Dry EEG Electrodes. *Sci. Rep.* **2018**, *8*, 14708. [CrossRef] [PubMed]
32. Sani, O.G. Quick SSVEP: A Web-Based SSVEP Stimulation Interface. *Zenodo* **2020**. [CrossRef]
33. Peirce, J. Generating stimuli for neuroscience using PsychoPy. *Front. Neuroinform.* **2009**, *2*, 10. [CrossRef]
34. Ganin, I.P.; Shishkin, S.L.; Kaplan, A.Y. A P300-based Brain-Computer Interface with Stimuli on Moving Objects: Four-Session Single-Trial and Triple-Trial Tests with a Game-Like Task Design. *PLoS ONE* **2013**, *8*, e77755. [CrossRef]
35. Tomita, Y.; Gaume, A.; Bakardjian, H.; Maurice, M.; Cichocki, A.; Yamaguchi, Y.; Dreyfus, G.; Maurice, F.B. Concatenation Method for High-temporal Resolution SSVEP-BCI. In Proceedings of the International Conference on Neural Computation Theory and Applications, NCTA 2011, Paris, France, 24–26 October 2011; pp. 444–452.
36. Liu, A.; Liu, Q.; Zhang, X.; Chen, X.; Chen, X. Muscle Artifact Removal Toward Mobile SSVEP-Based BCI: A Comparative Study. *IEEE Trans. Instrum. Meas.* **2021**, *70*, 1–12. [CrossRef]
37. Safeea, M.; Bearee, R.; Neto, P. End-Effector Precise Hand-Guiding for Collaborative Robots. In *ROBOT 2017: Third Iberian Robotics Conference*; Ollero, A., Sanfeliu, A., Montano, L., Lau, N., Cardeira, C., Eds.; Springer: Cham, Switzerland, 2018; pp. 595–605.
38. OnRobot. Case Study: Hand Guiding with OnRobot Force Torque Sensor. Available online: https://onrobot.com/en/case-studies/hand-guiding-with-onrobot-force-torque-sensor (accessed on 1 January 2022).

machines

MDPI

Article

Decision Support Method for Dynamic Production Planning

Simona Skėrė [1,*] , **Aušra Žvironienė** [2] , **Kazimieras Juzėnas** [1] **and Stasė Petraitienė** [2]

[1] Faculty of Mechanical Engineering and Design, Kaunas University of Technology, 51424 Kaunas, Lithuania
[2] Faculty of Mathematics and Natural Sciences, Kaunas University of Technology, 51368 Kaunas, Lithuania
*   Correspondence: simona.bukantaite@ktu.lt; Tel.: +37-063096806

Abstract: Small and medium-sized engineering production companies face challenges that are related to unpredicted rapid changes of availability of the work force, materials and equipment. Those challenges are especially difficult to solve for companies focusing on unit or batch production and when they are collaborating with customers who require short lead times. A four-month observation was carried out in a metal processing company in Lithuania to understand the most common rising problems and developing solution for computerised decision support systems. It was discovered that the company needs a computerised "employee centred" system for the improvement of the allocation of tasks to employees. Such a need proved to be the most urgent one, especially during pandemics. An algorithm for the analysis and automated allocation of the employees' tasks has been developed and tested. The proposed algorithm is universal and may be applied in different SMEs for engineering production.

Keywords: production planning; decision support method; production engineering

check for
updates

Citation: Skėrė, S.; Žvironienė, A.;
Juzėnas, K.; Petraitienė, S. Decision
Support Method for Dynamic
Production Planning. *Machines* **2022**,
10, 994. https://doi.org/10.3390/
machines10110994

Academic Editors: Raul
D.S.G. Campilho and Francisco J.
G. Silva

Received: 14 October 2022
Accepted: 27 October 2022
Published: 29 October 2022

Publisher's Note: MDPI stays neutral
with regard to jurisdictional claims in
published maps and institutional affiliations.

1. Introduction

Every production company seeks the best performance to maintain competitiveness. However, each day, production faces different challenges. In few last years, the lives of manufacturing companies have become more dynamic than ever, and this was caused by pandemics, energetic issues, political desicions and other factors which could not be planned or predicted. These turbulences are more effective to SMEs (small and medium-sized enterprises), and so flexibility and having fast response times are vital for such companies [1,2]. To accomplish faster production times, the more efficient use of raw materials and having automated processes of Industry 4.0 and the digitalization of their systems helps [3–5]. As well, having faster delivery times, improved quality of the products, increased customization and financial factors such as cost reduction, revenue growth and increased productivity are key factors to start using tools of Industry 4.0 [6]. However, it might sound easily achievable, but small and medium-sized companies cannot afford smart ERP systems, cloud manufacturing or automated production lines, and as a result it is much harder to implement digitalization or automation procedures [7–9]. SMEs often limit themselves only with such tools such as Cloud computing or Internet of Things [10], virtual factory, smart manufacturing or digital factory applications [11–13]. In different literature studies, cyber physical systems, cloud-based freeware or knowledge management is described as optional help for the SMEs [14,15]. Using only part of the possible newest solutions, the SMEs cannot obtain the interconnectivity and integration that they need in these times [16]. Such companies as much more likely to produce unique and non-recurring orders, and thus, their production processes cannot be simplified or unified [17]. In [18], the advantages of mass customization are described as a strength of the SMEs. To control such processes, production planning and control (PPC) systems are used, but they should be transformed to smart production planning and control (SPPC) systems [19,20]. SPPC systems require real-time management systems, dynamic production planning and autonomous execution control [21]. Thus, there is a great need to help and improve the performance of the SMEs companies since in most economies, SMEs take precedence [22]. That is why this article focuses on the decision support algorithm which could be adapted to production, and especially, it could improve "employee centred" companies which are mostly small or medium-sized.

2. Methodology

2.1. Information about the Case Study Company

For this research, the main data were collected from a metal processing company. This company is in Lithuania and its total number of employees during the examination was 75. This is categorized as a medium-sized company. This company produces furniture components such as metal tube legs, brackets, frames for shelves, tables and others. In total, more than 500 different article numbers for their products are active because company's production is defined by its customers' needs. It accepts small or individual sample orders and also provides metal processing services such as turning, punching, cutting and finishing. Such production tasks are different each day, and flexibility, having a quick response and adaptability are key factors. This company is a part of a group of more than 20 companies, and it is a supplier to them as well as for other companies that are all over Europe. Since the final products of this company are sold to other production companies, any delay in their delivery is unaccepted because that would cause production problems in the other companies. In such business models, a precision and accuracy are highly required. At the moment, all of the equipment is serviced by employees, and no robotic or automated line is available.

In this company, the monitoring of the production lasted for 4 months—from January to April of 2022. This was conducted to understand and confirm the main emerging problems. The company works in three shifts, each of them having 18 employees that work during the production. Different data were gathered to have as much information for further investigations as possible—every shift, every operation and every product was included in this. As seen in Figure 1, the absence of workers was a leading problem, mainly due to isolation and seasonal flus. Another issue, which ended up being ranked in the second place, was a delay of the delivery of the components. A significant increase in this issue was noticed in March since this company lost the relationship with its main suppliers from Ukraine. The third problem was machinery failures. Since the company works in three shifts and for five days a week, the machinery should be in the best condition to provide an uninterrupted workflow. The last section was about the changes in orders, and more specifically the changes by the customers. This issue was connected with the situation in Europe—the customers of this company are furniture producers, and they faced challenges with the supply of other components, thus, they cancelled and changed the dates for the open orders.

To create a workflow and test program, a specific product and its production was selected. It is a metal furniture leg (Figure 2). The furniture leg is made of a steel tube, a punched metal plate, a threaded stud and a plastic cap. These kind of furniture legs are lacquered or painted, and finally, they are packed in plastic bags according to the client, one by one or in kits.

There are 7 main technological operations to produce such furniture leg:

- Cutting operation (1);
- Punching operation (2);
- 1st welding operation (3);
- 2nd welding operation (4);
- Finishing operation (5);
- Assembling operation (6);
- Packaging operation (7).

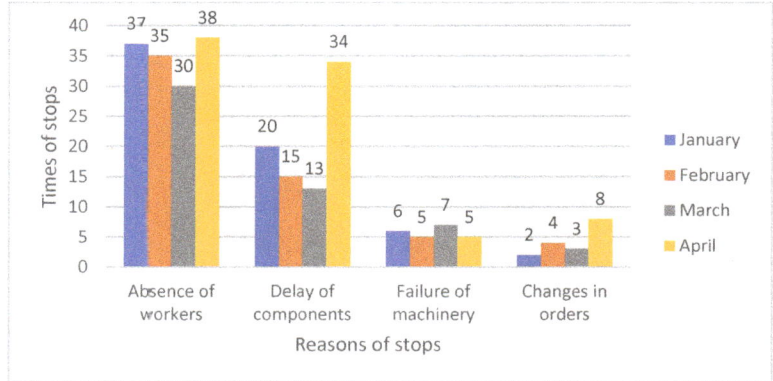

Figure 1. Observation results: issues of production stops.

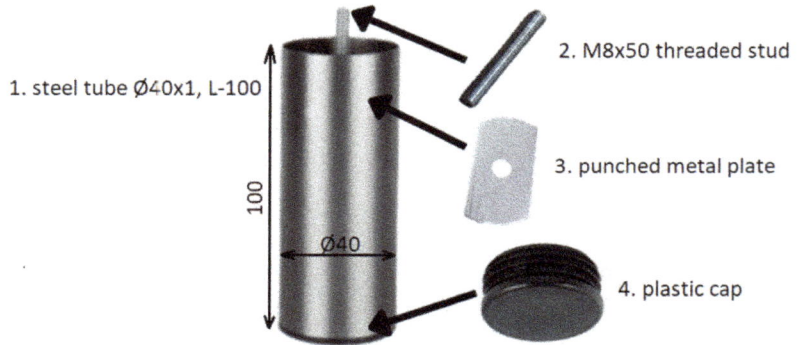

Figure 2. Steel metal furniture leg—D40 × L100 mm.

Each of these operations were performed by an employee, any absence caused a delay. Since production is rapidly changing, every employee had a main specific task to perform but they also had to learn how to cover different tasks. To gather data about the employees' skills and level of knowledge of the operations, matrix of skills was developed (Table 1). This helped to speed up a reaction time when a change was needed—making this process automatic is the main goal. As well, this matrix differentiated the salary of the workers—workers with more skills had a bigger coefficient. An employee can have 4 different levels of skills—from 0 to 1. In the investigated company, this ranking of skills is confirmed by the production manager, and it is based on the opinion about the abilities of each employee. Even though this might be subjective, it is the only way viable method that is available at the moment.

Table 1. Matrix of skills.

E. No.	1	2	3	4	5	6	7	Main Operation
1	0.5	0.5	0.75	0.75	0	1	1	Welding
2	0.5	0.5	0.75	0.75	0	1	1	Welding
3	0.5	0.5	0.75	0.75	1	0.75	1	Welding
4	0.25	0.5	0.75	0.75	0	0.75	1	Welding
5	0.25	0.25	0	0	0	1	1	Packaging
6	0	0	0	0	0	1	1	Packaging
7	0	0	0	0	0	1	1	Packaging

Table 1. *Cont.*

E. No.	1	2	3	4	5	6	7	Main Operation
8	0	0	0	0	0	1	1	Packaging
9	0	0	0	0	0	1	1	Assembling
10	0	0	0	0	0	0.75	1	Assembling
11	0	0	0	0	0	0.75	1	Assembling
12	0.5	0.5	0	0	0	1	1	Assembling
13	0.5	0.5	0	0	1	1	1	Finishing
14	0.5	0.5	0	0	1	0.75	1	Finishing
15	1	0.75	0	0	0	0.75	0.75	Punching
16	1	1	0	0	0	1	0.75	Punching
17	1	1	0	0	0	1	0.75	Cutting
18	1	1	0	0	0	1	1	Cutting

2.2. Workflow

Based on this conducted information, a workflow was created. The workforce algorithm is presented in Figure 3. The workflow reflects the situation in the observed factory, but it could be adapted to a different company by changing the questions.

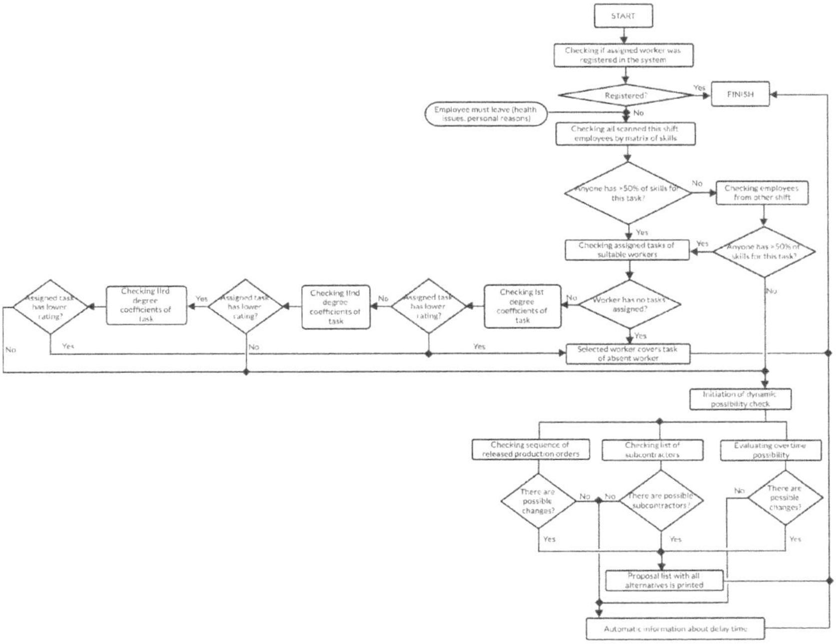

Figure 3. Workflow of presence of employees.

The information about whether an employee came to work or left the production site was received automatically because everyone must scan a personal card. The system was notified about employees personal data such as their name, surname and the time when the check-in and check-out was performed. If the system did not get notification about an employee, it automatically initiates further steps wherein the matrix of skills was needed. The algorithm checked whether someone on that shift had more than 50% of the skills of the absent employee.

However, even if there is a worker who is skilled enough, the system must run a three-stage evaluation process to evaluate which task is more important—the one that employee was performing at the time or the one which has no assigned employee. To run this evaluation, different factors should be evaluated.

This specific company has selected 5 factors of production importance:
- Delivery date;
- The need of this task (technological operation) for further production processes;
- Quantity;
- Clients ranking;
- Extra requirements (i.e., it is sample order for large quantities; parts should be sent to subcontractor; etc.).

Table 2 presents the values of the factors that evaluate the importance of each indicator. The closer to 1 that is value is, the more important it is according to this specific company.

Table 2. Values of selected factors.

Factor	Value	Degree
Delivery date (f_1)	0.8	1
The need of this task for further processes (f_2)	0.9	1
Quantity (f_3)	0.75	2
Clients ranking (f_4)	0.7	2
Any extra addition requirement (f_n)	0.2	3

The equation for the calculation of each coefficient value V is as follows:

$$V = \sum_n^1 f_n \cdot v_n, \tag{1}$$

where, v_n is the value of factor for the specific task and f_n is the factor value itself. Both of the values are confirmed by the production manager, and they are based on an expert opinion.

However, in real life, another situation could appear when there is no available or skilled enough employee to make this production reconfiguration. Then, the algorithm initiates the Dynamic decision support (DDS) module. In this module, the main additional external information is evaluated to give a proposal. For example, in case there was no available employee to cover the needed task, then the DDS module would check the possible changes in production orders, available subcontractors, or overtime possibilities. It is important for a company to import all of the data such as contacts, working hours, or lead times from the subcontractors so that the system can automatically check whether there is a solution or not. Figure 4 presents the DDS module for machinery workflow.

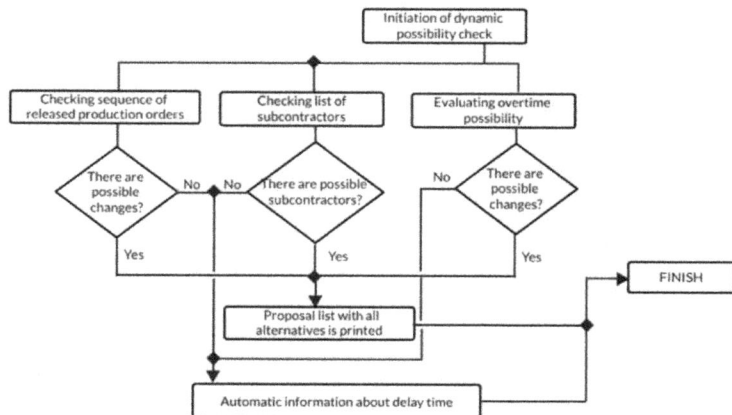

Figure 4. Dynamic Decision Support Module for presence of employees.

2.3. Workflow Check

To ensure that the workflow is usable and adaptable to real life situations, the tests were performed in Excel program (Figure 5). The test was based on the data from the company and it recreated a situation that occurred.

Matrix E	Matrix M	Matrix T	TN	ON	EN	Is EN=1?	E*M*T				
Machinery	Material	Task	Order	Operation	Employee	Employee	Can employee work?	Is task performed?	Result	If Result is 0, additional check:	Final result
1	1	1	1	1	1	1	1	1	2		1 = there is an available employee
1	0	1	2	3	2	1	0	0	1		1 = there is an available employee
0	1	1	3	4	3	1	0	0	1		1 = there is an available employee
1	1	1	4	4	4	0	1	0	0		0 = we need to cover this employee
1	1	1	5	6	5	1	1	1	2		2 = no further steps
1	1	1	6	6	6	1	1	1	2		2 = no further steps
0	1	1	6	6	7	1	0	0	1		1 = there is an available employee
1	1	1	6	6	8	1	1	1	2		2 = no further steps
1	1	1	5	7	9	1	1	1	2	If J & H columns = 0, then no	2 = no further steps
1	0	1	7	10	0	0	0	0	0	further steps, if no - looking	0 = no further steps
0	1	1	6	7	11	1	0	0	1	for an employee	1 = there is an available employee
1	1	1	6	7	12	1	1	1	2		2 = no further steps
1	1	1	5	5	13	1	1	1	2		2 = no further steps
1	1	1	6	5	14	1	1	1	2		2 = no further steps
1	1	1	7	2	15	1	1	1	2		2 = no further steps
1	1	1	8	2	16	1	1	1	2		2 = no further steps
1	0	1	7	1	17	1	0	0	1		1 = there is an available employee
1	1	1	8	1	18	1	1	1	2		2 = no further steps

Figure 5. Results of the algorithm testing.

The "Matrix E", "Matrix M" and "Matrix T" columns represent the availability of equipment, materials, and task, respectively—if any of these are missing, then the employee cannot work even if he is at work. Each employee has an assigned task (operation) and an order number ("TN" and "ON" columns). An employee number is provided in "EN" column. The values in "Is EN = 1?" column is used to check whether an employee is at work. The results in "Result" column can be as follows:

- "1" means that an employee is in reserve and is available for any work because there is no possibility for them to work due to lack of materials, equipment or other reasons;
- "2" means that everything has gone according to the plan;
- "0" has two meanings: either nothing should be done because there is no employee and there is no task, and either there is a need to look for someone to cover this task. The replacement is needed in case when values in H column is equal to "1".
- The next step is to find someone with the required skills to cover the task. For this, the matrix of skills is used. When the company has an employee who could cover the task, there is a need to check if the initially planned task is more or less important. To do that, a three-stage evaluation process is initiated.

After this, we tested and proved that algorithm was working and gave correct results, and then, further tests were made using Matlab. In order to start the program, some information was required. A manual data upload methos was only used in this test model because the idea is that information about equipment, materials and employees will be added in the system in real time.

2.4. Mathematical Transformation

To create a method that is based on previous tests, a mathematical transformation was performed. Matrix $P1$ shows the initial day plan:

$$P1 = (p1_{ij}), \ i = \overline{1, \ k}, \ j = \overline{1, \ 6}, \tag{2}$$

In this case, $k = 18$ since there are 18 employees.

$$P1 = \begin{pmatrix} 1 & 1 & 1 & 1 & 3 & 1 \\ 1 & 0 & 1 & 2 & 3 & 2 \\ 0 & 1 & 1 & 3 & 4 & 3 \\ 1 & 1 & 1 & 4 & 4 & 4 \\ 1 & 1 & 1 & 5 & 6 & 5 \\ 1 & 1 & 1 & 5 & 6 & 6 \\ 0 & 1 & 1 & 6 & 6 & 7 \\ 1 & 1 & 1 & 6 & 6 & 8 \\ 1 & 1 & 1 & 5 & 7 & 9 \\ 1 & 0 & 1 & 5 & 7 & 10 \\ 0 & 1 & 1 & 6 & 7 & 11 \\ 1 & 1 & 1 & 6 & 7 & 12 \\ 1 & 1 & 1 & 5 & 5 & 13 \\ 1 & 1 & 1 & 6 & 5 & 14 \\ 1 & 1 & 1 & 7 & 2 & 15 \\ 1 & 1 & 1 & 8 & 2 & 16 \\ 1 & 0 & 1 & 7 & 1 & 17 \\ 1 & 1 & 1 & 8 & 1 & 18 \end{pmatrix} \tag{3}$$

The corresponding elements of the columns of the matrix $P1$ can take on certain values:

$$P1_{i,1} = \begin{cases} 0, equipment\ is\ available, \\ 1, equipment\ is\ not\ available; \end{cases} \tag{4}$$

$$P1_{i,2} = \begin{cases} 0, materials\ are\ missing, \\ 1, aterials\ are\ in\ stock; \end{cases} \tag{5}$$

$$P1_{i,3} = \begin{cases} 0, no\ task\ is\ assigned, \\ 1, task\ is\ assigned; \end{cases} \tag{6}$$

The assigned task number is presented in the fourth column. In this case, $m = 8$ since there are 8 different production orders that run through this shift:

$$P1_{i,4} = \overline{1,\ m}, \tag{7}$$

The operation number is in the 5th column. In this case, $n = 7$ since there are 7 different production operations:

$$P1_{i,5} = \overline{1, n}, \tag{8}$$

The last column describes the number of employees. In this case, $k = 18$:

$$P1_{i,6} = \overline{1,\ k}, \tag{9}$$

The first check of the plan is whether all of the employees have come to work. The data are entered by simulating the real situation. A new matrix S is created:

$$S = (s_{ij}),\ i = \overline{1,\ k},\ j = \overline{1,\ 4}, \tag{10}$$

$$S = \begin{pmatrix} 1 & 1 & 1 & 2 \\ 1 & 0 & 0 & 1 \\ 1 & 0 & 0 & 1 \\ 0 & 1 & 0 & 0 \\ 1 & 1 & 1 & 2 \\ 1 & 1 & 1 & 2 \\ 1 & 0 & 0 & 1 \\ 1 & 1 & 1 & 2 \\ 1 & 1 & 1 & 2 \\ 0 & 0 & 0 & 0 \\ 1 & 0 & 0 & 1 \\ 1 & 1 & 1 & 2 \\ 1 & 1 & 1 & 2 \\ 1 & 1 & 1 & 2 \\ 1 & 1 & 1 & 2 \\ 1 & 1 & 1 & 2 \\ 1 & 0 & 0 & 1 \\ 1 & 1 & 1 & 2 \end{pmatrix} \qquad (11)$$

The corresponding elements of the columns of the matrix S can have certain values:

$$S_{i,1} = \begin{cases} 0, employee\ is\ absent, \\ 1,\ employee\ is\ working; \end{cases} \qquad (12)$$

Another check should define if the employee is able to start a new operation, has all of the materials, the equipment is properly working, and they have an initial task that has been allocated to them. To do that, the data from matrix $P1$ are needed:

$$S_{i,2} = P1_{i,1} \cdot P1_{i,2} \cdot P1_{i,3} = \begin{cases} 0, employee\ can\ not\ work, \\ 1,\ employee\ can\ work; \end{cases} \qquad (13)$$

The third column of matrix S shows whether the planned task is performed. Even if $S_{i,2} = 1$, but $S_{i,1} = 0$, it is not possible to perform a task as the assigned employee is not at work:

$$S_{i,3} = S_{i,1} \cdot S_{i,2} = \begin{cases} 0,\ task\ is\ not\ performed, \\ 1, task\ is\ performed; \end{cases} \qquad (14)$$

According to the values that are received in first three columns, the result column appears:

$$S_{i,4} = S_{i,1} + S_{i,3} = \begin{cases} 0, \\ 1, \\ 2, \end{cases} \qquad (15)$$

The result 1 means that employee is in reserve, the result 2 means that no changes are needed and if the result is equal to 0, an additional check is needed to see if $S_{i,2}$ is equal to 0 (no actions required) or 1 (replacement is needed).

Then, matrix C was created:

$$C = \begin{pmatrix}
1 & 2 & 0,50 & 0,50 & 0,75 & 0,75 & 0,00 & 1,00 & 1,00 \\
2 & 1 & 0,50 & 0,50 & 0,75 & 0,75 & 0,00 & 1,00 & 1,00 \\
3 & 1 & 0,50 & 0,50 & 0,75 & 0,75 & 1,00 & 0,75 & 1,00 \\
4 & 0 & 0,25 & 0,50 & 0,75 & 0,75 & 0,00 & 0,75 & 1,00 \\
5 & 2 & 0,25 & 0,25 & 0,00 & 0,00 & 0,00 & 1,00 & 1,00 \\
6 & 2 & 0,00 & 0,00 & 0,00 & 0,00 & 0,00 & 1,00 & 1,00 \\
7 & 1 & 0,00 & 0,00 & 0,00 & 0,00 & 0,00 & 1,00 & 1,00 \\
8 & 2 & 0,00 & 0,00 & 0,00 & 0,00 & 0,00 & 1,00 & 1,00 \\
9 & 2 & 0,00 & 0,00 & 0,00 & 0,00 & 0,00 & 1,00 & 1,00 \\
10 & 0 & 0,00 & 0,00 & 0,00 & 0,00 & 0,00 & 0,75 & 1,00 \\
11 & 1 & 0,00 & 0,00 & 0,00 & 0,00 & 0,00 & 0,75 & 1,00 \\
12 & 2 & 0,50 & 0,50 & 0,00 & 0,00 & 0,00 & 1,00 & 1,00 \\
13 & 2 & 0,50 & 0,50 & 0,00 & 0,00 & 1,00 & 1,00 & 1,00 \\
14 & 2 & 0,50 & 0,50 & 0,00 & 0,00 & 1,00 & 0,75 & 1,00 \\
15 & 2 & 1,00 & 0,75 & 0,00 & 0,00 & 0,00 & 0,75 & 0,75 \\
16 & 2 & 1,00 & 1,00 & 0,00 & 0,00 & 0,00 & 1,00 & 0,75 \\
17 & 1 & 1,00 & 1,00 & 0,00 & 0,00 & 0,00 & 1,00 & 0,75 \\
18 & 2 & 1,00 & 1,00 & 0,00 & 0,00 & 0,00 & 1,00 & 1,00
\end{pmatrix} \tag{16}$$

The corresponding elements of columns of the matrix C can take on certain values:

$$C_{i,1} = \overline{1, k}, \tag{17}$$

$$C_{i,2} = S_{i,4}, \tag{18}$$

$$C_{i,j} = \begin{cases} 0; \\ 0.25; \\ 0.5; \\ 1 \end{cases}, \tag{19}$$

$j = \overline{3, 9}$ because there are 7 operations, and each employee has specific knowledge that can be used for the task. It represents the information from Table 1.

As in this specific example, there were two absent employees—No. 4 and No. 10, but only the employee No. 4 needed a replacement. Firstly, the program scans the employees with $C_{i,2} = 1$ because they are free to the cover task that was initially planned for employee No. 4. They need to have enough skills for operation No. 4, which was the operation that was assigned to an absent employee. In this example, two employees are available—No. 2 and No. 3. If there were no one available from $C_{i,2} = 1$ list, then the $C_{i,2} = 2$ list would be checked. The one that would have lower task ranking position would cover the task.

In this specific case, there are 8 different tasks. Each of them has two Ist degree values, two IInd degree values and IIIrd degree values, and thus, matrix OW is created:

$$OW = \begin{pmatrix}
1 & 0.10 & 0,11 & 0,05 & 0,08 & 0,82 & 0,21 & 0,13 & 0,82 \\
2 & 0,12 & 0,13 & 0,21 & 0,21 & 0,09 & 0,25 & 0,42 & 0,09 \\
3 & 0,81 & 0,05 & 0,55 & 0,37 & 0,00 & 0,86 & 0,92 & 0,00 \\
4 & 0,15 & 0,12 & 0,15 & 0,07 & 0,00 & 0,27 & 0,22 & 0,00 \\
5 & 0,75 & 0,65 & 0,24 & 0,89 & 0,00 & 1,40 & 1,13 & 0,00 \\
6 & 0,88 & 0,50 & 0,18 & 0,89 & 0,00 & 1,38 & 1,07 & 0,00 \\
7 & 0,79 & 0,47 & 0,49 & 0,41 & 0,75 & 1,26 & 0,90 & 0,75 \\
8 & 0,20 & 0,39 & 0,58 & 0,25 & 0,00 & 0,59 & 0,83 & 0.00
\end{pmatrix} \tag{20}$$

The corresponding elements of the columns of the OW matrix can take on certain values:

$$OW_{i,1} = \overline{1, m}, \tag{21}$$

$$OW_{i,2} = \overline{0,1}, \tag{22}$$

This is factor f_1 from Table 1.

$$OW_{i,3} = \overline{0,1}, \tag{23}$$

This is factor f_2 from Table 1.

$$OW_{i,4} = \overline{0,1}, \tag{24}$$

This is factor f_3 from Table 1.

$$OW_{i,5} = \overline{0,1}, \tag{25}$$

This is factor f_4 from Table 1.

$$OW_{i,6} = \overline{0,1}, \tag{26}$$

This is factor f_5 from Table 1.

$$OW_{i,7} = OW_{i,2} + OW_{i,3}, \tag{27}$$

This is the overall first-degree tasks ranking coefficient.

$$OW_{i,8} = OW_{i,4} + OW_{i,5}, \tag{28}$$

This is the overall second-degree tasks ranking coefficient.

$$OW_{i,9} = OW_{i,6}, \tag{29}$$

This is the overall third-degree production tasks ranking coefficient. Now the condition is being checked. If

$$OW_{x,7} > OW_{i,7}, \tag{30}$$

where x is absent employee and i is employee which covers this task and whose $S_{i,4} = 2$, then, the decision is that the employee i changes to the x employee. Otherwise, 2nd degree check is initiated and if

$$OW_{x,8} > OW_{i,8}, \tag{31}$$

then, the employee i changes to the x employee. Otherwise, the DDS is initiated.

In this part, the mathematical background is presented which is needed to transform this method to the Matlab program and test it further. The tests showed that algorithm is correct and solves the described problem—the absence of employees. The received results from tests in the Matlab program are presented in the following section.

3. Results

All of received the data are up to date so that the program with the rearranged production plan can adapt in a short time. Since the investigated production company works in three shifts, the production manager is only available 8 hours per day, so this kind of software gives a solution and changes the existing production plan when it is based on real situations. The created method includes the selection of the working plan; the working day could be divided into two or one production plans for the employees. This is important since the production often has quantities which are only planned for half of the shift.

In the first case, the machine of the 1st operation is not working, and the company has all of the materials. All of the employees are at work. The visual result of the initial data is shown in Figure 6. Each operation has its own colour—red represents the cutting operation, green represents the punching operation, the 1st welding operation is dark blue, the 2nd welding operation is light blue, the finishing one is purple, the assembly one operation is yellow and the packaging one is black. If for any reason operation could not be performed, then the box is not filled with colour. The outline of the box shows the operation that was assigned. If there is a star-shaped box, it means that the employee is absent and the outline colour represents the operation which was assigned to this employee.

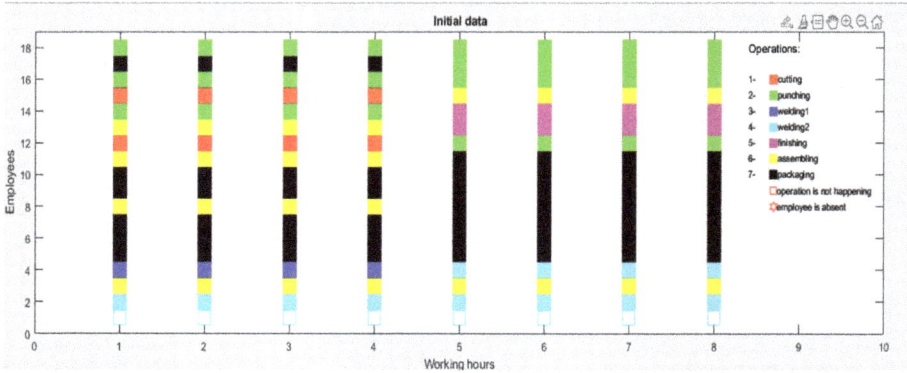

Figure 6. Initial data, 1st case.

The results are that the production plan stays as it was in the beginning of this case analysis. The first employee will remain in reserve. Figure 7 shows the visual presentation of the results.

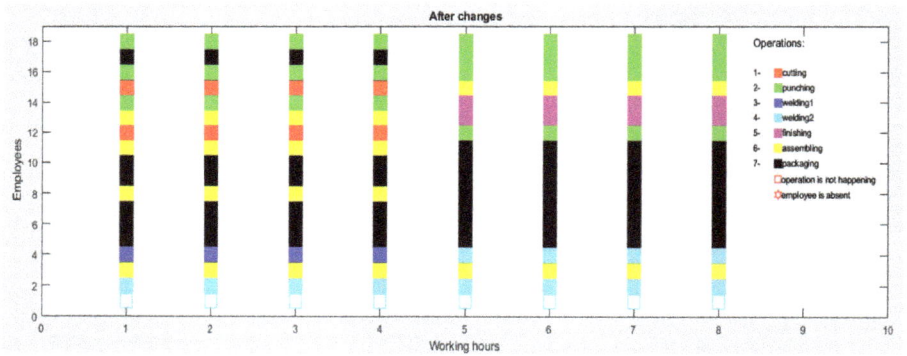

Figure 7. Data after changes, 1st case.

When the program starts, as mentioned, it requests for the data about the equipment, materials and employees that are available at the moment and the daily plan of production. For the second example, the initial data show that two machines stopped working, the 1st task was not performed due to lack of materials, the working plan is 8 h and two employees were absent (Figure 8). The program immediately gives a solution that employees No. 9 and No. 10 must be changed. Three other employees could perform the tasks of the absent ones—employees No. 5, No. 17 and No. 18. The system first gives the proposal to use those employees who were not able to work due to a lack of machinery and materials. The system proposes that they perform the operations instead of the absent ones, and only in the case that these are not suitable for the operation, it suggests someone from employee list that has tasks which has been already allocated.

```
Command Window
  Number of machines forced to stop due to critical condition:2
  Number of operations forced to stop due to lack of materials:1
  Working plan: 1 - 8 hours continuously; 2 - changing each 4 hours:1
  Number of absent employees:2
  ...........................................
  9 employee must be changed
  5 employee changes 9
  17 employee changes 9
  18 employee changes 9

   Select who changes absent 9 employee 17
  ---------------------------------------
  10 employee must be changed
  5 employee changes 10
  18 employee changes 10

   Select who changes absent 10 employee 5
  ---------------------------------------
fx >>
```

Figure 8. Entered information, 2nd case.

To automatically select one of the employee of several that are available, an additional coefficient was created—the employee effectiveness coefficient. As this program seeks to optimize the production, changing the employees at random is not be the best solution. Thus, an employee effectiveness coefficient will be calculated by evaluating:

- The time period that is required for an employee to start doing a specific task (technological operation)—even if the employee has enough skills to cover task, maybe this task was performed by them a long time ago, and therefore, remembering and preparing for the task will take more time when they are compared to some other employee;
- The productivity of the employee in the certain technological operation—at the same time, a different person will perform different amount of valuable tasks, and this should be evaluated to reach maximum performance;
- The quality—the information of rejected parts/scrap from production is collected so that everyone has the data about the percentage of the inappropriate parts. It is better to select an employee who has a better quality coefficient for a specific product.

The visual presentation of the data that were entered, and the selection of the employee is shown in Figure 9.

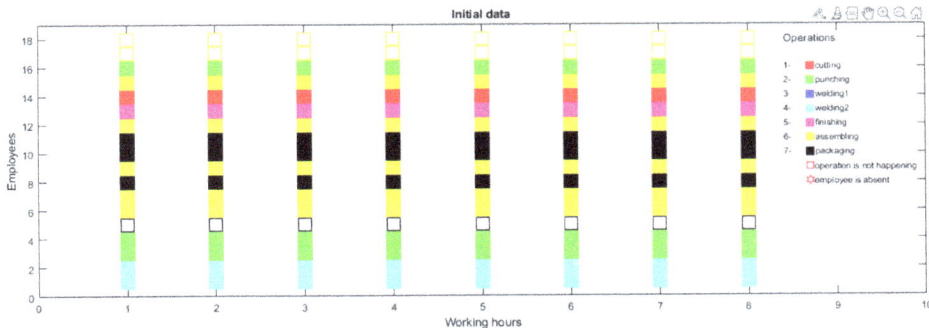

Figure 9. Initial data, 2nd case.

As it is shown in Figure 10, the selected employee No. 17 covers the task that was previously planned for the employee No. 9. Employee No. 9 had to perform an assembly operation (indicated by yellow colour) so the employee No. 17 is coloured in yellow. The same applies to the absent employee No. 10—the employee No. 5 covers the packaging operation (indicated by black colour). So as a result, only employee No. 18 stays in reserve.

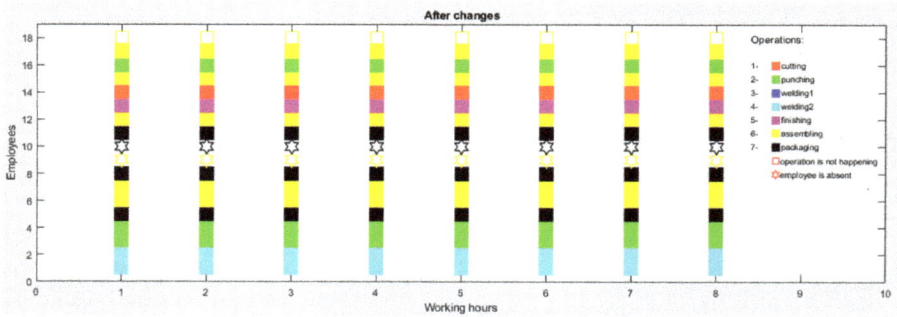

Figure 10. Data after changes, 2nd case.

There is a need to test a situation wherein all of the machines are working, and all of the materials are in place. In that case, a simulated situation is when no employees are in reserve, and thus, the selection for an absent employee is made from someone who has been allocated the task. For the selection, the task rating is initiated, and the production tasks importance coefficients are used. The input data for this specific situation are shown in Figure 11. The program selects the optimal change of the employees.

```
Command Window
  Number of machines forced to stop due to critical condition:0
  Number of operations forced to stop due to lack of materials:0
  Working plan: 1 - 8 hours continuously; 2 - changing each 4 hours:2
  Number of absent employees:1
  ****************************************
  8 employee must be changed
  There is noone free who can change absent employee
  Optimal replacement 5
  -------------------------------------------
  ****************************************
  8 employee must be changed
  There is noone free who can change absent employee
  Optimal replacement 5
  -------------------------------------------
fx >>
```

Figure 11. Entered information, 3rd case.

Since the working plan for this specific case is 4 hours, the program finds an optimal replacement for the first and for the second part of the day (shift). In this case, it is employee No. 5 who was completing a packaging operation. According to the plan, employee No. 8 was also assigned to the packaging operation. The initial plan is shown in Figure 12.

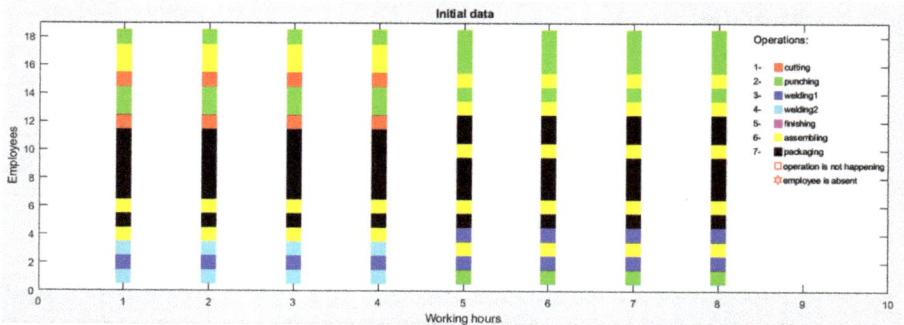

Figure 12. Initial data, 3rd case.

That means that the system checked the possible employees and selected the one that was working with least necessary operation. There were several other employees who could cover this task because based on the skills matrix, the packaging operation is commonly known to most of the employees. That is why task ranking coefficients need to be carefully selected by the authorities so the program could automatically solve such cases without additional questions being asked (additional data input is required). As a result, the task of employee No. 5 will not be processed (Figure 13).

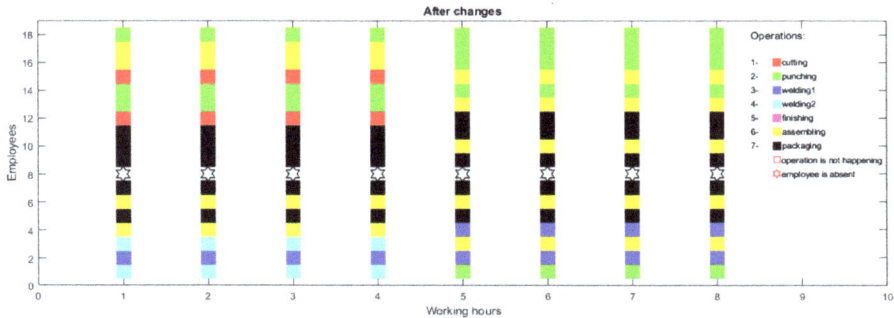

Figure 13. Data after changes, 3rd case.

All of these automatic changes in production not only provide an immediate response for the situation that occurs at that time, but it can also be used for proposals of future improvements. The collected data could give a proper result to develop further strategic actions. This is a future research topic to develop how this method could work not only in a daily background, but also provide sufficient data for management. The displayed data could present which technological equipment or operations need upgrading or to plan the development of the employees' skills.

While modelling possible situations, the program was activated 200 times. As the company has three shifts and the program is activated every time, it obtains an update on the equipment, materials or employees 200 times, which could possibly be achieved in approximately 2 months. This test presented the results of how often each employee did not come to work, and these data are presented in Figure 14.

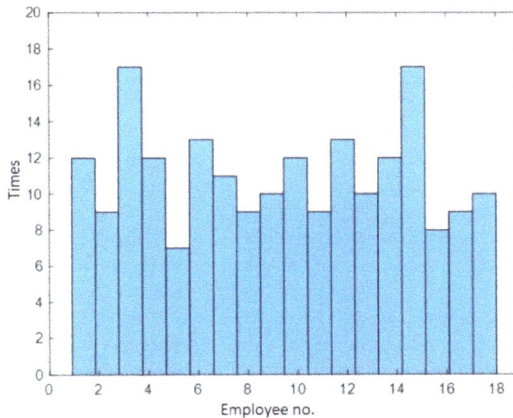

Figure 14. Number of absences after 200 program activations.

From this test, the company could find that employees No. 3 and No. 15 were absent 17 times out of 200 times. Employee No. 3 is a welder so there are only few employees with such skills, and possibly, they have been assigned welding operation tasks. The data that indicate that in two months, this employee was absent for almost 10% of working days could indicate a need to make strategical changes. The additional generated statistics present the frequency of the replacements of each employee (Figure 15).

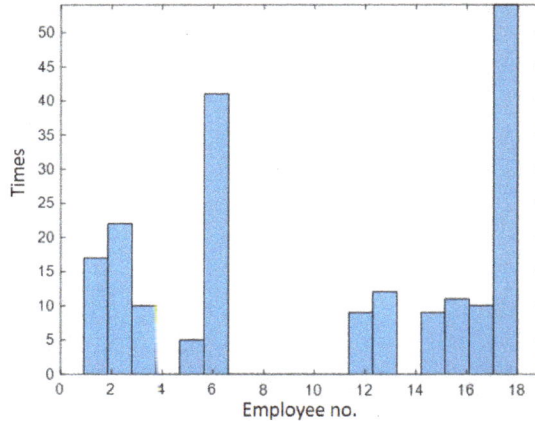

Figure 15. Number of changes after 200 program activations.

These data shows that employee No. 18 was involved in about 25% of all of the cases. Originally, this employee as assigned to cutting operation tasks. However, seven employees have not changed their tasks with another employee's during these test runs. The reasons for this should be checked, and if they cannot change theirs with other employee due to them having a lack of knowledge in another operations, this could show a need of trainings. This would reduce the load of employee No. 18, and the cutting operation (which is the starting operation in the production process) could proceed without interruptions. As well, it is possible that this employee was covering other tasks because he was in reserve or did not have the equipment or materials. In such a case, different solutions could be carried out. Having these statistics still cannot predict the occurrence of future stops or absences. Thus, the necessity of such a decision support method is needed. Each day has a different case, and this proves the dynamic of production and existing problem of SME companies that without advanced ERP systems, no other accessible solution can be provided.

4. Conclusions and Discussion

There is not a single day without unpredicted cases in the production industry, and this might not be a huge problem for large companies with ERP systems, mass production and advanced technologies. However, small and medium-sized companies face these problems differently. Production in such business is based on them having small batches and a wide range of products. The manufacturing is based on the work of the employees, and they do not have an accessible program to control the situation. That is why this article presents a decision support method for an "employee centred" company.

The case study in a medium-sized metal processing company was made, and the observations of 4 months gave a solid background to start the development of such a method. Having access to real life data presented the possibility to check the created method and obtain sufficient data. This also presented the ability to make future recommendations and proposals.

The created algorithm requires several coefficients—the production task importance coefficients, the data about the skills and the employee effectiveness coefficient will be added. The algorithm solves unpredicted situations and either gives a straight command to the workers or presents possible solutions.

In the future, this algorithm will be adapted for replanning equipment, the reconfiguration of the task sequence in case of machinery failure, the lack of materials, etc. This method is universal and could be easily used for equipment and materials in the same way it was used for employees. The statistics might be used to make strategical decisions about the improvement to the machinery, the training of employees, a change in suppliers, etc. After the investigation, a recommendation for the companies is to pay more attention to factors which are confirmed only by the production manager. To avoid subjective assessments, automatization should be discussed for coefficients such as the level of skills. The higher levels can be only gained when the automatic quality control for specific task exceeds the target value. A step before automatization could lead to a periodical competence verification which should be carried out to understand the existing level of skills. In this case, the evaluation would be connected to a specific value, but not to the personal opinion.

Author Contributions: Conceptualization, S.S. and K.J.; methodology, K.J.; software, S.P.; validation, S.S.; formal analysis, A.Ž.; investigation, S.S.; resources, S.S. and S.P.; data curation, S.P.; writing—original draft preparation, S.S.; writing—review and editing, S.S.; visualization, S.S.; supervision, K.J.; project administration, S.S. All authors have read and agreed to the published version of the manuscript.

Funding: This research received no external funding.

Data Availability Statement: Not applicable.

Conflicts of Interest: The authors declare no conflict of interest.

References

1. Wadhwa, R.S. Flexibility in manufacturing automation: A living lab case study of Norwegian metalcasting SMEs. *J. Manuf. Syst.* **2012**, *31*, 444–454. [CrossRef]
2. Spena, P.R.; Holzner, P.; Rauch, E.; Vidoni, R.; Matt, D.T. Requirements for the Design of Flexible and Changeable Manufacturing and Assembly Systems: A SME-survey. *Procedia CIRP* **2016**, *41*, 207–212. [CrossRef]
3. Moeuf, A.; Pellerin, R.; Lamouri, S.; Tamayo-Giraldo, S.; Barbaray, R. The industrial management of SMEs in the era of Industry 4.0. *Int. J. Prod. Res.* **2018**, *56*, 1118–1136. [CrossRef]
4. Li, W.; Liu, K.; Belitski, M.; Ghobadian, A.; O'Regan, N. E-Leadership through Strategic Alignment: An Empirical Study of Small- and Medium-sized Enterprises in the Digital Age. *J. Inf. Technol.* **2016**, *31*, 185–206. [CrossRef]
5. Issa, A.; Lucke, D.; Bauernhansl, T. Mobilizing SMEs Towards Industrie 4.0-enabled Smart Products. *Procedia CIRP* **2017**, *63*, 670–674. [CrossRef]
6. Colotla, I.; Fæste, A.; Heidmann, A.; Winther, A.; Høngaard Andersen, P.; Duvold, T.; Hansen, M. Winning the Industry 4.0 Race—How Ready Are Danish Manufacturers? BCG & Innovationsfonden: Copenhagen, Denmark, 2016.
7. Urbach, N.; Röglinger, M. Introduction to Digitalization Cases: How Organizations Rethink Their Business for the Digital Age. *Digit. Cases. Manag. Prof.* **2018**, 1–12. [CrossRef]
8. Stich, V.; Zeller, V.; Hicking, J.; Kraut, A. Measures for a successful digital transformation of SMEs. *Procedia CIRP* **2020**, *93*, 286–291. [CrossRef]
9. Zaverzhenets, M.; Łobacz, K. Digitalising and visualising innovation process: Comparative analysis of digital tools supporting innovation process in SMEs. *Procedia Comput. Sci.* **2021**, *192*, 3805–3814. [CrossRef]
10. Ingaldi, M.; Ulewicz, R. Problems with the Implementation of Industry 4.0 in Enterprises from the SME Sector. *Sustainability* **2020**, *12*, 217. [CrossRef]
11. Hao, Y.; Helo, P.; Shamsuzzoha, A. Virtual factory system design and implementation: Integrated sustainable manufacturing. *Int. J. Syst. Sci. Oper. Logist.* **2016**, *5*, 116–132. [CrossRef]
12. Stoldt, J.; Trapp, T.U.; Toussaint, S.; Süße, M.; Schlegel, A.; Putz, M. Planning for Digitalisation in SMEs using Tools of the Digital Factory. *Procedia CIRP* **2018**, *72*, 179–184. [CrossRef]
13. Mittal, S.; Khan, M.A.; Purohit, J.K.; Menon, K.; Romero, D.; Wuest, T. A smart manufacturing adoption framework for SMEs. *Int. J. Prod. Res.* **2020**, *58*, 1555–1573. [CrossRef]
14. Tobón Valencia, E.; Lamouri, S.; Pellerin, R.; Dubois, P.; Moeuf, A. Production Planning in the Fourth Industrial Revolution: A Literature Review. *IFAC-Pap.* **2019**, *52*, 2158–2163. [CrossRef]
15. Amaral, A.; Peças, P. SMEs and Industry 4.0: Two case studies of digitalization for a smoother integration. *Comput. Ind.* **2021**, *125*, 1–25. [CrossRef]

16. Bär, K.; Herbert-Hansen, Z.N.L.; Khalid, W. Considering Industry 4.0 aspects in the supply chain for an SME. *Prod. Eng.* **2018**, *12*, 747–758. [CrossRef]
17. Centobelli, P.; Cerchione, R.; Esposito, E. Efficiency and effectiveness of knowledge management systems in SMEs. *Prod. Plan. Control* **2019**, *30*, 779–791. [CrossRef]
18. Safar, L.; Sopko, J.; Bednar, S.; Poklemba, R. Concept of SME Business Model for Industry 4.0 Environment. *TEM J.* **2018**, *7*, 626–637. [CrossRef]
19. Cañas, H.; Mula, J.; Campuzano-Bolarín, F.; Poler, R. A conceptual framework for smart production planning and control in Industry 4.0. *Comput. Ind. Eng.* **2022**, *173*, 108659. [CrossRef]
20. Zheng, C.; Qin, X.; Eynard, B.; Bai, J.; Li, J.; Zhang, Y. SME-oriented flexible design approach for robotic manufacturing systems. *J. Manuf. Syst.* **2019**, *53*, 62–74. [CrossRef]
21. Saad, S.; Bahadori, R.; Jafarnejad, H.; Putra, M. Smart Production Planning and Control: Technology Readiness Assessment. *Procedia Comput. Sci.* **2021**, *180*, 618–627. [CrossRef]
22. Mittal, S.; Khan, M.A.; Romero, D.; Wuest, T. A critical review of smart manufacturing & Industry 4.0 maturity models: Implications for small and medium-sized enterprises (SMEs). *J. Manuf. Syst.* **2018**, *49*, 194–214. [CrossRef]

machines

Review

Environmental Risk Assessment and Management in Industry 4.0: A Review of Technologies and Trends

Janaína Lemos [1], Pedro D. Gaspar [1,2,*] and Tânia M. Lima [1,2]

[1] Department of Electromechanical Engineering, University of Beira Interior, 6201-001 Covilhã, Portugal
[2] C-MAST—Centre for Mechanical and Aerospace Science and Technologies, 6201-001 Covilhã, Portugal
* Correspondence: dinis@ubi.pt

Abstract: In recent decades, concern with workers' health has become a priority in several countries, but statistics still show that it is urgent to perform more actions to prevent accidents and illnesses related to work. Industry 4.0 is a new production paradigm that has brought significant advances in the relationship between man and machine, driving a series of advances in the production process and new challenges in occupational safety and health (OSH). This paper addresses occupational risks, diseases, opportunities, and challenges in Industry 4.0. It also covers Internet-of-Things-related technologies that, by the real-time measurement and analysis of occupational conditions, can be used to create smart solutions to contribute to reducing the number of workplace accidents and for the promotion of healthier and safer workplaces. Proposals involving smart personal protective equipment (smart PPE) and monitoring systems are analyzed, and aspects regarding the use of artificial intelligence and the data privacy concerns are also discussed.

Keywords: occupational risk assessment; Industry 4.0; internet of things; smart PPE

Citation: Lemos, J.; Gaspar, P.D.; Lima, T.M. Environmental Risk Assessment and Management in Industry 4.0: A Review of Technologies and Trends. *Machines* **2022**, *10*, 702. https://doi.org/10.3390/machines10080702

Academic Editors: Raul D. S. G. Campilho and Francisco J. G. Silva

Received: 26 July 2022
Accepted: 16 August 2022
Published: 17 August 2022

Publisher's Note: MDPI stays neutral with regard to jurisdictional claims in published maps and institutional affiliations.

1. Introduction

According to the International Labour Organization [1], occupational injury is a personal injury, disease, or death that results from an occupational accident. Occupational accidents, in turn, are unexpected occurrences, including acts of violence, arising out of or in connection with work and resulting in one or more workers incurring personal injury, disease, or death. Occupational diseases are acquired through personal exposure to environmental risks, such as physical, chemical, and biological agents in situations above the tolerance limits imposed by legislation or applicable standards. These diseases are caused or aggravated by specific activities, and are characterized when the causal link is established. between damage to the worker's health and exposure to certain work-related risks. Occupational diseases occur after various years of exposure, and in some cases, they can arise even after the worker is no longer in contact with the causative agent [2].

Many countries have prioritized concerns about workers' health in recent decades, but statistics show an urgent need to take further action to prevent accidents and illnesses related to work. Worldwide, about two million people die every year because of work-related illnesses or work-related accidents. Many work-related accidents and diseases are not reported, because in several countries there are no adequate data collection systems. Even in countries that adopt sufficient methods for this purpose the number of reported accidents often does not reflect reality, due to the presence of informal workers [3,4].

In addition, the incidence of fatalities in the workplace varies considerably between developed and developing countries. Insufficient OSH services contribute to the occurrence of accidents and deaths in low- and middle-income countries [5]. In terms of economic sectors, agriculture, forestry, mining, and construction have the highest death rates. Companies with fewer than 50 employees have a higher incidence of serious and fatal injuries [6]. In general, migrant workers are more susceptible to informal, abusive, and dangerous work, because the types of work they accept is often affected by lower levels of education [7].

With respect to methods and systems to prevent occupational diseases and accidents, as new technologies are added to workplaces, new risks are identified as well as new opportunities. Especially in the last decade, the term Industry 4.0 became more popular, referring to a new paradigm that has revolutionized factories by inserting and integrating several different technologies. Industry 4.0 technologies have impacted OSH by providing new possibilities in environmental risk monitoring and preventing accidents. Workers' health conditions can also be monitored in real-time. However, aside from new opportunities, new concerns have emerged, especially related to the types of activity commonly performed in Industry 4.0 workplaces. For example, the availability of jobs characterized by sedentary postures or interactions with robots has grown. In this scenario, illnesses related to a sedentary lifestyle and accidents because of interactions with robots are likely to become more and more common. In view of developments to date, it has become necessary for companies to adapt their OSH policies and seek appropriate solutions to this new reality [8,9].

This paper addresses occupational risks and diseases reported in Industry 4.0, as well as opportunities and challenges. Technologies and devices for use in risk assessment in Industry 4.0 are described, and studies that have successfully applied these technologies are analyzed, especially regarding the use of artificial intelligence and data privacy concerns. In addition, this work indicates some directions for addressing data privacy in IoT and Industry 4.0, and comments on issues within this new context.

2. Occupational Risks and Diseases in Industry 4.0

An occupational risk factor is an agent that can cause damage to a worker's health. The potential risk factor is called hazard. Occupational risk is the combination of the probability of an adverse effect (damage) on the worker's health and the severity of this damage, assuming that there is exposure within the work environment [10].

Examples of common occupational diseases include occupational asthma [10,11], vibration-related diseases [12–14], noise related diseases [15], pulmonary fibrosis [16], bronchopulmonary pleural fibrosis and damage caused by the inhalation of asbestos dust [17], and occupational cancer [18].

As mentioned above, in Industry 4.0 workplaces the presence of new technologies brings new opportunities and new risks. In addition to common occupational diseases, the nature of work in Industry 4.0 has the potential to contribute to the increasing frequency of other diseases, including mental disorders and diseases related to sedentary behavior. In Industry 4.0, several workers can often be involved in creative value-added tasks, while routine activities, as well as certain dangerous tasks, are often performed by robots. This scenario, along with early and continuous risk analysis and management based on various technologies, could make workplaces safer. On the other hand, semi-skilled employees could lose workplace opportunities because of potential difficulties in performing more complex tasks. At the same time, the use of digital tools to continuously monitor the performance of employees may become common, which could result in privacy invasion and psychological pressure [19,20].

In addition, the risks related to interactions between humans and machines have increased and greater connectivity makes it possible to work anywhere at any time. This scenario brings benefits such as flexibility, but also has the potential to impact individuals' work–life balance, which may in turn be harmful to mental health [21]. According to [20], depression is very common in workplaces compared to other mental disorders, and affects workers by reducing productivity, diminishing job retention, and increasing the risk of accidents at work. Another issue related to Industry 4.0 is the existence of many sedentary jobs, such as computer-based work. High levels of sedentary posture are associated with an increased risk of cardiovascular disease and type 2 diabetes, several cancers including lung and breast, and mental disorders such as depression. In addition, poor lighting conditions in workplaces (for example, store warehouses, since online commerce has been growing) can cause severe headaches and discomfort. Insufficient lighting makes it difficult to perceive the depth, shape, speed, and proximity of objects, and related accidents may often occur [22].

3. Organizational Culture as a Key Factor in OSH

According to [23], the occurrence of occupational diseases and accidents causes significant losses in companies' reputation and decreases their productivity. For example, a worker who becomes aware of a colleague's illness may become discouraged and start to produce less or may look for another job opportunity with better OSH conditions. To combat or significantly minimize these problems, it is necessary to perform preventive actions. The management of a company has an obligation to foresee, organize, and coordinate the organization of work, providing methods for preventing incidents and accidents in the workplace, through the effective management of occupational risks [24].

Risk perception depends on a variety of factors, including values and educational level [25,26]. Environments where workers feel pressured and overworked are in general quite prone to accidents. In addition, unqualified workers are generally more susceptible to accidents, because they often perform dangerous tasks. The low education of these workers tends to affect perception of risks present in the work environment and may make it difficult to understand the issues addressed in the health and safety training provided by companies. This issue demands special attention from professionals who plan and train these workers, to make sure that the topics covered are really understood [27].

Aiming to ensure the effectiveness of measures to prevent illnesses and accidents in the work environment, it is necessary that managers remain continuously engaged with the objective of promoting actions focused on the safety and well-being of workers. Improvements within a company should not happen only after an unwanted event has occurred, because this type of approach often means workers fail to take proper precautions after a time and even forget about them completely [23].

In this context, the ISO 45001:2018—Occupational health and safety management systems–Requirements with guidance for use—is a standard that aims to provide guidelines to assist organizations in improving OSH performance and preventing work-related injuries and illnesses. This standard is applicable to any organization, regardless of its size or type [28].

4. Technologies and Trends in OSH

This section covers concepts related to smart PPE, IoT, and Industry 4.0, describing works that have successfully applied these technologies in the construction of equipment and systems. Later in this section, devices and communication protocols for IoT and possibilities involving machine learning are addressed, representing alternatives for building systems similar to those described here.

4.1. Smart Personal Protective Equipment

According to [29], if an activity carried out by a worker involves a risk that cannot be reduced or eliminated by collective, technical, or organizational means, the use of personal protective equipment (PPE) allows that person to perform their activities without risk or with reduced risk of suffering injuries. In recent years the term 'smart PPE' has become more common. Every piece of smart PPE can interact with the environment and/or react to environmental conditions. This type of equipment combines traditional PPE with an electronic aspect, such as sensors, data transfer modules, or batteries. Sensors are used to monitor real-time hazardous factors for workers. In addition, the use of computer-based systems can facilitate OSH functions related to risk identification and management [30].

Aiming to assure that no new risks are added by the inclusion of electronic devices, tests must be performed designed for traditional PPE and related to electrical safety, such as surface temperature and battery safety. However, there are still no standards available for smart PPE, and standardization bodies must formulate requirements and procedures for testing this type of equipment. In Europe, there are some initial standardization projects in progress. Some of the challenges for the development of smart personal protective equipment are reliability, privacy, security, ergonomics, acceptance by users, applicable certifications, market surveillance, recycling, and the avoidance of additional risks [30].

4.2. Industry 4.0 Related Technologies and the Internet of Things (IoT) in OSH

Industry 4.0 can be defined as the Fourth Industrial Revolution, and encompasses a broad system of advanced technologies that are changing production and business models around the world. Industry 4.0 is related to the integration of the manufacturing process, aiming at continuous improvement, and avoiding waste [8,9].

The term Internet of Things (IoT), in turn, was introduced in the late 1990s by Kevin Ashton, a researcher at the Massachusetts Institute of Technology (MIT), referring to the connection of different objects to exchange data with other devices and systems over the Internet. IoT aims to supply a network infrastructure with interoperable communication protocols and software to connect this variety of devices. The term industrial IoT (IIoT) is related to the application of IoT technology in industrial environments [31].

IoT has been used in many OSH applications, including monitoring physiological variables of workers engaged in dangerous activities, as well as for sensors and alarm systems to prevent a variety of accidents. For example, Li and Kara [32] presented a methodology for monitoring factory conditions including temperature and air quality, by using wireless sensor networks and IoT. According to Awolusi et al. [33], wearable systems have been employed in construction sites to collect data to detect environmental conditions, and for determining whether people are close to danger. The authors described how gyroscopes can verify the rotation of different parts of the body, while ultrasonic sensors can monitor muscle contractions. Described below are proposals for OSH that use Industry 4.0 and/or IoT-related technologies.

Aqueveque et al. [34] proposed a device to measure physiological variables including the electrocardiogram and respiratory activity of miners working at high altitudes. The proposed system's noninvasive sensors are embedded in a T-shirt. The device can monitor heart rate and respiration rate, and exchanges data with a central monitoring station.

Yu et al. [35] presented a wearable system involving physiological sensors embedded into firefighters' garments, assessing their physiological state by evaluating data collected from the sensors The data was sent to the command center and the system evaluated the gravity of the risk scenario, sending messages, for example, to instruct that the action should be canceled because it is too dangerous. All collected data and messages were sent to the cloud.

Wu et al. [36] presented a hybrid wearable sensor network system for IoT-connected safety and health monitoring applications for outdoor workplaces. A local server processed raw sensor signals, displaying the environmental and physiological data, and triggered an alert if any emergency circumstance was detected. Temperature, humidity, Ultraviolet (UV) radiation, CO_2, heart rate, and body temperature were measured by the wearable sensors. The gateway pre-processed the sensor signals, displayed the data, and triggered alerts when emergency occurred. An IoT cloud server was used for data storage, web monitoring, and mobile applications.

Marques and Pitarma [37] proposed a real-time indoor quality monitoring system using a sensor to measure particulate matter (PM), temperature, humidity, and formaldehyde. The system included a mobile application for data consultation and notifications, and served a dataset to plan changes for improving indoor quality. The dataset can also support clinical diagnostics and correlate health problems with living environment conditions.

Balakreshnan et al. [38] proposed a system to check the safety of workers in the vicinity of machines. The solution used artificial intelligence and machine vision to identify use of safety glasses in areas where there are risks to the eyes, and can also detect the lack of other equipment. The system can initiate different control actions when safety violations occur.

Sanchez et al. [39] proposed a smart PPE using a sensor network located on a helmet and a belt, to monitor the worker and their environment. The system monitored biometrics risks and can detect external impact, shock, luminosity, gases, and environmental temperature, and provided real-time recommendations. Data were observable by the user on a tablet or a mobile phone. The device incorporated a flashlight that activated automatically if the worker was in poorly lighted areas, and a loudspeaker to assist the detection of audible alarms.

Márquez-Sánchez et al. [40] presented a system for the detecting anomalies in workplaces using a helmet, a belt, and a bracelet. Intelligent algorithms are applied to collected data through edge computing, in which processing takes place closer to the data source, providing faster services. The system early predicts and notifies anomalies detected in working environment. Then, data is sent to the cloud, where deep learning models verify possible anomalies because of the training of the set of data inserted previously.

Shakerian et al. [41] the authors proposed and examined an assessment process to evaluate workers' bodily responses to heat strain, to continuously and non-intrusively collect and evaluate workers' physiological signals acquired from a wristband-type biosensor. The proposed process assesses heat strain exposure through the collective analysis of electrodermal activity, photoplethysmography, and skin temperature biosignals. The physiological signals are uploaded to a cloud server, decontaminated from noise, and the measurable metrics are extracted from the signals and interpreted as distinct states of workers' heat strain by employing supervised learning algorithms.

Kim et al. [42] proposed an IoT-based system to monitor construction workers' physiological data using an off-the-shelf wearable smart band. The platform was designed for construction workers performing at high temperatures, to collect a worker's physiological data through a wearable armband that consists of three sensors—photoplethysmography (for heart rate monitoring), a temperature sensor, and an accelerometer, which provides the current position of a worker. The acquired data reflect a worker's current physiological status, sent to the web and to a smartphone application for visualization.

Yang et al. [43] conducted a study to monitor the level of physical load during construction tasks, to assess ergonomic risk to an individual construction worker. By using an ankle-worn wearable inertial measurement unit to monitor a worker's bodily movements, the study investigated the feasibility of identifying various physical loading conditions by analyzing a worker's lower body movements. In the experiment, the workers performed a load-carrying task by moving concrete bricks. This study developed and evaluated a classification model to detect different physical load levels, using Bidirectional long short-term memory (Bi-LSTM).

Marques and Pitarma [44] presented a real-time acoustic comfort monitoring solution suitable for occupational usage. The system was designed to be easy to install and use, incorporating a device for ambient data collection called iSoundIoT, and including Web/mobile data access based on Wi-Fi communication. The solution includes a notification feature to alert people when poor acoustic comfort scenarios are verified, and continuous real-time data collection enabling the generation of reports containing sound level values and alerts.

Mumtaz et al. [45], motivated by the COVID-19 outbreak, proposed an IoT-based system for monitoring and reporting air conditions in real time with the data sent to a web portal and mobile app. The solution can monitor multiple air pollutants, including carbon dioxide (CO_2), particulate matter (PM) 2.5, nitrogen dioxide (NO_2), carbon monoxide (CO), and methane (CH_4), as well as temperature and humidity. The system generates alerts after detecting anomalies in the air quality. Various machine learning algorithms were employed to classify indoor air quality, and long short-term memory (LSTM) was applied for predicting the concentration of each air pollutant and predicting the overall air quality of an indoor environment.

Zhou and Ding [46] presented an IoT-based system to generate early warnings and alarms as dynamical safety barriers for different types of hazards on underground construction sites. Their solution was able to collect, analyze, and manage multisource information, automate monitoring and warning, and minimize the hazard energy coupling by using IoT. The data-sensing layer included an IoT reader, IoT tag with warning device, ultrasonic detector, and infrared access device, achieving about 1.5 m locating accuracy in underground workspaces. The portable warning device, designed with RFID-based positioning technology, was installed on the safety helmet. Each IoT tag consisted of a RFID chip and a wireless antenna, and stored information about the worker wearing it. In case of accident, the proposed system can be used also for investigation purposes.

Zhan et al. [47] proposed a monitoring system for cold storage based on Industrial IoT, to identify abnormal stationary and acquire the spatial-temporal information of workers in real time. In these workplaces, an abnormal stationary position is a sign of danger, such as falling or fainting. A deep neural network was applied to learn specific features involving location and vibration for anomaly detection. The Bluetooth low energy (BLE) and a log-distance path loss model were used to fulfill indoor localization to allow rapid responses to an incident on site. In addition, digital twin technology that mirrors physical objects in cyberspace can be used to enhance spatial-temporal traceability and cyber-physical visibility to enforce safety monitoring by managers. Cloud and edge computing can be used to improve overall computational efficiency and system responsiveness.

Campero-Jurado et al. [48] proposed a smart helmet prototype that monitored the conditions in the workers' environment and performed a near real-time evaluation of risks. The data collected by the sensors was sent to an AI-driven platform for analysis, where different intelligent models were evaluated by the authors. The design is intended to protect the operator from possible impacts, while monitoring the light, humidity, temperature, atmospheric pressure, presence of gases, and air quality. Alerts can be transmitted to the operator by means of sound beeps. For visualization of environmental data, through color codes an LED strip deployed on the helmet can notify the worker of anomalies in the environment.

A comparison between the proposals mentioned above can be found in Section 5, along with further discussion.

For the design of IoT systems and devices for OSH, such as those described above, various low-cost devices and free software allow the implementation and use of IoT-based systems by small and medium-sized companies. Some of these technologies are described below.

4.3. IoT Devices

According to Lacamera [49], embedded systems consist of a class of systems that run on an architecture based on microcontrollers, that offer constrained resources. A microcontroller or microcontroller unit (MCU) is a device made of a dedicated processor for the purpose of running a specific application, unlike general purpose computers. These devices are often designed to be inexpensive, low-resource, and low-energy consuming. These devices can be used in factories and for several IoT applications. They are often used as sensors, actuators, or smart devices and may form networks. Below, the Arduino and ESP32 platforms are described, which are each widely used in IoT applications.

Arduino is an open platform for prototyping, based on free software and low-cost hardware, where the programs are written in the simplified C++ language. Arduino Integrated Development Environment (IDE) is used to write code and upload it to the board. The hardware consists of an open hardware design with a microcontroller manufactured by the Atmel Microchip company. The boards are sold preassembled, but hardware design information is available for people who want to build or modify them [50]. There are various types of Arduino boards supporting different features, such as Wi-Fi [51], Bluetooth, Bluetooth Low Energy (BLE) [52] and Global System for Mobile Communication (GSM) [53].

ESP32 is a series of low-cost and low-power microcontrollers and is a system-on-a-chip (SoC) with integrated microcontroller, Wi-Fi, and Bluetooth. ESP32 is a dual-core system and be used as a standalone system or can serve as a slave device to a host microcontroller. ESP32 is commonly used for academic and industrial purposes, especially in IoT. It can be programmed by ESP-IDF, which is a framework developed by ESPRESSIF, or by the Arduino Integrated Development Environment (IDE), which is the easiest way to start writing code for this platform [54].

4.4. Protocols for IoT

Below are described two protocols widely used in IoT, the Constrained Application Protocol (CoAP) and the Message Queue Telemetry Transport (MQTT). According to Shelby et al. [55], Constrained Application Protocol (CoAP) is suitable for resource-constrained environments, including those with power-constrained devices, low-bandwidth links, and lossy networks. In this protocol, the network nodes interact through a request–response model and support the built-in discovery of services. CoAP is very similar to the client–server model of Hypertext Transfer Protocol (HTTP), the widely used protocol that allows contents to be requested and transmitted between browsers and web servers via the Internet. However, CoAP implementations can often act in client and server roles. A client sends a request using a method code on a resource (identified by a URI—Universal Resource Identifier) on a server. The server, in turn, sends a response with a response code. CoAP executes these interchanges asynchronously using User Datagram Protocol (UDP). The messages support optional reliability, and CoAP supports secure messages using Datagram Transport Layer Security (DTLS), described in [56].

By other hand, MQTT provides asynchronous communication between devices [57]. This protocol uses a message publishing and signature model, and was invented by the IBM company in the late 1990s. MQTT was originally designed to link oil pipeline sensors to satellites. It is a lightweight protocol that can be implemented on devices with many restrictions, such as low computational power, and in networks with limited bandwidth and high latency. These features make MQTT suitable for several applications in IoT; publish–subscribe is the standard model for exchanging messages in MQTT. MQTT comprises two entities: a broker and the clients, where the message broker is a server receiving messages from clients and then sending these messages to other clients, that can subscribe to any message topic. Clients must publish their messages on a topic and send the topic and the message to the broker. The broker then forwards the message to all clients who subscribe to that topic. Clients can connect to the broker through simple TCP/IP connections or encrypted TLS connections.

4.5. Machine Learning

As described by Abiodun et al. [58], machine learning (ML) is a branch of artificial intelligence (AI) that uses computers to simulate human learning. In ML, computers can autonomously modify their behavior based on their own experience (training). ML algorithms are classified based on the approach used in the learning process.

In supervised learning, the learning algorithm aims to predict how a given set of inputs conducts to the output. The algorithm receives labeled data and learns from this data. In unsupervised learning the algorithm does not receive labels. This type of algorithm is mainly focused on finding hidden patterns in data. Semi-supervised learning algorithms have an incomplete training set, often with many target outputs missing, from which they must learn. Finally, the algorithm used in reinforcement learning learns from the external feedback received in terms of punishments and rewards [59].

Below are described recommender systems [59–62], anomaly detection, [63,64] and long short-term memory (LSTM) [65–69], which have each been applied in a variety of systems, and more recently have been suggested for use in IoT, healthcare, and OSH solutions.

4.5.1. Recommender Systems

Recommender systems use artificial intelligence (AI) methods to serve users with item recommendations (filtered content). These systems try to predict a user's preference for an item, based on available information about items, users, and the interactions between them. These systems aim to retrieve only the most relevant information services from a large volume of data [59,60].

Traditional applications for recommender systems include movies, music, tourism, e-learning, and more recently, healthcare (Health Recommender Systems—HRS) [53]. In addition, using data obtained from IoT devices, such as smart wearable and smart PPE, recommender systems can extract information to be used in OSH, for example, to predict risks and try to predict the emergence of occupational diseases.

4.5.2. Anomaly Detection

During IoT data analysis, it is in general necessary to identify uncommon states within the systems being monitored by sensors. Defining maximum and minimum limits for sensor readings to identify problems may increase the number of false alarms and missed dangerous conditions. In this context, anomaly detection methods that have been largely applied in cybersecurity, financial surveillance, risk management, and healthcare, among other areas, can be useful in IoT applications, including OSH systems [63,64].

Anomalies can be defined as measurements or observations that do not reflect expected behavior. Considering the context of the IoT, an anomaly is related to the measurable consequences of an unexpected modification in a system which is outside its standard. Anomaly detection is the process of detecting measurements with relevant deviations from other data. Anomaly detection methods consider the combination of two or more variables to identify problems. Obstacles to the development of anomaly detection, especially for IoT/OSH, include the lack of datasets with real-world anomalies, and sensor readings that are often affected by significant noise [63,64].

4.5.3. Long Short-Term Memory (LSTM)

As described in [65], recurrent neural networks (RNNs) are artificial neural networks which handle sequential or time-series data. RNNs have "memory" since they use information from past inputs to induce current input and output. LSTM is an RNN, suitable for classifying, processing, and predicting time series with intervals of unknown length. The fact that LSTM is relatively insensitive to gap length is an advantage compared to traditional RNNs.

Traditional examples of applications for this type of deep learning algorithm include language translation and speech recognition. In addition, LSTMs have been used in a variety of solutions [66–69], such as machine health monitoring and air-pollution forecasting [70,71].

5. Discussion

It is important to develop solutions that allow daily monitoring of the health conditions of workers, and their exposure to occupational risks, for the reasons explained earlier in this work and because the data obtained can support studies by companies to identify problems and guide OSH policies.

The studies described above were compared regarding the use of artificial intelligence and the use of techniques to ensure data privacy. The comparison is presented in Table 1.

The data obtained from continuous monitoring of occupational health, risks, and environmental conditions can also support academic research. Such research may allow new relationships to be established in the long term between occupational hazards and the occurrence of certain diseases. Keeping an updated record of changes in the health conditions of each worker is also a fundamental part of the process, so that the technologies mentioned above can help companies more significantly in making long-term decisions. Reliable data obtained by companies can also guide changes in legislation [45,47,48].

In this context, the use of artificial intelligence and machine learning is essential to obtaining better results, by identifying within work environments which settings or conditions may be safer or more harmful to workers' health. This type of approach has the potential to reduce workers' long-term absences, as well as their early retirement. AI/ML can be used to identify dangerous conditions that could result in accidents and/or diseases; by training with large datasets obtained over long periods of time, AI/ML may identify trends and suggest changes in workplaces to make them safer. Various AI/ML techniques have been used in recent studies [38–41,43,45,47,48]. Despite the various approaches involving AI/ML, none of the works mentioned take into account the health history of workers, to generate personalized alerts for example. This is a point that can be explored in future research.

Table 1. Study comparison.

Title	Data Privacy	AI
Monitoring Physiological Variables of Mining Workers at High Altitude [34]	NO	NO
A wearable intelligent system for real time monitoring firefighter's physiological state and predicting dangers [35]	NO	NO
An Internet-of-Things (IoT) Network System for Connected Safety and Health Monitoring Applications [36]	YES	NO
mHealth: Indoor Environmental Quality Measuring System for Enhanced Health and Well-Being Based on Internet of Things [37]	NO *	NO
PPE Compliance Detection using Artificial Intelligence in Learning Factories [38]	YES	YES
Smart Protective Protection Equipment for an accessible work environment and occupational hazard prevention [39]	NO	YES
Intelligent Platform Based on Smart PPE for Safety in Workplaces [40]	NO	YES
Assessing occupational risk of heat stress at construction: A worker-centric wearable sensor-based approach [41]	NO	YES
Development of an IoT-Based Construction Worker Physiological Data Monitoring Platform at High Temperatures [42]	NO	NO
Deep learning-based classification of work-related physical load levels in construction [43]	NO	YES
A Real-Time Noise Monitoring System Based on Internet of Things for Enhanced Acoustic Comfort and Occupational Health [44]	NO	NO
Internet of Things (IoT) Based Indoor Air Quality Sensing and Predictive Analytic—A COVID-19 Perspective [45]	NO *	YES
Safety barrier warning system for underground construction sites using Internet-of-Things technologies [46]	NO	NO
Industrial internet of things and unsupervised deep learning enabled real-time occupational safety monitoring in cold storage warehouse [47]	NO	YES
Smart Helmet 5.0 for Industrial Internet of Things Using Artificial Intelligence [49]	NO	YES

* This is a solution for monitoring the environment, not a wearable item or PPE.

It is important to highlight that the challenges involved in implementing new technologies can vary significantly according to the activity. For example, construction sites are very dangerous places because workers are exposed to hazards that can be very hard to measure due to the way tasks are executed in this type of workplace [33].

With respect to data privacy, according to [72], data collected from wearable devices are transferred to a receiver through wireless networks, making data privacy a very critical issue for this type of device and making workers unwilling to use them. For example, workers may be very uncomfortable in sharing with employers their location information during rest periods. In the study conducted by Häikiö et al. [73] an anonymous online questionnaire was applied to construction workers to collect their opinions regarding IoT-based work safety. 4385 workers responded to the questionnaire. 49.7% were very (18.2%) or rather interested (31.5%) in using activity wristbands or other devices for monitoring their movement or physical activities in the workplace. Experienced professionals were less interested in using wearables than younger ones. In general, workers were more interested in sharing their data when they were sure it could help to preserve their health.

Systems for OSH often need to handle workers' personal data, which according to the General Data Protection Regulation (GDPR) must be anonymized [74]. Anonymized personal data is has gone through stages that ensure its disconnection from the person, for example, a document number may have some digits suppressed. In such a case, it would not be possible through technical or other means to find out who the data subject was. Anonymized data is no longer subject to the GDPR and is essential for expanding the use of IoT and artificial intelligence. However, in some applications anonymization is not feasible. For this purpose, pseudo-anonymized data that is subject to GDPR may be used. Pseudo-anonymization is treatment through which a data loses the possibility of association, directly or indirectly, with a person. Additional information may be kept separately in a controlled and safe environment, for example, under the responsibility of the company that develops and provides the application. If a system does not handle personal data, GDPR is not applicable. Data privacy was addressed only in [36,38]. However, it is important to note that not all solutions deal with personal information, as some are intended for monitoring environments. In these cases, it is understood that secure communication, despite being desirable, is not a priority.

According to Zamfir et al. [75], in respect to the IoT protocols described earlier in this paper, CoAP and MQTT communication can be secured by Transport Layer Security with digital certificates, as widely used in Internet applications. However, this approach may be costly for a large number of devices, and is often too heavy for IoT devices. In a simpler way, a pre-shared key (TLS-PSK) is an alternative. In this case, the messages are encrypted and signed using the shared key between the parties involved in communication. The same key is used for decryption and authentication of messages at the destination. It is recommended that the pre-shared key (PSK) is configured between each device and the server. Both approaches can be used to provide data privacy, especially when the applications handle sensitive information such as physiological data and location.

According to Maltseva [76], wearable devices' characteristics create multiple opportunities and can help to improve organizational performance. Wearable wristbands are very popular devices, which can continuously collect data such as heart-rate variability and can continue collecting data after working hours. These devices bring benefits and can help to identify health risks. However, extending the use of wearables after working hours causes confusion distinguishing work and rest. It is important to note that training individuals in a clearly and sufficient way is a key factor for success regarding the use of any technologies in the workplace. In addition, workers need to be aware that their data is being used to protect them from work-related diseases and that enough means are being used to keep that data safe.

When it comes to costs and people, the technical and organizational complexity of manufacturing processes have increased in Industry 4.0, and related technologies have imposed great challenges especially on small- and medium-sized enterprises (SMEs). Even with several options for free software and low-cost hardware, as mentioned before, more complex monitoring systems tend to be expensive because they demand continuous updating and maintenance service from the manufacturer. However, in Industry 4.0 the use of these technologies is likely to become increasingly common and, with the emergence of more manufacturers, prices will possibly become more affordable. Companies of all sizes are impacted by the availability of sufficiently qualified people to work within complex production systems. In this context, workers will need to spend some time in continuing education [77].

In addition to the great need to monitor physiological variables and environmental risks, as revealed in all the studies mentioned in this work, new occupational risks have emerged along with complexity in working environments, such as ergonomic and psychosocial risks, and those associated with the use of collaborative robots (cobots) [78]. Monitoring the use of PPE with the aid of computer vision, as implemented in [38], especially in high-risk activities such as operating machines and working with robots, is very important, mainly because unsafe actions can cause serious injury, amputation, or death.

Regarding psychosocial risks, the study by Verra et al. [79] presented a comparison of policies and practices in Europe for promoting health at work. It was identified that more than 70% of establishments in the European Union adopt preventive measures against direct physical damage, and more than 30% implement measures to avoid psychosocial risks. Psychosocial risks are often addressed in national policy, but they have not been addressed by most institutions. In the context of Industry 4.0, psychosocial risks deserve special attention because workers tend to be pressured towards greater productivity, and need to be constantly updated on new technologies, concepts, and tools. In addition, many workers feel obliged to respond to text messages and even solve problems outside work hours, jeopardizing their leisure and rest. Another point that deserves attention is that Industry 4.0 workplaces usually offer a variety of sedentary jobs, for example, information technology positions. As highly documented in the literature, a sedentary lifestyle is often associated with obesity and cardiovascular diseases. Certainly, monitoring psychosocial risks, risks related to sedentary working conditions, and the health conditions of workers in sedentary jobs without intruding on their personal lives are big issues, and bring significant challenges for the OSH sector in the context of Industry 4.0.

Finally, we identified the following points to be explored in future research:

- Use of AI to monitor the use of PPE, especially in dangerous activities.
- Monitoring psychosocial risks and risks related to sedentary working conditions.
- Consideration of workers' health history, together with data obtained from monitoring the work environment, to generate personalized alerts.
- Data privacy issues.

6. Conclusions

As mentioned above, Industry 4.0 has brought significant advances in the production process as well as several challenges for OSH. Various benefits arising from the integration of IoT-related technologies in OSH within this new context have been presented in this work. It is important to develop of solutions that allow daily monitoring of exposure to occupational risks and the health conditions of workers, because the data obtained can support more focused studies by companies and more assertively guide OSH policies. For example, artificial intelligence can contribute to building solutions that map existing problems and predict future problems.

Regarding privacy concerns, several studies have shown that data privacy is a critical issue in wearable technology development and that uncertainties around this topic can make workers especially reluctant to use wearable devices. In this context, it is important to highlight that training people in a clear and sufficient way is a key factor for success in the use of any workplace technology. In addition, workers need to be aware that the use of their health-related data may be important to protect them from work-related diseases, and that enough means will be used to keep that data safe. In this case, the agreement of workers is necessary and applicable laws and standards shall be adopted.

For future work, the authors are developing a system for individual environmental risk assessment based on IoT-related technologies. The device is intended to have sufficient energy autonomy to allow monitoring and communication for at least one working day. Issues related to the device's ergonomics and data privacy must be considered in the project, as well as durability and the viability of cost for industries of all sizes. The main goal is to contribute in the long run to reducing the incidence of occupational diseases resulting from exposure to harmful agents, by facilitating the visualization of data by organizations.

Author Contributions: Conceptualization, J.L., P.D.G. and T.M.L.; methodology, P.D.G. and T.M.L.; validation P.D.G. and T.M.L.; formal analysis, P.D.G.; investigation, J.L.; resources, P.D.G. and T.M.L.; data curation, J.L.; writing—original draft preparation, J.L.; writing—review and editing, P.D.G. and T.M.L.; supervision, P.D.G. and T.M.L. All authors have read and agreed to the published version of the manuscript.

Funding: This research received no external funding.

Data Availability Statement: The authors confirm that the data supporting the findings of this study are available within the article.

Acknowledgments: This work was supported in part by the Fundação para a Ciência e Tecnologia (FCT) and C-MAST (Centre for Mechanical and Aerospace Science and Technologies), under project UIDB/00151/2020.

Conflicts of Interest: The authors declare no conflict of interest.

References

1. Indicator Description: Occupational Injuries. Available online: https://ilostat.ilo.org/resources/concepts-and-definitions/description-occupational-injuries (accessed on 1 July 2021).

2. Teufer, B.; Ebenberger, A.; Affengruber, L.; Kien, C.; Klerings, I.; Szelag, M.; Griebler, U. Evidence-based occupational health and safety interventions: A comprehensive overview of reviews. *BMJ Open* **2019**, *9*, e032528. [CrossRef] [PubMed]
3. WHO/ILO: Almost 2 Million People Die from Work-Related Causes Each Year. Available online: https://www.who.int/news/item/16-09-2021-who-ilo-almost-2-million-people-die-from-work-related-causes-each-year (accessed on 15 July 2022).
4. ILO; WHO. *WHO/ILO Joint Estimates of the Work-Related Burden of Disease and Injury, 2000–2016*; Global Monitoring Report; ILO/WHO: Geneva, Switzerland, 2021.
5. Ncube, F.; Kanda, A. Current Status and the Future of Occupational Safety and Health Legislation in Low- and Middle-Income Countries. *Saf. Health Work.* **2018**, *4*, 365–371. [CrossRef] [PubMed]
6. Melchior, C.; Zanini, R.R. Mortality per work accident: A literature mapping. *Saf. Sci.* **2019**, *114*, 72–78. [CrossRef]
7. Ronda-Perez, E.; Gosslin, A.; Martínez, J.M.; Reid, A. Injury vulnerability in Spain. Examination of risk among migrant and native workers. *Saf. Sci.* **2019**, *115*, 36–41. [CrossRef]
8. Vaidya, S.; Ambad, P.; Bhosle, S. Industry 4.0—A Glimpse. *Procedia Manuf.* **2018**, *20*, 233–238. [CrossRef]
9. Yu, F.; Schweisfurth, T. Industry 4.0 technology implementation in SMEs—A survey in the Danish-German border region. *Int. J. Innov. Stud.* **2020**, *4*, 76–84. [CrossRef]
10. Atuação dos Industriais no Âmbito do Sistema da Indústria Responsável-SIR. Available online: https://www.act.gov.pt/(pt-PT)/crc/PublicacoesElectronicas/Documents/atuacaodosindustriaisnoambitodosistemadaindustriaresponsavel_SIR.pdf (accessed on 18 June 2022).
11. Kelly, K.; Poole, J. Pollutants in the workplace: Effect on occupational asthma. *J. Allergy Clin. Immunol.* **2019**, *143*, 2014–2015. [CrossRef]
12. Kociolek, A.; Lang, A.; Trask, C.; Vasiljev, R.; Milosavljevic, S. Exploring head and neck vibration exposure from quad bike use in agriculture. *Int. J. Ind. Ergon.* **2018**, *66*, 63–69. [CrossRef]
13. Lundström, R.; Noor Baloch, A.; Hagberg, M.; Nilsson, T.; Gerhardsson, L. Long-term effect of hand-arm vibration on thermotactile perception thresholds. *J. Occup. Med. Toxicol.* **2018**, *13*, 19. [CrossRef]
14. Gerhardsson, L.; Hagberg, M. Vibration induced injuries in hands in long-term vibration exposed workers. *J. Occup. Med. Toxicol.* **2019**, *14*, 21. [CrossRef]
15. Factsheet 57—O Impacto do Ruído no Trabalho. Available online: https://osha.europa.eu/pt/publications/factsheet-57-impact-noise-work (accessed on 18 July 2022).
16. Álvarez, R.; González, C.; Martínez, A.; Pérez, J.; Fernández, L.; Fernández, A. Guidelines for the Diagnosis and Monitoring of Silicosis. *Arch. Bronconeumol. Engl. Ed.* **2015**, *51*, 86–93. [CrossRef]
17. Kratzke, P.; Kratzke, R. Asbestos-Related Disease. *J. Radiol. Nurs.* **2018**, *37*, 21–26. [CrossRef]
18. World Cancer Report. 2014. Available online: https://publications.iarc.fr/Non-Series-Publications/World-Cancer-Reports/World-Cancer-Report-2014 (accessed on 23 July 2021).
19. Coenen, P.; Gilson, N.; Healy, G.; Dunstan, D.; Straker, L. A qualitative review of existing national and international occupational safety and health policies relating to occupational sedentary behavior. *Appl. Ergon.* **2017**, *60*, 320–333. [CrossRef]
20. Do, H.; Nguyen, A.; Nguyen, H.; Bui, T.; Nguyen, Q.; Tran, N.; Ho, C. Depressive Symptoms, Suicidal Ideation, and Mental Health Service Use of Industrial Workers: Evidence from Vietnam. *Int. J. Environ. Res. Public Health* **2020**, *17*, 2929. [CrossRef]
21. Leso, V.; Fontana, L.; Iavicoli, I. The occupational health and safety dimension of Industry 4.0. *La Med. Lav.* **2018**, *109*, 327–338. [CrossRef]
22. De Guzman, H. Microcontroller Based Automated Lighting Control System for Workplaces. In Proceedings of the 2018 IEEE 10th International Conference on Humanoid, Nanotechnology, Information Technology, Communication and Control, Environment and Management (HNICEM), Baguio City, Philippines, 29 November–2 December 2018; pp. 1–6.
23. Cardella, B. *Segurança no Trabalho e Prevenção de Acidentes: Uma Abordagem Holística*; Atlas: São Paulo, Brazil, 2016.
24. Kim, Y.; Park, J.; Park, M. Creating a Culture of Prevention in Occupational Safety and Health Practice. *Saf. Health Work* **2016**, *7*, 89–96. [CrossRef]
25. Leoni, T. What drives the perception of health and safety risks in the workplace? Evidence from European labour markets. *Empirica* **2010**, *37*, 165–195. [CrossRef]
26. Xia, N.; Wang, X.; Griffin, M.; Wu, C.; Liu, B. Do we see how they perceive risk? An integrated analysis of risk perception and its effect on workplace safety behavior. *Accid. Anal. Prev.* **2017**, *106*, 234–242. [CrossRef]
27. Swuste, P.; Groeneweg, J.; van Gulijk, C.; Zwaard, W.; Lemkowitz, S.; Oostendorp, Y. The future of safety science. *Saf. Sci.* **2020**, *125*, 104593. [CrossRef]
28. ISO 45001—Occupational Health and Safety. Available online: https://www.iso.org/publication/PUB100427.html (accessed on 12 July 2022).
29. Podgórski, D.; Majchrzycka, K.; Dąbrowska, A.; Gralewicz, G.; Okrasa, M. Towards a conceptual framework of OSH risk management in smart working environments based on smart PPE, ambient intelligence and the Internet of Things technologies. *Int. J. Occup. Saf. Ergon.* **2016**, *23*, 1–20. [CrossRef]
30. Smart Personal Protective Equipment: Intelligent Protection for the Future. Available online: https://osha.europa.eu/pt/publications/smart-personal-protective-equipment-intelligent-protection-future/view (accessed on 23 July 2022).
31. Khan, W.; Rehman, M.; Zangoti, H.; Afzal, M.; Armi, N.; Salah, K. Industrial internet of things: Recent advances, enabling technologies and open challenges. *Comput. Electr. Eng.* **2020**, *81*, 106522. [CrossRef]

32. Li, W.; Kara, S. Methodology for Monitoring Manufacturing Environment by Using Wireless Sensor Networks (WSN) and the Internet of Things (IoT). *Procedia CIRP* **2017**, *61*, 323–328. [CrossRef]
33. Awolusi, I.; Marks, E.; Hallowell, M. Wearable technology for personalized construction safety monitoring and trending: Review of applicable devices. *Autom. Constr.* **2018**, *85*, 96–106. [CrossRef]
34. Aqueveque, P.; Gutiérrez, C.; Rodríguez, F.; Pino, E.; Morales, A.; Wiechmann, E. Monitoring Physiological Variables of Mining Workers at High Altitude. *IEEE Trans. Ind. Appl.* **2017**, *53*, 2628–2634. [CrossRef]
35. Yu, B.; Wei, W.; Xianyi, Z.; Koehl, L.; Tartare, G. A wearable intelligent system for real time monitoring firefighter's physiological state and predicting dangers. In Proceedings of the 2015 IEEE 16th International Conference on Communication Technology (ICCT), Hangzhou, China, 18–21 October 2015. [CrossRef]
36. Wu, F.; Wu, T.; Yuce, M. An Internet-of-Things (IoT) Network System for Connected Safety and Health Monitoring Applications. *Sensors* **2019**, *19*, 21. [CrossRef]
37. Marques, G.; Pitarma, R. mHealth: Indoor Environmental Quality Measuring System for Enhanced Health and Well-Being Based on Internet of Things. *J. Sens. Actuator Netw.* **2019**, *8*, 43. [CrossRef]
38. Balakreshnan, B.; Richards, G.; Nanda, G.; Mao, H.; Athinarayanan, R.; Zaccaria, J. PPE Compliance Detection using Artificial Intelligence in Learning Factories. *Procedia Manuf.* **2020**, *45*, 277–282. [CrossRef]
39. Sanchez, M.; Sergio Rodriguez, C.; Manuel, J. Smart Protective Protection Equipment for an accessible work environment and occupational hazard prevention. In Proceedings of the 2020 10th International Conference on Cloud Computing, Data Science & Engineering (Confluence), Noida, India, 29–31 January 2020; pp. 581–585. [CrossRef]
40. Márquez-Sánchez, S.; Campero-Jurado, I.; Herrera-Santos, J.; Rodríguez, S.; Corchado, J.M. Intelligent Platform Based on Smart PPE for Safety in Workplaces. *Sensors* **2021**, *21*, 4652. [CrossRef]
41. Shakerian, S.; Habibnezhad, M.; Ojha, A.; Lee, G.; Liu, Y.; Jebelli, H.; Lee, S. Assessing occupational risk of heat stress at construction: A worker-centric wearable sensor-based approach. *Saf. Sci.* **2021**, *142*, 105395. [CrossRef]
42. Kim, J.; Jo, B.; Jo, J.; Kim, D. Development of an IoT-Based Construction Worker Physiological Data Monitoring Platform at High Temperatures. *Sensors* **2020**, *20*, 5682. [CrossRef]
43. Yang, K.; Ahn, C.; Kim, H. Deep learning-based classification of work-related physical load levels in construction. *Adv. Eng. Inform.* **2020**, *45*, 101104. [CrossRef]
44. Marques, G.; Pitarma, R. A Real-Time Noise Monitoring System Based on Internet of Things for Enhanced Acoustic Comfort and Occupational Health. *IEEE Access* **2020**, *8*, 139741–139755. [CrossRef]
45. Mumtaz, R.; Zaidi, S.M.H.; Shakir, M.Z.; Malik, U.; Malik, M.M.; Haque, A.; Mumtaz, S.; Zaidi, S.A.R. Internet of Things (IoT) Based Indoor Air Quality Sensing and Predictive Analytic—A COVID-19 Perspective. *Electronics* **2021**, *10*, 184. [CrossRef]
46. Zhou, C.; Ding, L. Safety barrier warning system for underground construction sites using Internet-of-Things technologies. *Autom. Constr.* **2017**, *83*, 372–389. [CrossRef]
47. Zhan, X.; Wu, W.; Shen, L.; Liao, W.; Zhao, Z.; Xia, J. Industrial internet of things and unsupervised deep learning enabled real-time occupational safety monitoring in cold storage warehouse. *Saf. Sci.* **2022**, *152*, 105766. [CrossRef]
48. Campero-Jurado, I.; Márquez-Sánchez, S.; Quintanar-Gómez, J.; Rodríguez, S.; Corchado, J. Smart Helmet 5.0 for Industrial Internet of Things Using Artificial Intelligence. *Sensors* **2020**, *20*, 6241. [CrossRef]
49. Lacamera, D. *Embedded Systems Architecture: Explore Architectural Concepts, Pragmatic Design Patterns, and Best Practices to Produce Robust Systems*; Packt Publishing Ltd.: Birmingham, UK, 2018.
50. Understanding the IEEE 802.11 Standard for Wireless Networks. Available online: https://www.juniper.net/documentation/en_US/junos-space-apps/network-director3.7/topics/concept/wireless-80211.html (accessed on 22 June 2022).
51. What Is Arduino? Available online: https://www.arduino.cc/en/Guide/Introduction (accessed on 15 February 2022).
52. Bluetooth Wireless Technology. Available online: https://www.bluetooth.com/learn-about-bluetooth/radio-versions (accessed on 10 January 2022).
53. Sauter, M. Global System for Mobile Communications (GSM). In *From GSM to LTE-Advanced: An Introduction to Mobile Networks and Mobile Broadband*; Wiley: Hoboken, NJ, USA, 2014; pp. 1–71. [CrossRef]
54. ESP 32. Available online: https://www.espressif.com/en/products/socs/esp32 (accessed on 6 January 2022).
55. Shelby, Z.; Hartke, K.; Bormann, C. RFC 7252—The Constrained Application Protocol (CoAP). Available online: https://tools.ietf.org/html/rfc7252 (accessed on 10 January 2022).
56. Rescorla, E.; Modadugu, N. RFC 6347—Datagram Transport Layer Security Version 1.2. Available online: https://tools.ietf.org/html/rfc6347 (accessed on 10 January 2022).
57. MQTT Version 5.0. MQTT. Available online: https://docs.oasis-open.org/mqtt/mqtt/v5.0/mqtt-v5.0.html (accessed on 13 November 2021).
58. Abiodun, O.; Jantan, A.; Omolara, A.; Dada, K.; Mohamed, N.; Arshad, H. State-of-the-art in artificial neural network applications: A survey. *Heliyon* **2018**, *4*, e00938. [CrossRef]
59. Portugal, I.; Alencar, P.; Cowan, D. The use of machine learning algorithms in recommender systems: A systematic review. *Expert Syst. Appl.* **2018**, *97*, 205–227. [CrossRef]
60. Lu, J.; Wu, D.; Mao, M.; Wang, W.; Zhang, G. Recommender system application developments: A survey. *Decis. Support Syst.* **2015**, *74*, 12–32. [CrossRef]

61. Alibabaei, K.; Gaspar, P.D.; Lima, T.; Campos, R.M.; Girão, I.; Monteiro, J.; Lopes, C.M. A review of the challenges of using deep learning algorithms to support decision-making in agricultural activities. *Remote Sens.* **2022**, *14*, 638. [CrossRef]
62. Tran, T.; Felfernig, A.; Trattner, C.; Holzinger, A. Recommender systems in the healthcare domain: State-of-the-art and research issues. *J. Intell. Inf. Syst.* **2021**, *57*, 171–201. [CrossRef]
63. Cook, A.; Mısırlı, G.; Fan, Z. Anomaly Detection for IoT Time-Series Data: A Survey. *IEEE Internet Things J.* **2020**, *7*, 6481–6494. [CrossRef]
64. Pang, G.; Shen, C.; Cao, L.; Van Den Hengel, A. Deep Learning for Anomaly Detection: A Review. *ACM Comput. Surv.* **2022**, *54*, 2. [CrossRef]
65. Manaswi, N. RNN and LSTM. In *Deep Learning with Applications Using Python*; Apress: Berkeley, CA, USA, 2018. [CrossRef]
66. Alibabaei, K.; Gaspar, P.D.; Lima, T. Modeling soil water content and reference evapotranspiration from climate data using Deep Learning methods. *Appl. Sci.* **2021**, *11*, 5029. [CrossRef]
67. Alibabaei, K.; Gaspar, P.D.; Lima, T. Crop yield estimation using Deep Learning based on climate big data. *Energies* **2021**, *14*, 3004. [CrossRef]
68. Alibabaei, K.; Gaspar, P.D.; Assunção, E.; Alirezazadeh, S.; Lima, T. Irrigation with a deep reinforcement learning model - Case study on a site in Portugal. *Agric. Water Manag.* **2022**, *263*, 107480. [CrossRef]
69. Alibabaei, K.; Gaspar, P.D.; Assunção, E.; Alirezazadeh, S.; Lima, T.M.; Soares, V.N.G.J.; Caldeira, J.M.L.P. Comparison of on-policy deep reinforcement learning A2C with off-policy DQN in irrigation optimization: A case study at a site in Portugal. *Computers* **2022**, *11*, 104. [CrossRef]
70. Zhao, R.; Yan, R.; Wang, J.; Mao, K. Learning to Monitor Machine Health with Convolutional Bi-Directional LSTM Networks. *Sensors* **2017**, *17*, 273. [CrossRef]
71. Huang, C.-J.; Kuo, P.-H. A Deep CNN-LSTM Model for Particulate Matter (PM2.5) Forecasting in Smart Cities. *Sensors* **2018**, *18*, 2220. [CrossRef]
72. Choi, B.; Hwang, S.; Lee, S. What drives construction workers' acceptance of wearable technologies in the workplace?: Indoor localization and wearable health devices for occupational safety and health. *Autom. Constr.* **2017**, *84*, 31–41. [CrossRef]
73. Häikiö, J.; Kallio, J.; Mäkelä, S.-M.; Keränen, J. IoT-based safety monitoring from the perspective of construction site workers. *Int. J. Occup. Environ. Saf.* **2020**, *4*, 1–14. [CrossRef]
74. Data Protection in the EU. Available online: https://ec.europa.eu/info/law/law-topic/data-protection/data-protection-eu_en (accessed on 25 July 2022).
75. Zamfir, S.; Balan, T.; Iliescu, I.; Sandu, F. A security analysis on standard IoT protocols. In Proceedings of the 2016 International Conference on Applied and Theoretical Electricity (ICATE), Craiova, Romania, 6–8 October 2016; pp. 1–6. [CrossRef]
76. Maltseva, K. Wearables in the workplace: The brave new world of employee engagement. *Bus. Horiz.* **2020**, *63*, 493–505. [CrossRef]
77. Erol, S.; Jäger, A.; Hold, P.; Ott, K.; Sihn, W. Tangible Industry 4.0: A Scenario-Based Approach to Learning for the Future of Production. *Procedia CIRP* **2016**, *54*, 13–18. [CrossRef]
78. Badri, A.; Boudreau-Trudel, B.; Souissi, A. Occupational health and safety in the industry 4.0 era: A cause for major concern? *Saf. Sci.* **2018**, *109*, 403–411. [CrossRef]
79. Verra, S.; Benzerga, A.; Jiao, B.; Ruggeri, K. Health Promotion at Work: A Comparison of Policy and Practice Across Europe. *Saf. Health Work.* **2019**, *10*, 21–29. [CrossRef]

Review

machines

MDPI

Human–Machine Relationship—Perspective and Future Roadmap for Industry 5.0 Solutions

Jakub Pizoń [1] and Arkadiusz Gola [2,*]

[1] Department of Enterprise Organization, Faculty of Management, Lublin University of Technology, ul. Nadbystrzycka 38, 20-618 Lublin, Poland
[2] Department of Production Computerization and Robotization, Faculty of Mechanical Engineering, ul. Nadbystrzycka 36, 20-618 Lublin, Poland
* Correspondence: a.gola@pollub.pl; Tel.: +48-81-538-4535

Abstract: The human–machine relationship was dictated by human needs and what technology was available at the time. Changes within this relationship are illustrated by successive industrial revolutions as well as changes in manufacturing paradigms. The change in the relationship occurred in line with advances in technology. Machines in each successive century have gained new functions, capabilities, and even abilities that are only appropriate for humans—vision, inference, or classification. Therefore, the human–machine relationship is evolving, but the question is what the perspective of these changes is and what developmental path accompanies them. This question represents a research gap that the following article aims to fill. The article aims to identify the status of change and to indicate the direction of change in the human–machine relationship. Within the framework of the article, a literature review has been carried out on the issue of the human–machine relationship from the perspective of Industry 5.0. The fifth industrial revolution is restoring the importance of the human aspect in production, and this is in addition to the developments in the field of technology developed within Industry 4.0. Therefore, a broad spectrum of publications has been analyzed within the framework of this paper, considering both specialist articles and review articles presenting the overall issue under consideration. To demonstrate the relationships between the issues that formed the basis for the formulation of the development path.

Keywords: human–machine relationship; human–machine collaboration; human-oriented manufacturing; Industry 5.0; human factor

Citation: Pizoń, J.; Gola, A. Human–Machine Relationship—Perspective and Future Roadmap for Industry 5.0 Solutions. *Machines* **2023**, *11*, 203. https://doi.org/10.3390/machines11020203

Academic Editors: Raul D.S.G. Campilho and Francisco J. G. Silva

Received: 10 December 2022
Revised: 16 January 2023
Accepted: 23 January 2023
Published: 1 February 2023

1. Introduction

With the advent of new revolutions and changes within manufacturing paradigms, the human–machine relationship becomes critical [1,2]. This change is taking place whether in technical, social, organizational, or ultimately legal and ethical dimensions [3]. The machine-human relationship has evolved over the centuries. Nevertheless, it was with the advent of machines that people began to delegate manual work (or that performed by humans) to machines. However, the delegation did not mean a complete handover of work to machines. Machines were and are still supervised by humans [4]. The degree of delegation was directly dependent on the technical level of the machine in question [5]. As technology changed over time, so too did the capabilities of machines become available [6]. Machines, over time, gained a new form of propulsion—from steam power to electricity). Simple electronic circuits with the ability to control machines were introduced [7]. Sensors were added, which allowed better monitoring as well as reporting of machine status [8]. Or, finally, machines were integrated into the Industrial Internet of Things network so that they could be constantly monitored [9]. The amount of data available on the machines and their condition would allow the use of advanced analytical methods, including artificial intelligence methods (deep learning) [10]. At the same time, this was only possible with the use of advanced information system solutions enabling the processing of multidimensional production data sets. In this respect, CPS and digital twin technologies are increasingly being used [11,12].

The implementation of these technologies made it possible that the supervision of machine operations not to require significant work [13]. To a certain extent, machines have become or are becoming autonomous. Consequently, the portion of work delegated to them can increase [14]. For this reason, the human–machine relationship is changing and is directly dependent on the level of technology [15]. The more machines can act autonomously, the more work is delegated to them. At the same time as technology was changing, production paradigms were also changing [16].

The human–machine relationship was dictated by human needs and what the market dictated at the time. While this relationship was difficult to demonstrate in ancient times, technology has been serving people since after the Second World War. Its changes are strictly dependent on people's needs. Hence, the question arises as to what the perspective of these changes and what trend accompanies these changes. This question constitutes a research gap, which the following article aims to fill. The article aims to identify the status of change and to indicate the direction of change in the human–machine relationship.

The task of the authors within the framework of this paper is to identify the perspective and development path (roadmap) for the human–machine relationship. To achieve this objective, a literature review was carried out to identify visible directions and perspectives of the human–machine relationship concerning the industrial revolutions and industry 5.0 [17]. A concept development plan was then drawn up, considering contemporary conditions as well as trends around robot-machine collaboration.

2. Industrial Revolutions

Regarding changes affecting production, it has been accepted to distinguish industrial revolutions [18]. The basis for distinguishing a revolution is to indicate the technological breakthrough that caused a rapid increase in production or development in the organization of the production process. Subsequent technological solutions introduced new possibilities, which historically formed the basis for distinguishing revolutions (Figure 1).

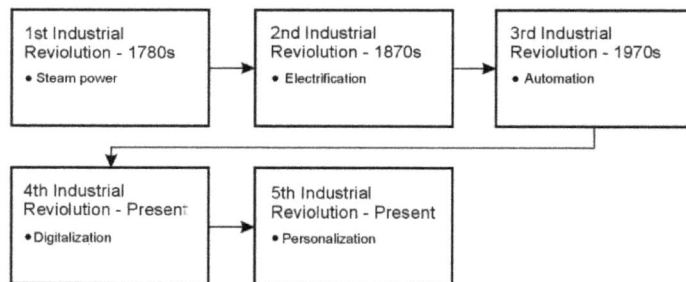

Figure 1. Industrial revolutions and their outcomes [19].

Electrification, which made it possible to transmit energy over long distances, separating where it was generated from where it was used. Or through automation, which has made it possible for machines to adapt to changing conditions and carry out repetitive work without human supervision. Or, finally, the digitalization that underpins Industry 4.0, which ensures the use of information technology in the continuous processing of real-time production data to seek optimization [20]. Personalization, on the other hand, forms the basis for distinguishing the next revolution referred to as Industry 5.0 [21]. Among the revolutions, the current ones will be detailed.

2.1. Industry 4.0

The industry 4.0 revolution was announced in 2011 at the Hannover Fair [22]. Industry 4.0 was created at the initiative of the German government as part of a strategic project to transform manufacturing factors with physical systems into manufacturing factors with cyber-physical systems [23,24]. Industry 4.0 revolution combines IoT, cloud computing, analysis of multidimensional data sets (Big data), artificial intelligence (AI), blockchain, and other high technologies into a high degree of automation and manufacturing [25].

It can offer the manufacturing environment opportunities for self-awareness, self-learning, autonomous decision-making, self-realization, and adaptation to production. Industry 4.0 integrates the production systems of various smart factories into a value chain in the form of a CPS, to obtain real-time data and make decisions for production. A CPS production system is characterized by a high degree of flexibility, adaptability, and agility [26].

2.2. Industry 5.0

Despite the enthusiasm of companies regarding Industry 4.0 pursuing the concept of the Smart Factory [27,28], both researchers and representatives of institutions have noted that this concept ignores a very important aspect related to productionthe human aspect. Industry 4.0, while focusing on the work of machines and their optimization, marginalizes the participation of people in the process implementation. This is understandable because of economic aspects or even because of employee safety. On the other hand, it should not be forgotten that it is people who are the source of disruptive innovations, i.e., solutions that have the effect of changing the manufacturing paradigm.

What is more, people can make associations of facts, technology features, and production conditions that lead to key improvements and competitive advantages at the manufacturing know-how level, in a way that is completely inaccessible to machines.

Here it is also very important to emphasize that machines have a very high capacity for analysis, if data is available. In the absence of aggregated data, there is no possibility to use it. On the other hand, people tire quickly, make mistakes, and need time to be trained before they can carry out a specific production. Nevertheless, they can make items not only guided by technical parameters of materials, equipment, and standards but above all by technical intuition. Given the above, it seems that the best solution is a synergy effect, a combination of the efficiency and reliability of machines together with the innovation of people expressed by the so-called 'human touch'. This is the basis for the announcement of the next industrial revolution, numbered 5.0.

Industry 5.0 is defined as an evolution that aims to harness the creativity of human experts working together with efficient, intelligent, and precise machines [29]. Saeid Nahavandi, on the other hand, points out that Industry 5.0 brings the human workforce back into the factory, where man and machine are paired to make processes more efficient, harnessing human brainpower and creativity by integrating workflows with intelligent systems [30].

Industry 5.0 brings human–machine collaboration to a new level. Industry 5.0 shifts the center of gravity of manufacturing from system-oriented manufacturing systems to human-oriented systems. This represents, in a way, a return to the roots of manufacturing, but aspects related to economics or sustainability remain important [31].

In addition, Industry 5.0 is a motivated change in sustainability in the area of manufacturing related to the use of creativity and human skills in collaboration with machines. At the same time, maintaining the cost criterion together with a high production volume to meet the needs reported by the market.

As a result, the customer will not have to give up his or her individual preferences in favor of cost-optimized production of the good in question—thus reducing the preferences resulting from mass production or mass customization. The prospect opens up for mass customization leading to the satisfaction of the needs of both parties, producer and consumer [32,33]. This approach to production implies the development of collaboration between machines and people. Industry 5.0 will revolutionize production systems around the world, taking tedium, drudgery, and repetitive tasks off the hands of workers. Intelligent robots and systems will infiltrate supply chains and factory floors to an unprecedented degree [30].

Importantly, Industry 5.0 is also a focus of the European Commission's commission studies. According to the commission, the strength of Industry 5.0 is a societal goal that goes beyond job creation and development to become a resilient provider of well-being by ensuring that manufacturing respects the limitations of the planet and puts the well-being of industrial workers at the center of the production process [34,35]. In addition, research conducted on the critical elements of Industry 5.0 for successful adaptation in manufacturing. The study suggests that Industry 5.0 enables smart manufacturing through the intelligent use of data by combining multiple factory data and advanced technologies, thereby producing more personalized products [36].

In addition, the literature review also identified five trends in which Industry 5.0 is heading, namely:

- Industry 5.0 in the context of assessing and optimizing the supply chain in manufacturing processes,
- Industry 5.0 in the context of business management, innovation, and digitalization,
- Industry 5.0 in the context of intelligent and sustainable manufacturing,
- Industry 5.0 transformation driven by IoT, Bigdata, and AI,
- Human–machine connectivity and coexistence [37].

Finding solutions to the management challenges posed by modern industrial revolutions will require governments, businesses, and individuals to make the right strategic decisions regarding the development and implementation of new technologies [16].

With new technological solutions, a whole new perspective is opened to meet the needs of individual customers on both large and small scales. Thus, particular revolutions have led to a change in manufacturing paradigms in terms of both volume and variety of products.

2.3. Main Differences between Industry 4.0 and 5.0

Industry 4.0 and Industry 5.0 have many common features. This is primarily due to the fact that Industry 5.0 is a development of Industry 4.0. The definition of Industry 5.0 is dominated by two basic visions. One of them concerns the technical approach (see e.g., [38]) while the second one focuses on biological aspects (see e.g., [39]). From a technical perspective, there are several elements that definitely differentiate them. The distinction between revolutions is based on questions of interest. On the one hand Industry, 4.0 focuses on digitization (i.e., the implementation of advanced data processing techniques and the implementation of advanced algorithms) mainly focusing on artificial intelligence in production. On the other hand, Industry 5.0 is described as a revolution paying special attention to the human aspect. This is due to the fact that in Industry 4.0 (if modern systems and a high level of automation are implemented) the domain of progress is innovation and creativity. Although modern advanced technologies (such as neural networks) are already able to create images or texts, the domain of production is innovation provided by people. So they are supposed to be a source of inspiration. Another field of interest is the manufacturing paradigms of both revolutions. Industry 4.0 is (thanks to vertical and horizontal integration) dedicated to the realization of mass customization ideas. On the other hand, Industry 5.0 (focusing on people) refers to new idea craft production paradigms using mobile collaborative robots. The main features that constitute the differences between the Industry 4.0 and Industry 5.0 paradigms are presented in Table 1.

Table 1. Main differences between Industry 4.0 and 5.0 paradigms [40].

	Industry 4.0	Industry 5.0
Objective	• Smart manufacturing (smart mass production, smart products, smart working, smart supply-chain), • Systems(s) optimization.	• Sustainability, • Environmental stewardship, • Human-Centricity, • Social benefit.
Systemic Approaches	• Real-time data monitoring, • Integrated chain that follows through the end of life-cycle phases.	• Utilization of technology ethically to advance human values and needs, • Socio-centric technological decisions, • 6R methodology and logistic efficiency design principles.
Human Factors	• Human reliability, • Human-computer interaction, • Repetitive movements.	• Employee safety and management, • Learning/training for employees.
Enabling Technologies and Concepts	• Cloud computing, • Internet of Things, • Big data and Analytics, • Cyber Security, • Digitization (simulation, digital twins, artificial intelligence, augmented, virtual or mixed technology), • Automation (advanced robotics, remote monitoring, autonomous robots, machine-to-machine communication), • Cyber-physical systems, • Horizontal and vertical integration (PLC, Supervisory Control and Data Acquisition (SCADA), Manufacturing Execution System (MES), Enterprise Resource Planning (ERP)), • Additive Manufacturing.	• Cloud computing, • Internet of Things, • Big data and Analytics, • Cyber Security, • Digitization (simulation, digital twins, artificial intelligence, augmented, virtual or mixed technology), • Human–machine interaction, • Multi-lingual speech and gesture recognition, • Tracking technologies for mental and physical occupational strain, • Collaborative robots • Bio-inspired safety and support equipment, • Decision support systems, • Smart grids, • Predictive maintenance.
Environmental Implications	• Systems are economic, • Waste prevention per data analytics, additive manufacturing and optimized systems, • Increased energy usage, • Extended product life cycle.	• Waste prevention and recycling, • Renewable energy sources, • Energy-efficient data storage, transmission, and analysis, • Smart and energy-autonomous sensors.

3. Production Paradigm Shift

Over the centuries, production methods have been defined by the technical solutions available, but also by product preferences (Figure 2). New technical solutions allowed for increased volume and variety (Figure 3), but this also involved changes in the organization and also entailed social changes.

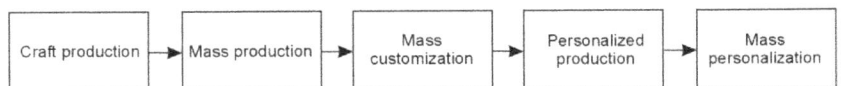

Figure 2. Production paradigm shift [40].

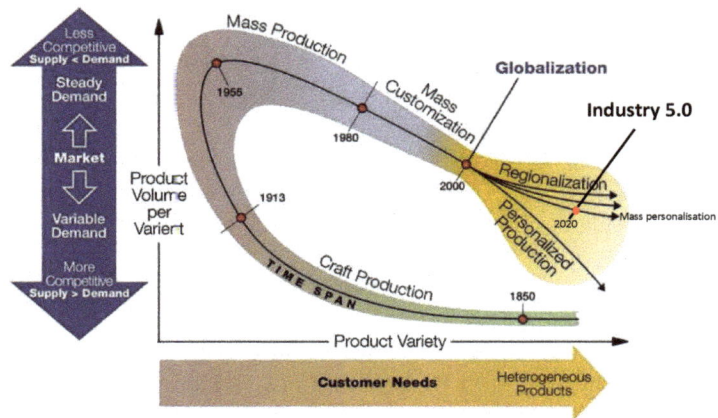

Figure 3. Shifting manufacturing paradigm and factors implying change (based on: [40]).

Artisanal production made it possible to meet the needs of selected customers. It was limited in terms of territory and resources. Products were targeted at the local market. Production was only possible with machinery available on the market. In this production, the overriding role was played by man. All elements of the technological process were made by hand and their quality depended on the skill, knowledge, and ability of the person making them.

With the advent of mass production, the role of man was reduced, but on the other hand, the scale increased. Goods were mass-produced, and the volume of production was for the local, national but also international markets. Specializations were introduced so that a given worker performed a specific task repetitively, and thus the quality of his work was repeatable. On the other hand, the introduced production tactics ensured that production became repetitive and predictable. On the other hand, it lacked variety, which is best expressed by Henry Ford's statement that the Ford Model T car produced was available in every possible color if it was black.

Mass customization, on the other hand, is a solution that has ensured product variety. The consumer can choose the product and customize it based on the possible options presented by the supplier. Thus, a customized product is already realized at the production stage. At the same time, to optimize costs, the base on which the product is made remains the same. With changing preferences and the desire to express individuality, the demand for personalized goods has emerged. The ability to meet the individual needs of individual customers on a large scale has come to be known as mass personalization [41,42]. In this way, the products provided, not only are better in terms of how they are produced, but also express the customer's needs. Given that mass customization is currently the dominant paradigm, it is crucial to understand the differences between the paradigms, as presented in Table 2.

It is also worth noting that mass personalization is the limiting case of mass customization. While both strategies are guided by a product price criterion consistent with mass production efficiency, the former (mass customization) targets a specific customer segment (individual customers), while the latter (mass personalization) targets segments dedicated to many different customers [43].

Table 2. Key differences between production paradigms (Source: based on [43]).

Paradigm / Features	Mass Production	Mass Customization	Personalized Production/ Mass Personalization
Production Aim	• Scale	• Scale • Scope	• Scale • Scope • Value
Desired properties of the product	• Quality • Cost	• Quality • Cost • Variety	• Quality • Cost • Variety • Efficacy
Role of Customer	• Buy	• Choose • Buy	• Design • Choose • Buy
Production System	• Dedicated manufacturing system	• Reconfigurable manufacturing system	• On-demand manufacturing system/ • Cobot-oriented manufacturing system

4. Human–Machine Relationship

With successive industrial revolutions and accompanying changes in the production paradigm, an evolution of the human–machine relationship is practically taking place, referred to in the literature as the 5C model: Coexistence, Cooperation, Collaboration, Compassion, and Coevolution (Figure 4) [44].

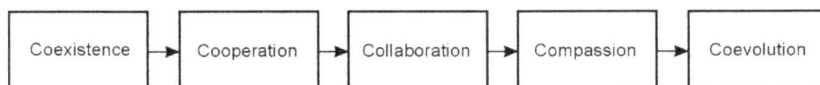

Figure 4. Human–machine 5C relationship model [44].

The different stages signify the successive change that took place in the relationship. At the same time, the successive stages did not exclude the previous ones. Evolution rather than revolution in this respect means that constant change opens new perspectives. Thus, the man-machine relationship is undergoing constant change, which now, due to unprecedented technological developments, is entering a completely new level. Therefore, an extremely important question is the direction in which it will change.

What is important is that the above changes took place over the revolutions indicated in the previous chapter. Therefore, it should be noted that during the first and second industrial revolutions, machines were the equipment of factories. Thus, they formed a 'cold' coexistence relationship in which machines were simple tools for humans or worked independently under supervision.

Reconfigured machines and production lines forming dynamic cooperative human–machine teams in integrated production processes is a characteristic of the third industrial revolution. In this case, humans, and machines, depending on the process, temporarily share a workspace and share some of their physical, cognitive, and computational resources. At the same time, it should be noted that they are not working on the same task at the same time.

In the next, already the fourth industrial revolution, intelligent machines collaborate with humans in a shared workspace, with a specific goal of completing tasks through synchronized interactive joint actions of all parties within a common team identity [45].

In contrast, at the level of the fifth industrial revolution, workers together with robots (cobots) form teams to carry out production tasks [17]. Central to this relationship is how workers feel in this environment, how decisions are made, who makes them, and how such teams are formed. In the case of mixed human-cobot teams, there can be trust challenges. Particularly in the engineering industry, where the work requires not only good qualifications but the ability to work, many challenges arise. Methods and ways of developing trust in collaboration with the machine are needed. So that the worker knows that the machine is working towards the same goal and is not in danger from the machine. So that the worker can communicate effectively with the machine and understands the decision-making process when working with the machine. Depending on the manufacturing system, it is determined who ultimately makes the decisions.

Today, due to the development of advanced information technology and the transition to the Industry 4.0 era, the human–machine relationship is turning towards the human. Therefore, human-centered production systems need to be characterized by bidirectional empathy, proactive communication, and collaborative intelligence to establish reliable human–machine co-evolution relationships and thus lead to high-performance human–machine teams. Thus, on the one hand, the evolution of relationships and on the other hand a series of challenges [18] that need to be met to realize mass personalization (Figure 5).

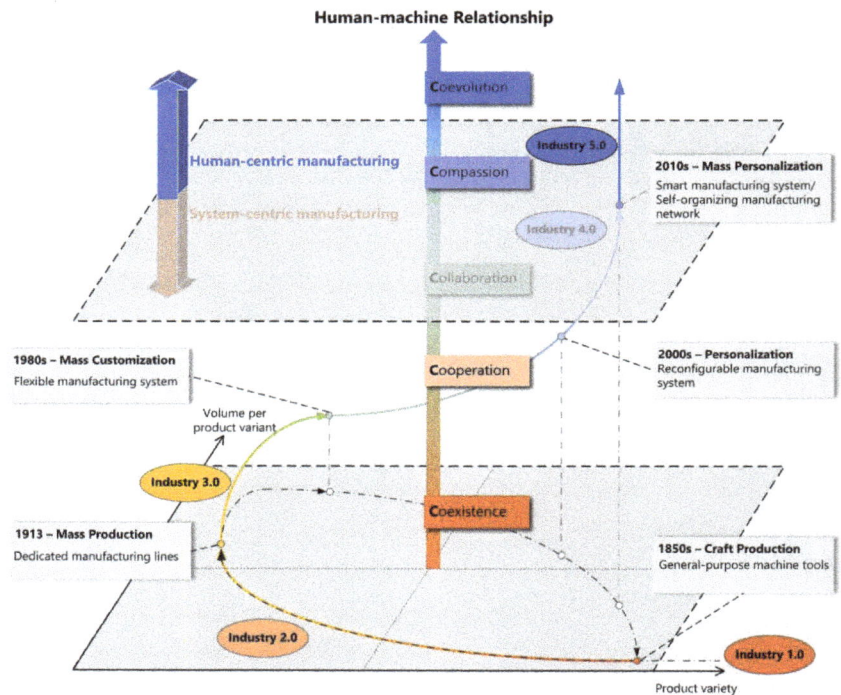

Figure 5. Evolution of the human–machine relationship towards human-centric production [27].

To fully characterize the human–machine relationship, further characteristics for the types distinguished within the 5Cs are indicated below.

4.1. Human–Machine Coexistence

This type of relationship envisages that man and machine are in the same environment and share the same space. Within this case, one speaks of monitored coexistence, that is when a robot and a human work closely together without the need for mutual contact or coordination, thus requiring continuous opportunities for the robot controller to avoid obstacles [46].

Furthermore, the coexistence of machines and humans for the enterprise realizes the goal of balancing automation/productivity and flexibility/capability. For example, using the human–machine interface, the operator is associated with a set of behavioral roles as a supervisor of multiple semi-automated production processes. The proposed model can be used to design manufacturing systems at different levels of enterprise architecture, particularly at the machine level of the manufacturing system, where operators interact with semi-automated machines to realize the goal of human-enhanced automation [47].

The coexistence of machines and humans in the production space creates the risk of collisions. Therefore, there is a need to adopt appropriate policies that safeguard production efficiency using safety features such as emergency stops [48].

4.2. Human–Machine Cooperation

Human–machine cooperation is defined as a group of agents in a collaborative situation when two conditions are present. The first occurs when the agents pursue goals that may conflict with those of others, at the level of their own goals, sub-goals, processes, or resource. The second, on the other hand, is that each agent seeks to manage these interferences to facilitate its tasks and those of others on a common task [49].

Some research work has applied the principles of interpersonal cooperation to the dynamic sharing of tasks between human operators and automated systems. These shared tasks are decision-making tasks (e.g., conflict detection, problem-solving, diagnosis, or image analysis and retrieval) and may involve several organizational configurations of the human–machine system. In this mode of human-robot interaction, the human operator and the robot are placed at the same decision-making level. The two agents work together to achieve a common goal by initiating an interactive dialogue. In the approach presented in this thesis, this dialogue can take different forms depending on the collaborative situation. This dialogue is carried out using various forms of collaboration [50].

While considering the human–machine collaboration relationship, it should be noted that the inclusion of the human in this type of relationship, even if it integrates the machine, can be justified as contributing to solving part of the automation problems in terms of human–machine relationships [2].

4.3. Human–Machine Collaboration

Technological advances increasingly envisage the use of robots interacting with humans in everyday life. Human-robot collaboration (HRC) is an approach that explores the interaction between a human and a robot, while pursuing a common goal, at a cognitive and physical level [51].

The human–machine collaboration relationship is a key means of manufacturing. The system that results from it performs surveillance, prognostics, and health management is related to the safety and sustainability of manufacturing [52,53].

Human–machine collaboration has great potential for making risky decisions. Machines could be more helpful in gathering information and assessing uncertainty and communicating key information to human decision-makers to save cognitive resources. Moreover, human decision-makers could debate their judgments with the help of a machine and reduce emotional influences [54].

As machines become increasingly intelligent and can perform more complex functions, a new relationship between humans and automation is emerging. This relationship is changing from 'master-servant' to 'master-collaborator' and requires a different approach to system design, human–machine information exchange and interface, and the imposition of additional requirements on the machine [55].

Current technological trends are enabling more and more physical contact between humans and machines. Humans and machines act together and communicate with each other not only through gestures and speech but mainly through the haptic channel. This current phase of human–machine interaction can be referred to as human–machine collaboration [1].

This has resulted in the development of a new type of robot—cobot, collaborative robot— which is geared towards human collaboration [56]. This term describes a robot capable of working and collaborating with humans. These collaborative robots will be aware of human presence and will therefore take care of safety and risk criteria. They will be able to notice, understand and feel not only the human but also the goals and expectations of the human operator. Similar to a learner, cobots will observe and learn how to perform a specific task. Once learned, the task will be performed by the cobots as it would be performed by a human [57].

4.4. Human–Machine Compassion and Coevolution

In human-centered production, given the successes in cognitive science and personalized AI, it is conceivable that empathic machines that sense human emotions, needs, and preferences could provide situational assistance to humans in addition to situational cooperation.

In such circumstances, based on reciprocity, humans will willingly monitor and take care of the 'health' of empathic machines. At the same time, this machine's health will be expressed in terms of quantitative measures related to workload, level of task fluctuation, etc. Such optics represent a whole new chapter in the human–machine relationship—human–machine empathy. It is also worth adding that intimate human–machine interactions will ultimately enable the growth of both human and machine capabilities, leading to continuous human–machine co-evolution in the future [45]. This mutual co-evolutionary development may lead to new forms of relationships no longer focusing on competition but on creating a better new future for machines and humans [58].

5. Perspectives and Future Roadmap for Industry 5.0 Solutions

5.1. Research Process

The development perspective and roadmap are issues that are conditioned by a broad spectrum of issues Therefore, to be able to speak of a perspective, a view is needed, which is formulated by analyzing several approaches to the issues in question. Unlike in the case of specialized issues, one can only speak of a perspective and a roadmap when the entire context is known.

Therefore, to define a perspective and be able to indicate a roadmap for development, it is necessary to examine a wide range of different scientific publications on the issues addressed in this article. For this reason, the author's attention has mainly been focused on both review articles, presenting the overall issue under consideration, and specialized articles. Therefore, it was considered that the review would refer to publications that contain references to the issues addressed within this paper.

The review focused on the renowned databases Web of Science and Scopus (Figure 6). These databases were considered reference databases, as their scope includes publications cited and produced worldwide. Based on the Scopus and Web of Science databases, a search was carried out, using as keywords those topics that relate to the human–machine relationship as well as Industry 5.0. Therefore, the following were included in the query: human–machine relationship, human–machine coexistence, human–machine cooperation, human–machine collaboration, human–machine compassion, and human–machine coevolution or Industry 5.0. In addition, to keep the context up-to-date, the research was carried out regarding the last five years.

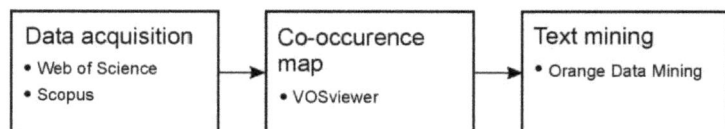

Data acquisition	Co-occurence map	Text mining
• Web of Science • Scopus	• VOSviewer	• Orange Data Mining

Figure 6. Research process.

This resulted in 9533 publications from the Web of Science database and 8430 records from the Scopus database. Furthermore, since the human–machine relationship is often indicated interchangeably with human–machine interaction, this word was also adopted as a keyword implemented in the analysis.

This was followed by a technical analysis using the data processing methodology of the VOSViewer software [59]. The analysis used a method of counting the type of term weights using full counting, which counts the occurrences of a given term in all processed documents. Based on the occurrences, it counts the weight of the relationship between the given terms. The result was a co-occurrence map of the terms with their assigned relationship strength. In the next step, text mining was carried out using Orange Data Mining analysis [60] indicating the weight of the given words related to the topics to demonstrate the topics through topic modeling with Latent Dirichlet Allocation.

5.2. Results

The results extracted from the Web of Science database (Figure 7) will be presented first. Both title and abstract were technically analyzed. The demonstrated study, based on the analysis criteria adopted, showed several clusters. The results showed that, despite the oriented query, the topics indicated in the query were not the dominant content of the analyzed publications. Furthermore, due to the broad spectrum of the inquiry, the analysis showed both technical issues (red) but also application-specific clusters (green, blue, orange, and mixed). It can be assumed that four main clusters were distinguished. For this analysis, the focus will be on the technology cluster.

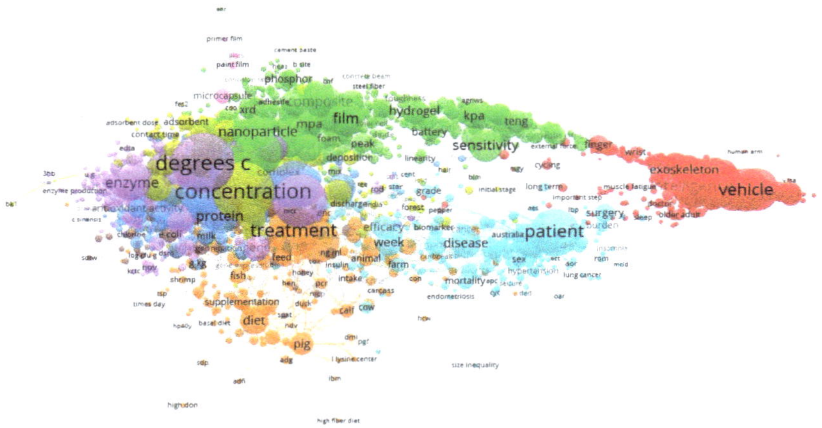

Figure 7. Term co-occurrence map based on provided publications Web of Science.

In the technology cluster (Figure 8), it is notable that the keyword highlighted is: collaboration. This is significant insofar as the other 5C terms are missing at this level of significance. On the one hand, this is due to the study's time horizon of 5 years. On the other hand, however, it confirms the thesis that the 5th industrial revolution is indeed taking place today and that man-machine collaboration is a topical and widely described issue. In the following section, key relationships are highlighted in the terms of greatest importance.

The first keyword is collaboration. The dependency associated with this word shows a possible link to the issues of automation, trust, or intelligence. This indicates that machine-to-machine collaboration is only possible with intelligence and automation, however, trust is needed for collaboration to take place. Trust is a key factor for the development perspective (Figure 9).

Based on the collaboration dependency (Figure 10), it is then indicated which elements make up its components. In this, intelligence is closely linked to data (big data) and the issue of collaboration.

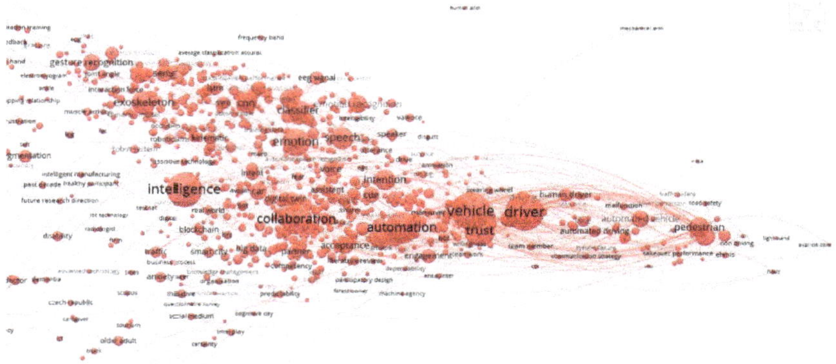

Figure 8. Technology cluster.

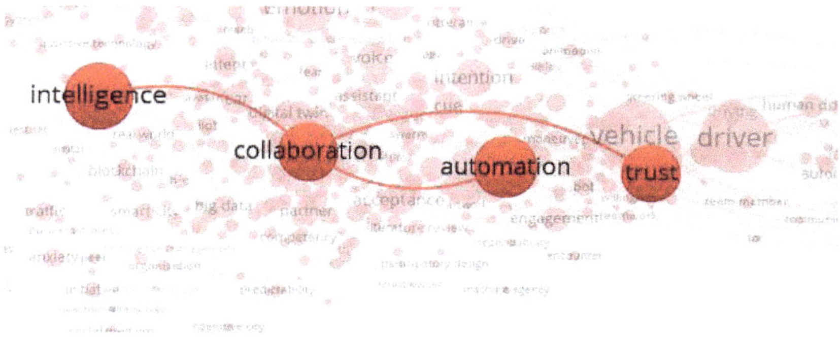

Figure 9. Collaboration dependency.

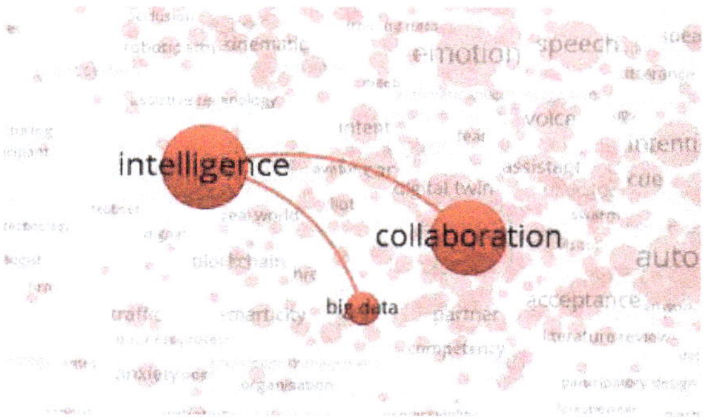

Figure 10. Intelligence dependency.

In a further step, attention was turned to the issue of automation. Here, two new issues arise vehicle and driver (Figure 11). Thus, cooperation in automation must concern the control of different vehicles, but there is a need to develop a specific control agent.

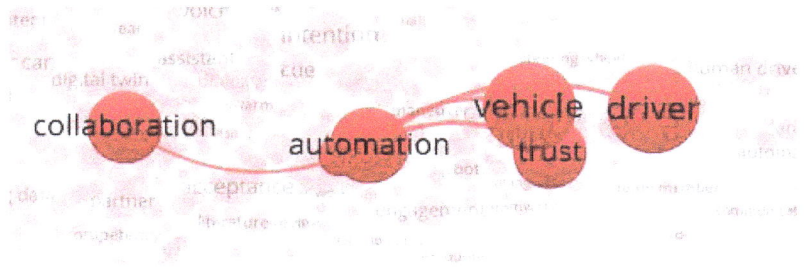

Figure 11. Automation dependency.

The issue of trust relates to issues of collaboration but touches on issues of acceptance, errors, automated vehicles, passers-by, or communication interfaces (Figure 12).

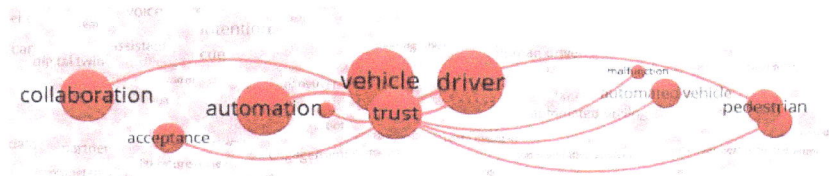

Figure 12. Trust dependency.

Vehicle dependency introduces human driver or road user issues (Figure 13). Thus, this dependency indicates the perspective of placing human–machine cooperation in terms of the environment. It returns to the issue of intention detection.

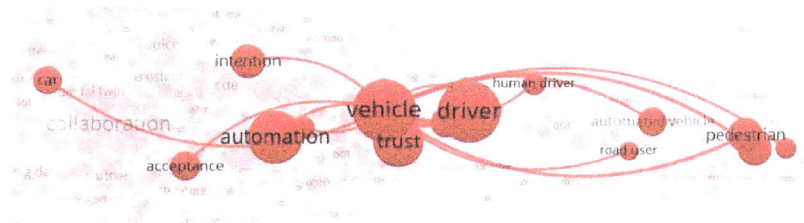

Figure 13. Vehicle dependency.

In addition to the previously indicated ones, driver dependency points out that it depends on both the emotions and the vehicle in which it is managed (Figure 14). Furthermore, attention is drawn to issues of engagement.

It is also worth describing here, the issue of emotions, which often identifies personality, which is closely related to the human being. From the perspective of identified relationships, the emotion is closely linked to the tools for classifying it, recognizing it or expressing it in speech (Figure 15).

In the next step, a volume analysis of the Scopus database documents was carried out (Figure 16). The analysis resulted in as many as eight clusters. This indicates a much greater fragmentation of the issues described from the perspective of the realized query. Given this, the dependencies obtained are not so clearly visible. Hence, clusters will be presented as dependencies to the highlighted terms with the highest weight—thus forming the largest perspective.

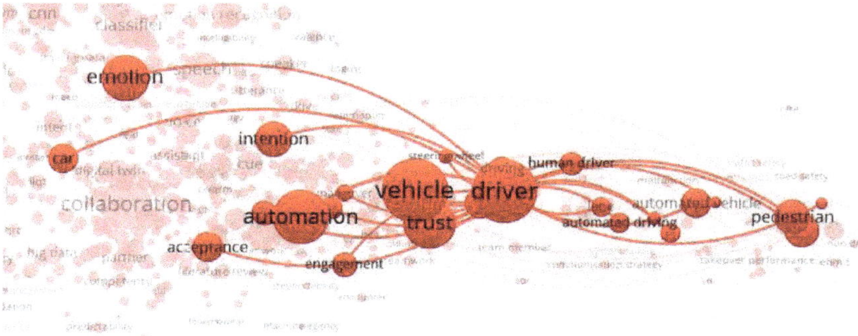

Figure 14. Driver dependency.

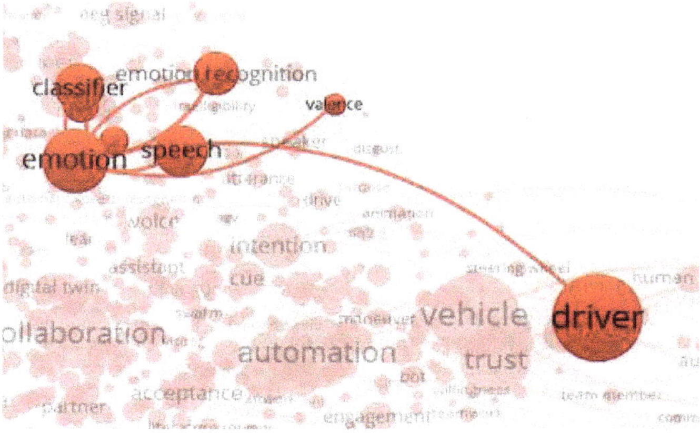

Figure 15. Emotion dependency.

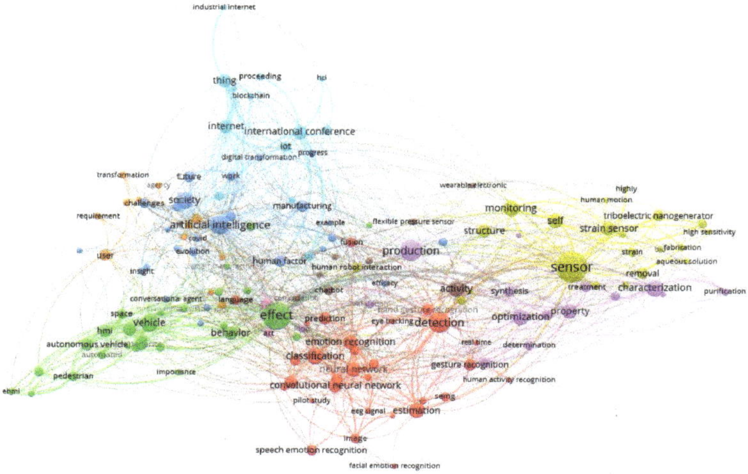

Figure 16. Term co-occurrence map based on provided publications Scopus.

One of the key dependencies highlighted within the analysis, linking to different scopes, is sensors (Figure 17). Here, several strong connections to issues such as artificial intelligence, monitoring, sensing, estimation, or neural networks can be identified. However, there is no explicit indication here of the issues present in the 5C model.

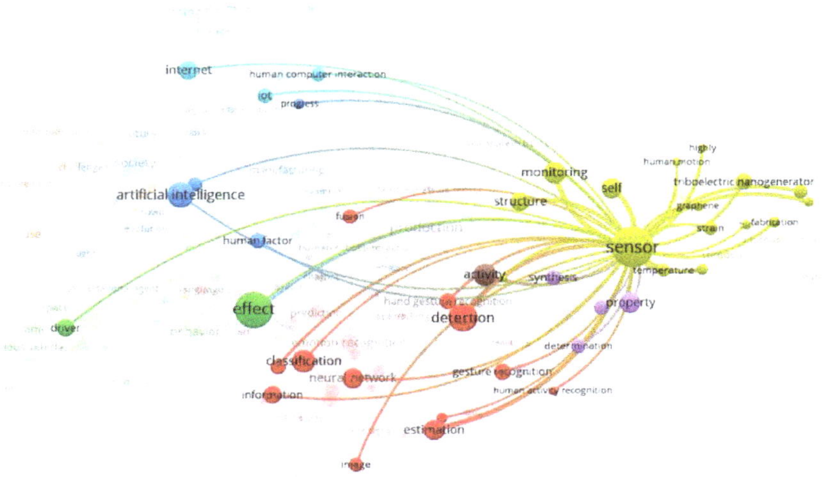

Figure 17. Sensor dependency.

Then there is the effect dependency (Figure 18), which relates to a broad spectrum of issues while linking the most important ones, such as artificial intelligence, sensors, and vehicles. This dependency is a good illustration of the range of tools used to obtain the causal result—the effect—of the work carried out. The issue of effect is crucial in managerial terms. Tools and technologies are used to achieve technical, social, and, above all, economic results.

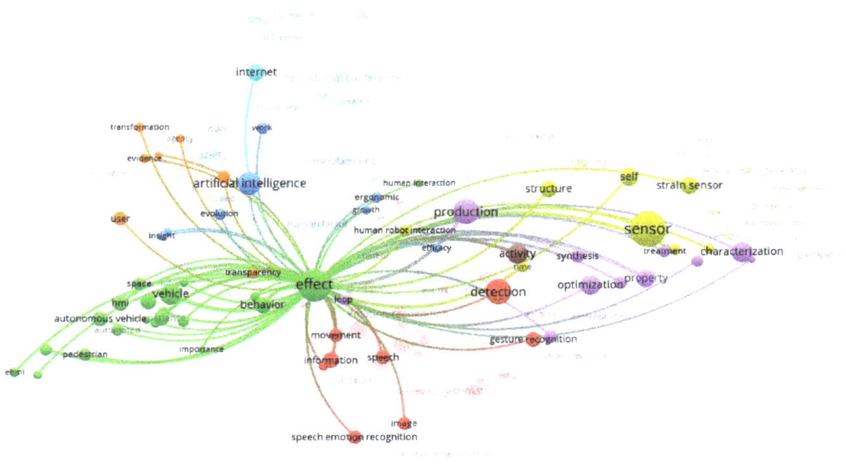

Figure 18. Effect dependency.

Another highlighted dependency relates to manufacturing (Figure 19). Here, one can see the publications focus both on technical issues, but also on application issues for optimization.

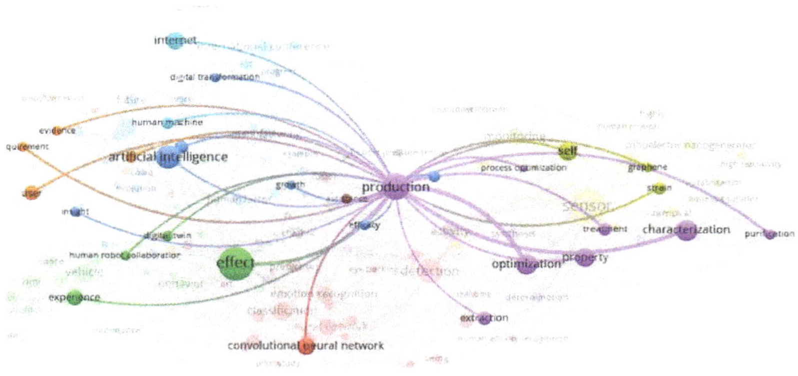

Figure 19. Production dependency.

Further, the dependency on artificial intelligence (Figure 20) brings together many terms from different scopes. It can be pointed out that it constitutes a kind of glue for all the indicated dependencies that appear in the analyzed literature.

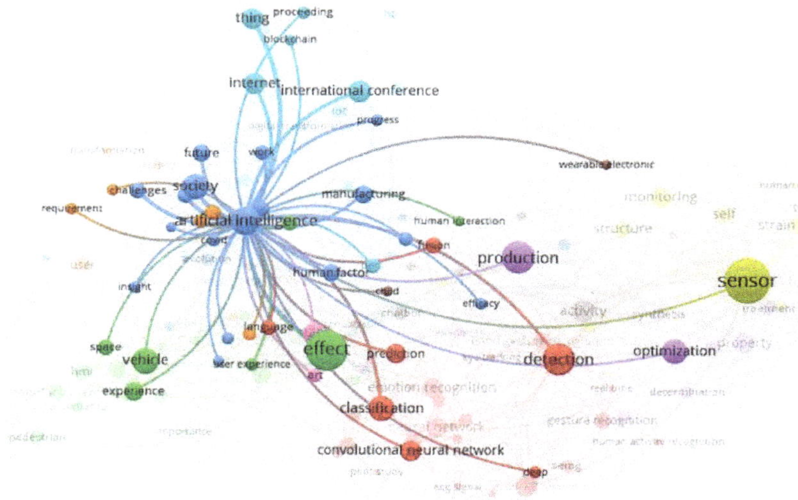

Figure 20. Artificial intelligence dependency.

Similar to the dependency on artificial intelligence, the issue of detection (Figure 21) is widely discussed within the articles. This demonstrates that it is crucial for machine-human collaboration that the machine can discover that there is a machine, product, or human in its environment.

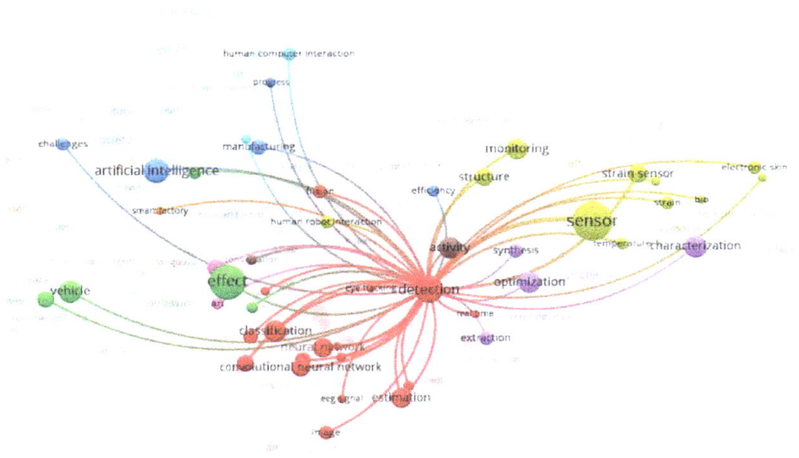

Figure 21. Detection dependency.

The demonstrated text mining analysis of the topics that dominate the analyzed Web of Science publications identified the following topics—the 10 topics with the highest weighting were assumed to be extracted:

1. *human, use, system, machin, interact, base, model, control, result, robot*
2. *human, use, machin, c, studi, robot, result, interact, industri, method*
3. *system, sensor, high, human, control, user, model, detect, applic, oper*
4. *sensor, system, model, high, human, use, method, devic, learn, data*
5. *robot, model, industri, control, sensor, data, method, base, machin, propos*
6. *industri, control, robot, human, technolog, product, propos, machin, model, research*
7. *robot, industri, human, c, data, control, sensor, technolog, studi, model*
8. *model, use, user, interact, c, control, design, interfac, industri, high*
9. *use, control, system, design, robot, user, base, human, c, interact*
10. *control, use, interact, user, robot, industri, model, c, human, task*

Similar results were obtained for the Scopus data analysis:

1. *human, machin, interact, base, system, interfac, robot, design, use, control*
2. *base, human, system, machin, use, recognit, sensor, control, network, design*
3. *system, base, industri, recognit, design, sensor, control, machin, human, network*
4. *use, base, recognit, industri, emot, network, learn, interact, neural, sensor*
5. *industri, interact, robot, system, recognit, use, human, collabor, applic, interface*
6. *interact, robot, machin, system, industri, interfac, human, vehicl, design, autom*
7. *robot, industri, interact, control, human, machin, collabor, interfac, design, intellig*
8. *design, interfac, vehicl, autom, system, robot, human, use, drive, control*
9. *sensor, high, flexibl, base, use, pressur, wearabl, strain, recognit, model*
10. *autom, design, machin, human, vehicl, intellig, drive, robot, model, artifice*

Given this, the perspective, despite the clear presence of keywords, is decisively about the human aspect—so emphasized by Industry 5.0. This confirms the chosen research direction and illustrates well the current shape of the human–machine relationship.

5.3. Roadmap

Moving towards human–machine relationship-oriented manufacturing, technologies must be made to work for people in a trustworthy and friendly way, as people are essential in creating personalised products of high value. Philosophical, social and ethical issues and theories must drive the development and implementation of assistive technologies in the real world [45]. This provides the source for the many challenges that the new industrial revolution brings with it. Industry 5.0 represents the next revolution in the development of manufacturing systems, which builds on the achievements of previous ones.

The identified dependencies were defined based on the literature review carried out by the authors. As a result of the analysis, the dependencies were extracted and put into a framework of three stages (Figure 22) forming a roadmap.

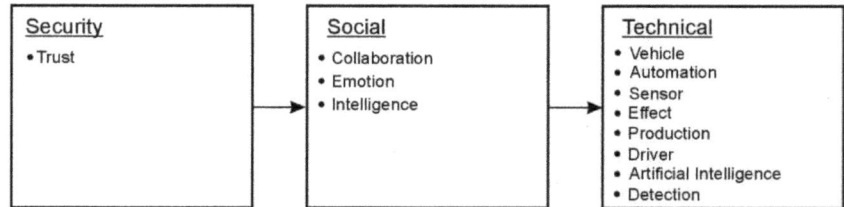

Figure 22. Human–machine relation roadmap.

The identified dependencies were used to formulate the roadmap. The proposed sequence was determined by the authors considering both their experience of working with companies and the literature review carried out.

The first stage of the roadmap concerns the security dimension. In this dimension, it is necessary to develop protocols to secure the data processed by the machine. A key focus should be on issues of remote access to manage cobots. These inherently mobile devices will be controlled remotely. Interception of cobot control could end up losing the integrity of the entire manufacturing process. This raises a new threat of intercepting the management of the process in progress.

The next stage is formed by social dependencies. This dimension is directly related to the working environment. The key here is how the employees concerned feel in this environment, how decisions are made, who makes them, and how such teams are formed. In the case of mixed human-cobot teams, there can be challenges related to trust. Particularly in the engineering industry, where the work requires not only good qualifications but the ability to work, many challenges arise. Methods and ways of developing trust in collaboration with the machine are needed. So that the worker knows that the machine is working towards the same goal and is not in danger from the machine. So that the worker can communicate effectively with the machine and understands the decision-making process when working with the machine. Depending on the manufacturing system, it is determined who ultimately makes the decisions.

Because of the answers formulated in the previous stages, the task of the technical stage is to implement solutions in line with social and safety considerations. Solutions constructed based on clearly defined boundaries can be the starting point for further work. In addition, it is worth noting the direct benefits of information technology solutions. It will be necessary to provide new learning datasets for collaborative algorithms (image recognition, self-awareness, performance of technological tasks), to develop algorithms that perform safety functions of collaborative work with a person.

In summary, the roadmap for the development of the human–machine relationship should mainly focus on how to facilitate this process for workers. True collaboration will only happen when people can trust the machine. The machine, on the other hand, will be able to identify the intentions of the people it is working with. Therefore, this dimension will need answers to the questions of how to introduce effective training models. How do we convince people to work with cobots? Only after answering these questions will it be necessary to focus on technical issues to answer other important questions such as how to create a cobot that works with people.

6. Discussion

The results presented represent a development pathway for the human–machine relationship illustrated in the literature. The path shown can be seen as a practical development framework.

The article provides direction and formulates the next steps on the development path for the human–machine relationship. By doing so, the systematization of the issue has been realized. What is most important, the path is a signpost both for researchers, but also for specialists in the areas of economics, technology, and industry, who are developing product development paths or developing strategies related to Industry 5.0. The demonstrated relationships should be considered when formulating long-term management strategies in the area of planning the development of enterprise manufacturing systems. The roadmap proposed in the article is a forecast based on the author's experience and the literature review carried out.

Of course, the issue described, for differently defined criteria, may show differences in the relationships shown. At the same time, it should be noted that the article pays attention to the development perspective and roadmap. Thus, the perspective is intended to illustrate the direction of change and provide information. The roadmap, on the other hand, is an expectation of the next development steps with a defined choreography. The issues analyzed are characterized by dynamics related to changing trends, but also the economic situation. Therefore, it can be interpreted differently in terms of the experience of different researchers. Which is understandable and appropriate for this type of study and the methodology adopted.

At the same time, in the opinion of the authors, the work carried out has made it possible to capture those relationships that illustrate contemporary transformations. Therefore, the results and conclusions of the research can be useful.

7. Summary and Conclusions

The fifth industrial revolution uses technological advances to bring the human aspect back into production. To harness the creativity and the potential inherent in workers. This has a definite impact on how the human–machine relationship is shaped. This change in the relationship involves a series of dependencies and challenges of social and legal nature.

Unlike previous revolutions, Industry 5.0 redefines the man-machine relationship by restoring the human aspect to production. This practically reverses the vector of relationships, which hitherto focused on the development of technical solutions specific to machines. The collaborative relationship dominates, with the issue of compassion and coevolution still an issue of the future. At the core is the issue of trust and automation. Therefore, such a major shift in the manufacturing paradigm, which has so far been concerned with the implementation of economically motivated new technical solutions (machines, equipment) in favor of minimizing 'expensive' human labor, will have a significant impact on how goods are produced. This shift towards sustainability in tipping systems represents an open development perspective. It holds the promise of a new shared future, where social needs and responsibility for the goods provided are the ultimate goals of manufacturing. At the same time, it opens many questions in terms of the development roadmap. Different parts require the development of methods, techniques, and the development of good practices to fully understand the potential that lies in the human–machine relationship, which will be the subject of further research.

Author Contributions: Conceptualization, J.P. and A.G.; methodology, J.P. and A.G.; validation, A.G.; formal analysis, A.G.; investigation, J.P.; resources, J.P.; data curation, J.P.; writing—original draft preparation, J.P.; writing—review and editing, A.G.; visualization, J.P.; supervision, A.G.; project administration, A.G.; funding acquisition, A.G. All authors have read and agreed to the published version of the manuscript.

Funding: This research received no external funding.

Data Availability Statement: Not applicable.

Conflicts of Interest: The authors declare no conflict of interest.

References

1. Inga, J.; Ruess, M.; Robens, J.H.; Nelius, T.; Kille, S.; Dahlinger, P.; Thomaschke, R.; Neumann, G.; Matthiesen, S.; Hohmann, S.; et al. Human-machine symbiosis: A multivariate perspective for physically coupled human-machine systems. *Int. J. Hum. Comput. Stud.* **2023**, *170*, 102926. [CrossRef]
2. Hoc, J.M. From human—Machine interaction to human—Machine cooperation. *Ergonomics* **2000**, *43*, 833–843. [CrossRef] [PubMed]
3. Johannsen, G. Human-machine systems research for needs in industry and society. *IFAC Proc. Vol.* **2001**, *34*, 1–9. [CrossRef]
4. Kłosowski, G.; Kulisz, M.; Lipski, J.; Maj, M.; Bialek, R. The use of transfer learning with very deep convolutional neural network in quality management. *J. Eur. Res. Stud.* **2021**, *24*, 253–263. [CrossRef]
5. Javaid, M.; Haleem, A.; Singh, R.P.; Suman, R. An integrated outlook of cyber–physical systems for Industry 4.0: Topical practices, architecture, and applications. *Green Technol. Sustain.* **2023**, *1*, 100001. [CrossRef]
6. Rymarczyk, T.; Król, K.; Zawadzki, A.; Oleszek, M.; Kłosowski, G. An intelligent sensor platform with an open architecture for monitoring and controlling cyber-physical. *Przegląd Elektrotechniczny* **2021**, *3*, 141–145. [CrossRef]
7. Świć, A.; Gola, A. Economic analysis of casing parts production in a flexible manufacturing system. *Act Prob. Econ.* **2013**, *141*, 526–533.
8. Borucka, A.; Mazurkiewicz, D. Production Process Stability: The Advantages of Going Beyond Qualitative Analysis. *Lect. Notes Mech. Eng.* **2023**, 143–148. [CrossRef]
9. Bocewicz, G.; Wójcik, R.; Sitek, P.; Banaszak, Z. Towards digital twin-driven performance evaluation methodology of FMS. *Appl. Comput. Sci.* **2022**, *18*, 5–18. [CrossRef]
10. Bartezzaghi, E. The evolution of production models: Is a new paradigm emerging? *Int. J. Oper. Prod. Manag.* **1999**, *19*, 229–250. [CrossRef]
11. Pizoń, J.; Cioch, M.; Kański, Ł.; García, E.S. Cobots Implementation in the era of Industry 5.0 using modern business and management solutions. *Adv. Sci. Technol. Res. J.* **2022**, *16*, 166–178. [CrossRef]
12. Petrillo, A.; Felice, F.; De Cioffi, R.; Zomparelli, F. Fourth Industrial Revolution: Current Practices, Challenges, and Opportunities. In *Digital Transformation in Smart Manufacturing*; Books on Demand: Norderstedt, Germany, 2018; pp. 1–20. [CrossRef]
13. Kuryło, P.; Wysoczański, A.; Cyganiuk, J.; Dzikuć, M.; Szufa, S.; Bonarski, P.; Burduk, A.; Franowsky, P.; Motyka, P.; Medyński, D. Selected determinants of machines and devices standardization in designing automated production processes in Industry 4.0. *Materials* **2023**, *16*, 312. [CrossRef]
14. Jasiulewicz-Kaczmarek, M.; Antosz, K.; Zhang, C.; Waszkowski, R. Assessing the barriers to Industry 4.0 implementation from a maintenance perspective—Pilot study results. *IFAC-PapersOnLine* **2022**, *55*, 223–228. [CrossRef]
15. Pohl, R.; Oehm, L. Towards a new mindset for interaction design-understanding prerequisites for successful human-machine cooperation using the example of food production. *Machines* **2022**, *10*, 1182. [CrossRef]
16. Doyle-Kent, M.; Kopacek, P. Industry 5.0: Is the manufacturing industry on the cusp of a new revolution? *Lect. Notes Mech. Eng.* **2020**, 432–441. [CrossRef]
17. Stączek, P.; Pizoń, J.; Danilczuk, W.; Gola, A. A digital twin approach for the improvement of an autonomous mobile robots (AMR's) operating environment—A case study. *Sensors* **2021**, *21*, 7830. [CrossRef]
18. Pizoń, J.; Gola, A. The Meaning and Directions of Development of Personalized Production in the Era of Industry 4.0 and Industry 5.0. In *Innovations in Industrial Engineering II.*; Springer: Cham, Switzerland, 2023; p. 279569. [CrossRef]
19. Tay, S.I.; Lee, T.C.; Hamid, N.Z.A.; Ahmad, A.N.A. An overview of industry 4.0: Definition, components, and government initiatives. *J. Adv. Res. Dyn. Control Syst.* **2018**, *10*, 1379–1387.
20. Lalik, K.; Flaga, S. A Real-Time Distance Measurement System for a Digital Twin Using Mixed Reality Goggles. *Sensors* **2021**, *21*, 7870. [CrossRef]
21. Anczarski, J.; Bochen, A.; Głąb, M.; Jachowicz, M.; Caban, J.; Cechowicz, R. A Method of Verifying the Robot's Trajectory for Goals with a Shared Workspace. *Appl. Comput. Sci.* **2022**, *18*, 37–44. [CrossRef]
22. Gola, A.; Pastuszak, Z.; Relich, M.; Sobaszek, Ł.; Szwarc, E. Scalability analysis of selected structures of a reconfigurable manufacturing system taking into account a reduction in machine tools reliability. *Eksplaot. Niezawodn.* **2021**, *23*, 242–252. [CrossRef]
23. Alvarez-Aros, E.L.; Bernal-Torres, C.A. Technological competitiveness and emerging technologies in industry 4.0 and industry 5.0. *An. Acad. Bras. Cienc.* **2021**, *93*, 20191290. [CrossRef] [PubMed]
24. Bordel, B.; Alcarria, R.; Robles, T. Prediction-Correction Techniques to Support Sensor Interoperability in Industry 4.0 Systems. *Sensors* **2021**, *21*, 7301. [CrossRef]
25. Mon, A.; Del Giorgio, H.R. Analysis of Industry 4.0 Products in Small and Medium Enterprises. *Procedia Comput. Sci.* **2022**, *200*, 914–923. [CrossRef]
26. Lu, Y. The Current Status and Developing Trends of Industry 4.0: A Review. *Inf. Syst. Front.* **2021**, 1–20. [CrossRef]
27. Lu, Y. Industry 4.0: A survey on technologies, applications and open research issues. *J. Ind. Inf. Integr.* **2017**, *6*, 1–10. [CrossRef]
28. Mrugalska, B.; Wyrwicka, M.K. Towards Lean Production in Industry 4.0. *Procedia Eng.* **2017**, *182*, 466–473. [CrossRef]

29. Maddikunta, P.K.R.; Pham, Q.-V.; Prabadevi, B.; Deepa, N.; Dev, K.; Gadekallu, T.R.; Ruby, R.; Liyange, M. Industry 5.0: A survey on enabling technologies and potential applications. *J. Ind. Inf. Integr.* **2021**, *26*, 100257. [CrossRef]

30. Nahavandi, S. Industry 5.0—A Human-Centric Solution. *Sustainability* **2019**, *11*, 4371. [CrossRef]

31. Jafari, N.; Azarian, M.; Yu, H. Moving from Industry 4.0 to Industry 5.0: What Are the Implications for Smart Logistics? *Logistics* **2022**, *6*, 26. [CrossRef]

32. Bauer, D.; Stock, D.; Bauernhansl, T. Movement Towards Service-orientation and App-orientation in Manufacturing IT. *Procedia CIRP* **2017**, *62*, 199–204. [CrossRef]

33. Ozdemir, M.; Verlinden, J.; Cascini, G. Design methodology for mass personalisation enabled by digital manufacturing. *Des. Sci.* **2022**, *8*, e7. [CrossRef]

34. Breque, M.; De Nul, L.; Petridis, A. *Industry 5.0 Towards a Sustainable, Human-Centric and Resilient European Industry*; Publications Office of the European Union: Luxembourg City, Luxembourg, 2021.

35. Dixson-Declève, S.A. *Transformative Vision for Europe. ESIR Policy Brief No. 3*; Publications Office of the European Union: Luxembourg City, Luxembourg, 2021.

36. Javaid, M.; Haleem, A. Critical components of industry 5.0 towards a successful adoption in the field of manufacturing. *J. Ind. Integr. Manag.* **2020**, *5*, 327–348. [CrossRef]

37. Akundi, A.; Euresti, D.; Luna, S.; Ankobiah, W.; Lopes, A.; Edinbarough, I. State of Industry 5.0-Analysis and Identification of Current Research Trends. *Appl. Syst. Inn.* **2022**, *5*, 27. [CrossRef]

38. Van Oudenhoven, B.; Van de Calseyde, P.; Basten, R.; Demerouti, E. Predictive maintenance for Industry 5.0: Behavioural inquiries from a work system perspective. *Int. J. Prod. Res.* **2022**; *preprint*. [CrossRef]

39. Demir, K.A.; Döven, G.; Sezen, B. Industry 5.0 and human robot co-working. *Procedia Comput. Sci.* **2019**, *158*, 688–695. [CrossRef]

40. Lu, Y.; Xu, X.; Wang, L. Smart manufacturing process and system automation—A critical review of the standards and envisioned scenarios. *J. Manuf. Syst.* **2020**, *56*, 312–325. [CrossRef]

41. Sikhwal, R.K.; Childs, P.R.N. Towards Mass Individualisation: Setting the scope and industrial implication. *Des. Sci.* **2021**, *7*, e16. [CrossRef]

42. Aheleroff, S.; Mostashiri, N.; Xu, X.; Zhong, R.Y. Mass Personalisation as a Service in Industry 4.0: A Resilient Response Case Study. *Adv. Eng. Inform.* **2021**, *50*, 101438. [CrossRef]

43. Hu, S.J. Evolving Paradigms of Manufacturing: From Mass Production to Mass Customization and Personalization. *Procedia CIRP* **2013**, *7*, 3–8. [CrossRef]

44. Kumar, A. From mass customization to mass personalization: A strategic transformation. *Int. J. Flex. Manuf. Syst.* **2007**, *19*, 533–547. [CrossRef]

45. Lu, Y.; Zheng, H.; Chand, S.; Xia, W.; Liu, Z.; Xu, X.; Wang, L.; Qin, Z.; Bao, J. Outlook on human-centric manufacturing towards Industry 5.0. *J. Manuf. Syst.* **2022**, *62*, 612–627. [CrossRef]

46. Magrini, E.; Ferraguti, F.; Ronga, A.J.; Pini, F.; De Luca, A.; Leali, F. Human-robot coexistence and interaction in open industrial cells. *Robot Comput. Integr. Manuf.* **2020**, *61*, 101846. [CrossRef]

47. Cochran, D.S.; Arinez, J.F.; Collins, M.T.; Bi, Z. Modelling of human–machine interaction in equipment design of manufacturing cells. *Enterp. Inf. Syst.* **2016**, *11*, 969–987. [CrossRef]

48. Ishigooka, T.; Yamada, H.; Otsuka, S.; Kanekawa, N.; Takanashi, J. Symbiotic Safety: Safe and Efficient Human-Machine Collaboration by utilizing Rules. In Proceedings of the 2022 Design, Automation & Test in Europe Conference & Exhibition (DATE), Online, 14–23 March 2022; pp. 280–281. [CrossRef]

49. Habib, L.; Pacaux-Lemoine, M.P.; Millot, P. A method for designing levels of automation based on a human-machine cooperation model. *IFAC-PapersOnLine* **2017**, *50*, 1372–1377. [CrossRef]

50. Zieba, S.; Polet, P.; Vanderhaegen, F. Using adjustable autonomy and human–machine cooperation to make a human–machine system resilient—Application to a ground robotic system. *Inf. Sci.* **2011**, *181*, 379–397. [CrossRef]

51. Semeraro, F.; Griffiths, A.; Cangelosi, A. Human–robot collaboration and machine learning: A systematic review of recent research. *Robot Comput. Integr. Manuf.* **2023**, *79*, 102432. [CrossRef]

52. Wang, T.; Li, J.; Kong, Z.; Liu, X.; Snoussi, H.; Lv, H. Digital twin improved via visual question answering for vision-language interactive mode in human–machine collaboration. *J. Manuf. Syst.* **2021**, *58*, 261–269. [CrossRef]

53. Simmler, M.; Frischknecht, R. A taxonomy of human–machine collaboration: Capturing automation and technical autonomy. *AI Soc.* **2021**, *36*, 239–250. [CrossRef]

54. Xiong, W.; Fan, H.; Ma, L.; Wang, C. Challenges of human—Machine collaboration in risky decision-making. *Front. Eng. Manag.* **2022**, *9*, 89–103. [CrossRef]

55. Trujillo, A.C.; Gregory, I.M.; Ackerman, K.A. Evolving Relationship between Humans and Machines. *IFAC-PapersOnLine* **2019**, *51*, 366–371. [CrossRef]

56. Simões, A.C.; Pinto, A.; Santos, J.; Pinheiro, S.; Romero, D. Designing human-robot collaboration (HRC) workspaces in industrial settings: A systemic literature review. *J. Manuf. Syst.* **2022**, *62*, 28–43. [CrossRef]

57. Pizoń, J.; Gola, A.; Świć, A. The Role and Meaning of the Digital Twin Technology in the Process of Implementing Intelligent Collaborative Robots. In *Lecture Notes in Mechanical Engineering*; Springer: Cham, Switzerland, 2022; pp. 39–49. [CrossRef]

58. Santosuosso, A. About coevolution of humans and intelligent machines: Preliminary notes. *BioLaw J. Riv. Di BioDiritto* **2021**, 445–454. [CrossRef]

59.   Van Eck, J.N.; Waltman, L. VOSviewer Manual; 2022. Available online: https://www.vosviewer (accessed on 6 December 2022).
60.   Orange Data Mining—Data Mining. Available online: https://orangedatamining.com (accessed on 6 December 2022).